Localized Dynamics of Thin-Walled Shells

Monographs and Research Notes in Mathematics

Series Editors:
John A. Burns, Thomas J. Tucker, Miklos Bona, Michael Ruzhansky

About the Series

This series is designed to capture new developments and summarize what is known over the entire field of mathematics, both pure and applied. It will include a broad range of monographs and research notes on current and developing topics that will appeal to academics, graduate students, and practitioners. Interdisciplinary books appealing not only to the mathematical community, but also to engineers, physicists, and computer scientists are encouraged.

This series will maintain the highest editorial standards, publishing well-developed monographs as well as research notes on new topics that are final, but not yet refined into a formal monograph. The notes are meant to be a rapid means of publication for current material where the style of exposition reflects a developing topic.

Inverse Scattering Problems and their Application to Non-Linear Integrable Equations
Pham Loi Vu

Spectral Methods Using Multivariate Polynomials On The Unit Ball
Kendall Atkinson, David Chien, and Olaf Hansen

Glider Representations
Frederik Caenepeel, Fred Van Oystaeyen

Lattice Point Identities and Shannon-Type Sampling
Willi Freeden, M. Zuhair Nashed

Summable Spaces and Their Duals, Matrix Transformations and Geometric Properties
Feyzi Basar, Hemen Dutta

Spectral Geometry of Partial Differential Operators (Open Access)
Michael Ruzhansky, Makhmud Sadybekov, Durvudkhan Suragan

Linear Groups: The Accent on Infinite Dimensionality
Martyn Russel Dixon, Leonard A. Kurdachenko, Igor Yakov Subbotin

Localized Dynamics of Thin-Walled Shells
Gennadi I. Mikhasev, Petr E. Tovstik

For more information about this series please visit: https://www.crcpress.com/Chapman--HallCRC-Monographs-and-Research-Notes-in-Mathematics/book-series/CRCMONRESNOT

Localized Dynamics of Thin-Walled Shells

Gennadi I. Mikhasev
Belarusian State University

Petr E. Tovstik
Saint Petersburg State University

CRC Press
Taylor & Francis Group
Boca Raton London New York

CRC Press is an imprint of the
Taylor & Francis Group, an **informa** business
A CHAPMAN & HALL BOOK

First edition published 2020
by CRC Press
6000 Broken Sound Parkway NW, Suite 300, Boca Raton, FL 33487-2742

and by CRC Press
2 Park Square, Milton Park, Abingdon, Oxon, OX14 4RN

Library of Congress Cataloging-in-Publication Data

Names: Mikhasev, G. I., author. | Tovstik, P. E., author.
Title: Localized dynamics of thin-walled shells / Gennadi I. Mikhasev and Petr E. Tovstik.
Description: First edition. | Boc Raton : CRC Press, 2020. | Series: Chapman & Hall/CRC monographs and research notes in mathematics | Includes bibliographical references and index.
Identifiers: LCCN 2019060133 | ISBN 9781138069749 (hardback) | ISBN 9781315115467 (ebook)
Subjects: LCSH: Shells (Engineering)--Vibration--Mathematical models. | Thin-walled structures--Vibration--Mathematical models. |
Localized waves. | Wave packets. | Asymptotic expansions.
Classification: LCC TA660.S5 M487 2020 | DDC 624.1/7762015118--dc23
LC record available at https://lccn.loc.gov/2019060133

ISBN: 978-1-138-06974-9 (hbk)
ISBN: 978-1-315-11546-7 (ebk)

Typeset in CMR
by Nova Techset Private Limited, Bengaluru & Chennai, India

Contents

Preface

The study of vibrations of thin shells is an extensive scientific problem which includes free and forced vibrations, vibrations of shells in contact with a liquid or gas, random oscillations, parametric vibrations, propagation of elastic waves, and others. This book is dedicated to the following two aspects of the problem: *localized* vibrations and propagation of localized waves.

The phenomenon of localization of vibrations in thin-walled structures is undesirable, and sometimes dangerous, because it is associated with the concentration of stresses in the structure, which can lead to loss of stability, accumulation of material fatigue, the appearance of cracks, and other destructive phenomena.

In our book, vibrations which are concentrated in small neighborhoods of some fixed or moving lines or points on the shell middle surface and decay exponentially with distance from these lines or points are called localized vibrations. Fixed lines or points in the vicinity of which localization of free oscillations takes place are called the *weakest* ones. For the first time, the term *weakest line* was introduced in

- P.E. Tovstik. Two-dimensional problems of buckling and vibrations of the shells of zero Gaussian curvature. *Soviet Physics Doklady*, 28(7): 593–594, 1983

in the study of the localized buckling and vibrations modes of thin elastic cylindrical and conical shells. The reasons leading to the localization of both buckling and free vibrations modes are often the same. They are related to variability of the shell Gaussian curvature and/or of the shell thickness, or to non-homogeneity of the material properties and/or initial stress-strain state, or to complex shape of the shell edges, etc. Localization in the neighborhood of an edge may be related to specific feature of the boundary conditions on the edge.

Localized parametric vibrations of a thin shell under external forces periodically varying in time can be easily excited near the *weakest* line (if such a line on the middle surface exists). Inhomogeneity of periodical loading may also result in localization in zones which are most pre-stressed.

Travelling vibrations concentrated in the vicinity of moving lines or points on the middle surface are called here the *wave packets* (WP) with centers at these lines or points. Such type of running not-stationary vibrations may be also readily excited near the weakest lines or points. Localized waves in thin shells can be also treated as the response of a shell to short-time concentrated loading.

For any type of localized vibrations, it is assumed that one or both sizes of the localization region are small compared with the dimensions of the entire middle surface. Therefore, to study the localized dynamics of a thin shell, methods of asymptotic integration of the governing equations can be used. The aim of the authors is to demonstrate the asymptotic methods capable of dealing with a wide class of problems on the localized dynamics of thin-walled structures, to construct asymptotic solutions and give the qualitative analysis of these solutions as well.

The book contains nine chapters. Chapter 1 (Introduction) presents a brief overview on two-dimensional (2D) theories of thin shells as well as hypotheses underlying them. Various asymptotic approaches used for solving boundary-value problems in the shell theory are also given. The state of the art of problems on localized vibrations and waves in thin shells is briefly discussed.

Chapter 2 is devoted to the governing equations of the 2D theories of thin shells. The equations based on the Kirchhoff-Love or Timoshenko-Reissner hypotheses are given. The equivalent single layer models for laminated shells based on the generalized Timoshenko hypotheses are also presented.

Chapter 3 is focused on localized vibrations of plates and shells of revolution. At first, the simplest problems on edge vibrations of rectangular and circular plates are considered. Then the asymptotic algorithms to construct eigenmodes localized near an edge as well as near a parallel lying strictly within a shell of revolution are proposed.

Chapter 4 is devoted to the asymptotic approach allowing constructing eigenmodes localized in the neighborhood of the weakest generatrix on the middle surface of a cylindrical or conical shell. The method is applied to study free localized vibrations of non-circular cylindrical and conical shells with arbitrary (not necessarily plane) edges.

Chapter 5 is dedicated to localized parametric vibrations of single layer cylindrical/ conical shells and laminated thin cylinders under pulsing axial forces and pressure. The asymptotic method developed in Chapter 4 in combination with the multiple scale method with respect to time is utilized to construct solutions localized near the weakest generatrix.

Chapters 6 and 7 present the asymptotic method for constructing solutions in the form of wave packets (WPs) travelling in the circumferential direction in cylindrical and conical shells. The detail analysis of the influence of geometrical parameters of a shell, the initial conditions as well as of external forces, including not-stationary ones, on the dynamic behavior of WPs is given.

In Chapter 8, infinitely-long shells of revolution subjected to the initial conditions localized in the vicinity of a parallel are considered. Applying the asymptotic method developed in Chapter 6, solutions of the governing equations are constructed in the form of the superposition of packets of bending, longitudinal and torsional waves propagating in the axial direction.

Chapter 9 generalizes the method stated in Chapters 6-7 to the case of two-dimensional WPs. Solutions are constructed in the form of bending waves localized in the neighborhood of moving points on the mid-surface of a shell of an arbitrary shape.

In part the book revises and supplements the results published in the Russian-language monograph

- G.I. Mikhasev, P.E. Tovstik. *Localized Vibrations and Waves in Thin Shells. Asymptotic Methods.* FIZMATLIT, Moscow, 2009,

while some outcomes are being published for the first time.

Finally, we have to thank our pupils and followers. The first author is grateful to R. Ataev, I.V. Avdoshka E.A. Korchevskoy and S.P. Kuntsevich, and the second author expresses special gratitude to Z.G. Ershova, S.B. Filippov, D.N. Ivanov, A.V. Krotov, A.I. Molchanov, N.V. Naumova and A.V. Zelinskaya, the results of which formed the basis of some sections. We also thank T.P. Tovstik who supported us with some numerical calculations and graphic design.

Gennadi I. Mikhasev
Minsk, Belarus

Petr E. Tovstik
Saint Petersburg, Russia
November, 2019

Authors

Gennadi I. Mikhasev is an expert in the theory of non-stationary localized dynamics of thin shells and in the mathematical modelling in biomechanics. He is a co-author of two monographs in mechanics of thin shells, two handbooks in bio-mechanics, and of around 100 papers. He is the head of the department of Bio- and Nanomechanics in the Belarusian State University, as well as a part-time researcher in the laboratory of Theoretical and Applied Mechanics.

Petr E. Tovstik is a specialist in the theory of asymptotic methods and its applications to the shell vibrations and buckling and to the various branches of mechanics. He is a co-author of ten monographs and of around 150 papers. He is the head of department of Theoretical and Applied Mechanics of the Saint Petersburg State University, as well as a part-time researcher in the Institute for Problems in Mechanical Engineering RAS.

Chapter 1

Introduction

This chapter has the introductory character and consists of five sections. Section 1.1 contains a brief overview of different two-dimensional theories for thin elastic shells, including the classical theory based on the Kirchhoff-Love hypotheses, and the first- and higher-order shear deformation theories as well. Section 1.2 is devoted to the phenomenon of localization of free vibrations in thin-walled structures. Reasons for the appearance of localized modes, such as features in boundary conditions, presence of a free or partially clamped edge, availability of an interface boundaries, inhomogeneity of the shell parameters as well as of external influences, are discussed. In Sections 1.3, some general information on parametric vibrations of thin shells is presented, the reasons leading to the localization effect of parametrically excited vibrations in thin shells are also discussed. In Section 1.4, classification of the localized waveforms of motion called wave packets is given. Finally, Section 1.5 contains a brief overview of the asymptotic methods used in studying the localized dynamics of thin-walled shells, with special attention being given to the modified Maslov-WKB method.

1.1 Two-dimensional theories in the dynamics of thin shells: Brief overview

The choice of a suitable theory for analyzing the stress-strain state of an elastic shell depends on many factors. However, if a shell is sufficiently thin so that [173] $h/R < 1/20$, where h, R are the thickness and the characteristic size of the shell, respectively, and strains and stresses in the cross section may be neglected, then two-dimensional (2D) theories can be applied to give a satisfactory result in solving many problems.

In dynamical problems, the second of the above requirements implies in particular that high frequency vibrations with formation of a large number of short waves in the normal direction cannot be described in the framework of any 2D theory. Such type of localized high frequency vibrations of plates or shells (which should be considered as 3D bodies) [99] will not be studied in our book, although the approach used in [99] and to be applied here is the same. We shall be concerned with the localized vibrations of thin shells which are accompanied by the formation of waves only in two directions at the middle surface. The accuracy of the 2D theory for correctly predicting such oscillations and waves (under some assumptions concerning the wave's length) in a thin shell is quite sufficient.

There are a huge number of 2D theories of thin plates and shells, which are based on different assumptions and hypotheses (see, for example, [8, 12, 50, 51, 86, 105, 119, 126, 161, 168, 173, 198–200, 202, 203]). Following Qu et al. [191], they may be classified as the classical shell theory (CST), the first-order shear deformation theory (FSDT), and the higher-order deformation theory (HSDT). The simplest of them is the CST proceeding from the Kirchhoff-Love hypotheses [105,125]. These kinematic hypotheses, which were originally formulated by Kirchhoff for thin plates, read as follows:

- straight line normal to the undeformed middle surface remains straight and normal to the deformed middle surface;

- the elements of the midplane during the deformation have no dilatation.

These assumptions ignore the in-plane displacement and presuppose only the deflection w and imply the following kinematic constrains: no transverse shear and no thickness changes. Love introduced additionally the in-plane displacements for the middle surface of the shell [126]. Thus, the original CST for thin isotropic shells leads to three differential equations with respect to the in-plane displacements u_1, u_2 and the transverse displacement (deflection) w.

Later, many variations of improved CSTs appeared. Depending on additional assumptions for strain-displacement, constitutive and balance equations, they may be classified as theories [158] by Ambartsumyan [8], Donnel [50], Flügge [57], Mushtari and Galimov [168], Love [126], Mindlin [161],

Novozhilov [173], Reissner [202], Sanders [205], Vlasov [241], etc. All above approaches, not depending on modification, again lead to three equations with respect to three unknowns.

The CSTs fail for shells pliable to shear (such as sandwiches with soft core, laminates with different material properties for layers, and functionally graded shells as well). Such theories can be carefully exploited for the static analysis of thin laminates [178, 179]. As for dynamic problems, their application for thin laminated shells is limited by low-frequency range [191] and result in errors up to 30% under the analysis of high-frequency vibrations [199].

In the so-called FSDTs proposed by Reissner [203], Lurie [129], Mindlin [161], etc., and actually being the development of the Timoshenko's beam theory [219], the deflection w is independent of the normal coordinate z, while the in-plane components u_1, u_2 of the middle surface are assumed to be linear functions of z. The last assumption allows us to take into account the two components of transverse shear strains that turn out to be constant along the shell thickness. Thus, the FSDT results in five coupled differential equations with respect to five unknowns. Often, this original version of the FSDT for thin shells is called the Timoshenko-Reissner theory [70, 166, 167, 231].

The main drawback of both the CSTs and the FSDTs is that they do not permit one to satisfy the traction-free boundary conditions at the shell surfaces. To improve these models, Reissner [202] and Mindlin [161] introduced the shear correction factors. However, they are difficult to calculate for an arbitrary shell because they are affected by the geometric and physical parameters of a shell, loading and boundary conditions as well [27, 46, 214, 240, 244]. All subsequent modifications of the original FSDTs for isotropic shells as well as their extensions for layered ones (see the references in [158]) gave only minor corrections, but did not solve the problem involving traction-free boundary conditions.

Further improvements of 2D theories for thin shells, including laminated ones, were related to quadratic, cubic or higher-order expansions at least of the in-plane displacements u_1, u_2. These theories called HSDTs (see among many others [7,80,121,209]) also results in five equations, but in contrast to FSDTs, they comply with the traction-free boundary conditions on the top and bottom surfaces. Here, we draw attention to the so-called equivalent single layer (ESL) model for laminated shells proposed by Grigolyuk and Kulikov [80] that will be used below to study localized vibrations of cylindrical laminates. This model is based on introducing the generalized kinematic hypotheses of Timoshenko for the in-plane displacements. Complying with boundary conditions at free face surfaces, this theory satisfies in addition the stress continuity conditions at the interface surfaces. In the particular case, when the dynamic stress-strain state of a shell has a large variability at least in one direction at the middle surface, the system of five equations is reduced to three differential equations with respect to the displacement, stress and shear functions, χ, Φ, Θ. It is of interest to note that these simplified coupled equations, on the one hand, coincide with similar equations derived by Tovstik and Tovstik [227] from the

3D theory of elasticity and, on the other hand, unify the well-known equations of the Mushtary-Donell-Vlasov theory.

To date, there are other high accurate theories like the layer-wise theories accounting for the zig-zag effects. However, these approaches being extremely complicated and highly costly in terms of computations are applied for calculation of moderately thick multi-layered shells.

Completing the short overview of the existing 2D theories for thin shells, we refer the readers to the following survey articles [4, 39–41, 79, 96–98, 188, 190, 196, 197, 206, 213, 220] and the books by Qatu [189], Reddy [199], Gorshkov et al. [74], Altenbach et al. [5], Mikhasev and Altenbach [158].

The error of any of the 2D theories for thin isotropic shells depends on the type of a problem under consideration. In statics, it has an order of the relative thickness of a shell [172], and in the dynamical problems, it is additionally influenced by the wave length, the natural frequency or the variability in time [65, 67, 70, 110] as well as the rate of attenuation of the waves amplitude in the case of localized vibrations. High frequency vibrations with formation of extremely short waves as well as highly concentrated oscillations are not considered here. The detailed estimate of an error of various equations used in the book will be discussed in Chapter 2.

1.2 Localized modes in the dynamics of thin-walled structures

The phenomenon of localization of free vibrations is well-known both in physics and in mechanics [49]. In the physics of condensed matter, it was first discovered in 1958 by Anderson [10] while studying oscillations of various lattice structures. He observed that irregularities in the distribution of electrons can vary vibration characteristics and result in the mode localization. This phenomenon was called *Anderson's localization* [87].

In mechanics, the early extensive studies of the mode localization were performed for discrete one-dimensional systems such as open or closed chains of pendulums or beams (see, e.g., in [87, 182, 183, 185, 245]). Pierre [182] has demonstrated that nearly periodic structures such as assemblies of coupled vertically aligned beams are highly sensitive to small disorder among the component systems and under certain conditions possess strongly localized modes. Studies of free vibrations of periodic continuous systems like chains of beams, multi-span beams, weakly taut strings resting on a Winkler foundation, cables with lumped masses, and large space structures (see among many others in [2, 48, 88, 128, 184]) have also confirmed that localization of free vibrations occurs as a rule in disordered structural systems consisting of weakly coupled component systems.

The pioneering works of Rayleigh [195] and Stoneley [192] in the field of continuum mechanics and solids gave a powerful impetus for further numerous studies of edge and interfacial vibrations and waves in elastic bodies, including plates and shells.

When considering thin-walled structures (plates and shells), one needs to distinguish the following basic reasons causing the mode localization:

- features in boundary conditions,
- presence of a free or partially clamped edge,
- availability of a surface dividing a plate or a shell into two parts with different physical properties,
- variability of geometrical or/and physical parameters, or/and external factors affecting the shell stress-strain state.

Consider briefly the above cases.

1.2.1 Mode localization induced by features in boundary conditions

The presence of points restricting deformations in thin elastic 2D structures can result in strong mode localization. Nowacki [174] was apparently the first who has studied similar vibrations of simply supported rectangular plates with internal constrained points. Gorman [72] and Gorman and Singal [73] analytically and experimentally investigated free vibrations of plates clamped only at symmetric points on the diagonals and revealed the effect of localization in the vicinity of these points. Rao et al. [194], Raju and Amba-Rao [193] and Utjes et al. [236] have numerically examined the localized vibrations of rectangular plates with point supports. Subsequent numerous studies on localized free vibrations of plates having point supports or more strong point restrictions were performed in different complicated settings on the basis of both the Kirchhoff theory and FSDTs accounting for transverse shear deformations (see the extended review in the recently published research paper [42]). These investigations have demonstrated phenomenon of mode localizations of both bending and in-plane vibrations of plates.

1.2.2 Edge vibrations of thin plates and shells

The phenomenon of edge vibrations in the case of plane stress-strain state was first observed experimentally by Shaw [207] when studying vibrations of a thick circular barium titanate disk. Analytic verification of this phenomena based on a plate theory was made by Gazis and Mindlin [58], and subsequently confirmed in other papers (see, e.g., [14, 76, 221]).

The edge flexural vibrations of a plate were first studied by Konenkov [111] in 1960, although the identical problem on the localized buckling of a plate was solved by Ishlinsky [91] much earlier, in 1954. Later, the results of Konenkov were rediscovered in [210, 217].

As a rule, the forms of both in-plane and flexural edge vibrations are as found in terms of an exponential function slowly decaying far from a free edge. For both types of edge vibrations, solutions describe the Rayleigh wave running along an edge. The natural frequency, below which there does not exist a solution oscillating along the edge, is called the cut-off frequency. Resonant frequencies arise below the limiting frequencies for the travelling wave and are weakly dependent on plate sizes.

In recent years, the edge phenomenon in plates was studied on the basis of different theories, accounting for anisotropy, lamination and other external factors (see, e.g., [60, 62, 81, 186, 204, 216, 248–250]).

The phenomenon of edge vibrations in shells is more complicated. Due to curvature, the localized forms of in-plane and bending vibrations turn out to be coupled, this leads to low level radiation damping of the in-plane shell edge waves. Considering a semi-infinite cylindrical shell, Kaplunov et al. [101] revealed three types of free localized shell vibrations:

- bending vibrations satisfying conditions

$$r = 2\varsigma - 1, \quad \frac{1}{2} \le \varsigma < 1, \tag{1.1}$$

- super-low-frequency semi-membrane vibrations obeying conditions

$$r = 2\varsigma - 1, \quad 0 \le \varsigma < \frac{1}{2}, \tag{1.2}$$

- extensional (in-plane) vibrations with

$$r = \varsigma, \quad \varsigma \ge 0, \tag{1.3}$$

where ς and r are orders of the wave number n in the circumferential direction and the dimensionless frequency parameter λ compared with a natural small parameter (relative thickness) $h_* = h/R$, respectively, which are introduced as

$$n \sim h_*^{\varsigma}, \quad \lambda = \frac{R^2 \rho \omega^2}{E} \sim h_*^{-2r}. \tag{1.4}$$

In (1.4), R is the radius of mid-surface, E is Young's modulus, ρ is the mass density, and ω is the radian frequency.

The amplitudes of bending and in-plane displacements for each type of vibrations satisfy the certain asymptotic relations depending on the variability of a solution in the axial direction (see in [101]). It is also pointed out that the first and third types of localized vibrations are similar to bending and extensional vibrations of a semi-strip for a large number of waves in the circumferential direction. In addition, the natural frequencies of bending vibrations are real, and the natural frequencies of in-plane vibrations turn out to be complex with asymptotically small imaginary parts. It is worth mentioning that apart from the classical case of a traction-free edge, the authors revealed three variants of weak boundary conditions (with *partially clamped* edges) corresponding to aforementioned types of localized vibrations.

To date, there are more recent studies on free edge vibrations taking into account the effect of anisotropy and more complicated shell shapes (see in [11, 20, 61, 83, 103]).

1.2.3 Interfacial vibrations in thin plates and shells

Interfacial vibrations are vibrations occurring on the boundary between two joined bodies with different elastic properties. Apparently, the first time this phenomenon for thin-walled structures was detected in papers [59, 252] in which the analogies of Stoneley's waves [192] in joined plates and half-strips were studied. In the papers [104, 239], using an asymptotic method, interfacial vibrations of joined shells of revolution were studied. Free interfacial vibrations of infinite closed and open cylindrical shells composed of two semi-infinite orthotropic elastic cylindrical shells with different elastic properties were examined in the recent papers [61, 82, 84]. Using the asymptotic approach, the authors of these studies derived the dispersion equations as well as formulas for eigenfrequencies, and revealed the asymptotic links between the dispersion equations of problems under consideration and analogous problems for joined rectangular plates and half-strips.

One needs to point out certain similarity between edge and interfacial vibrations of thin plates and shells: for both types of localized vibrations, the modes are found in the form of functions exponentially decaying far from an edge or interfacial surface. Finally we refer readers to the paper by Lawrie and Kaplunov [118] which provides a focused history and overview of such phenomenon as edge and interfacial vibrations in strips, rods, plates and shells. The mathematical background for studying these phenomenon in bodies with particular emphasis on structures such as plates and shells may be found in book by Vilde et al. [239].

1.2.4 Mode localization induced by inhomogeneity of geometrical and physical parameters

Natural modes of free vibrations for a thin shell are well known to be strongly affected by its shape. If for a circular cylindrical shell (with constant physical parameters and, particularly, clamped both edges) the mode vibrations cover the entire surface of the shell with waves extending from edge to edge, then for a convex shell, under certain conditions, it is possible for localization of waves to occur in the neighborhood of a curve being far from its edges [64]. For example, for an elongated ellipsoid of revolution, the vibration modes are localized near the diameter [141]. For some variants of boundary conditions, an ellipsoid of revolution possesses eigenmodes localized near an edge with *weak* boundary conditions.

For shells of zero Gaussian curvature, the localization of modes occurs as a rule in the vicinity of a generatrix [226, 233]. In what follows, this generatrix will be called the *weakest* one as it was proposed in paper [233] where the study of the localization phenomenon in cylindrical and conical shells was carried out. Here, one needs again to distinguish the localization inside a shell (for instance, in the problem on free vibrations of an elliptical cylinder) and in the neighborhood of a straight edge. In the first case, we have asymptotically multiple frequencies of natural vibrations, when four different modes of vibration correspond to eigenfrequencies differing very little from each other. Free vibrations of a shell in the form of a straight circular cylinder cover the entire middle surface. Localization inside a cylinder is possible if the symmetry is broken (cylinder is non-circular and/or edges are oblique and/or there are initial imperfections). Localization in the vicinity of an edge takes place only for weak clamping of this edge.

External influences, such as axial forces, pressure, temperature, magnetic or electric field, etc., may also cause the mode localization [37, 139, 140, 158]. So, in the book by Mikhasev and Altenbach [158] devoted to laminated shells, it was shown that the axial forces non-uniformly distributed on the shell edges result in localization of modes in the neighborhood of a generatrix which is the most compressed. Moreover, the paper [37] demonstrates the effect of localization of low-frequency free vibrations in a viscoelastic cylindrical shell non-uniformly heated in the circumferential direction. Another interesting example is related to sandwich shells containing a core made of polarized magnetorheological elastomer [139, 140]. It was revealed that applying magnetic field to such a shell distorts eigenmodes and, under some conditions, may result in their strong localization in the neighborhood of a generatrix with minimal storage modulus (the real part of the complex shear modulus).

For shells of negative Gaussian curvature, the modes of natural vibrations tend to propagate along characteristics. The waves reflected from the shell edges fill the entire middle surface; therefore, no localized modes were found for such shells. Shells of negative Gaussian curvature will not considered in this book.

1.3 Localized parametric vibrations of thin-walled structures

Parametric vibrations of a thin-walled structure occur when one or several of its parameters vary periodically as a result of external influence (forces, boundary conditions, temperature, magnetic field, etc.). For instance, an external force having even a small non-stationary component can generate in a thin shell periodically varying membrane stress resultants which excite not necessarily periodic vibrations and, under some conditions, lead to its dynamic instability [34].

Chelomey [43] is apparently the first who considered parametric vibrations of a thin shell. He studied the dynamic instability of a circular cylindrical shell under the action of non-stationary forces applied at the shell ends. Later, various problems on parametric vibrations of thin elastic cylinders under periodic axial and radial forces were considered by Markov [130], Oniashvili [176], Wenzke [243], Yao [246, 247], Vijayaraghavan and Evan-Ivanowski [238], Baruch [21] and many others. An accurate and general mathematical formulation of similar problems for thin isotropic cylindrical shells was given by Bolotin [34]. He derived the system of differential equations in variations describing dynamics of a thin shell in the neighborhood of membrane stress state periodically varying in time. Subsequently, parametric vibrations of thin shells under various complicating factors (in a non-linear setting, taking into account energy dissipation, anisotropy, reinforcement, transverse shears, temperature field, initial imperfections etc.), and for different loading schemes were investigated by many researchers (among many others, [28, 30–33, 36, 109, 113, 175, 187, 242]).

The problems on parametric vibrations are generally reduced to the Mathieu equation [133, 134] which leads to the well-known stability chart called *Strutt-Ince diagram* [90, 215]. The use of this diagram is associated with the calculation of coefficients in the Mathieu equation. The method for their determination depends on the shell shape, the loading scheme, and boundary conditions. For example, for a non-circular cylindrical shell under periodic axil forces, they are found in the form of expansions by the beam functions in the axial coordinate and the Fourier series in the circumferential direction. This approach is as a rule used for analysis of parametric vibrations with formations of waves covering the entire shell surface.

If a shell possesses natural modes localized near the weakest line or point, then under some conditions for the external periodic influence, it is easy to excite parametric oscillations in some localized region on the shell surface. The first studies on locally excited parametric vibrations were apparently performed in the papers [115, 151], where the authors considered non-circular thin elastic [151] and viscoelastic [115] cylinders subject to a non-uniform axial load consisting of static and weak periodic components. Low-frequency thermoparametric vibrations of an elastic non-circular cylindrical shell in a non-stationary temperature field were examined in [160]. Parametric instability of a thin conical shell subjected to a non-uniform pulsating pressure was investigated in [114]; the authors considered the problem in a general framework taking into account the variability of curvatures, generatrix length, and viscosity of a material as well. Recently, based on the ESL theory for laminated shells, Mikhasev and Ataev studied the effect of transverse shear on the main region of parametric instability for a thin laminated non-circular cylindrical shell under axial non-uniform axial forces periodically varying in time [136].

In the aforementioned papers dealing with localized parametric vibrations, the authors found the main region of dynamic instability and showed that it is invariant with respect to both the shell shape and the scheme of loading.

1.4 Localized waves in thin shells

In shell theory, non-stationary problems on localized vibrations propagating over the shell surface generally arise when studying the dynamic response of a shell to pulsed or transient loads non-uniformly distributed on the middle surface [201,208]. Non-stationary vibrations propagating in a shell occur also when a shell interacts with hydrodynamic pressure waves, and under unsteady diffraction of waves on a shell as well [78, 85, 162, 163, 171, 181]. Problems on localized waves running in the axial direction in long shells of revolution were often considered by studying of the response of semi-infinite cylindrical shells to short-time dynamic loads applied to the shell edge (see, among many others, [100, 112, 180]). Localized wave processes in a shell having a *weak* line or point may be caused by an external non-stationary influence such as growing pressure [155], axial force [15] or magnetic field [138]. Sometimes, unsteady waves in a shell are treated as a response to the given distribution of initial displacements and velocities [135].

In any case, the aforementioned problems are quite involved and in rare cases admit accurate solutions when a number of simplifying assumptions is introduced. For example, in studying waves running in a cylindrical shell in the circumferential direction, the assumption of load independence upon the longitudinal coordinate is introduced [85, 135, 162, 181, 201]. Under this assumption, a cylinder with finite dimensions is replaced by an infinitely long cylindrical shell, and the original two-dimensional (with respect to curvilinear coordinates) problem is reduced to a one-dimensional, the shell being actually replaced by a ring. One of the few papers where an attempt was made to construct exact solutions in problems on the non-stationary dynamics of a finite length cylindrical shell with boundary conditions taken into account is the paper by Sheng [208]. In this study and in several others, the governing equations were integrated by representing the displacements in the form of Fourier series in the circumferential coordinate.

One of the assumptions allowing the construction of an approximate analytical solution in shell dynamics is the assumption on strong localization of displacements and velocities in a running wave. Under this assumption, the entire surface of a shell is replaced by a small neighborhood of a moving line or a point at which the amplitude of waves is maximum. The approximate solution localized near this line or point is called the *wave packet* (WP) with a center on this line or point. In our book, we study three types of travelling WPs:

- WPs travelling in the circumferential direction in finite length cylindrical and conical shells;
- WPs running in the axial direction in infinitely long shells of revolution;
- WPs localized near moving points on the shell middle surface of an arbitrary shape.

In studying the first type of WPs, one arises an initial boundary-value problem with boundary conditions on the shell edges. In such a WP, waves extend from edge to edge and attenuate far from a generatrix being the WP center. If the shell edges lie in a plane perpendicular to the shell axis, then the problem is easily reduced to the one-dimensional (1D) one because it admits the separation of the axial coordinate [143], if not then the problem remains two-dimensional (2D). As shown in [150, 155], the required solution of the initial 2D problem can be found by asymptotic splitting into the sequence of 1D problems at a *moving generatrix*.

The second type of WPs is also reduced to the 1D problem, if the coefficients in the governing equations are not functions of the angular coordinate β [146, 149, 154]. In the case when even one parameter (geometrical or/physical) depends on β, the problem can be substituted by the sequence of 1D problems by expanding all functions into Fourier series with respect to the coordinate β, however such problems are not considered in this book. For the second type WP, waves are concentrated near a "moving parallel" on the shell surface. In this case, solutions are constructed in the form of a superposition of packets of bending, longitudinal and torsional waves travelling in the axial direction [145, 147–149, 154]. The behavior of running WPs strongly depends on the order of the wave number in the circumferential direction.

Finally, in studying the third type of WPs, the problem is reduced to analysing the behavior of points being the centers of 2D packets [153, 156], the solutions being like "quasi-photons" [16, 19] oscillating in the vicinity of moving points.

Due to the presupposed localization of solutions, the asymptotic method turns out to be a very effective mathematical tool in studies of the shell dynamics. A brief outline of these methods will be given in the next section.

1.5 Asymptotic methods in the analysis of localized dynamics of thin shells

Methods for the asymptotic integration of static two-dimensional equations of thin shell theory were mostly completely developed by Gol'denveizer [65,67].

In books [13,64], these methods were extended to the analysis of spectrum of free vibrations of thin shells. In some cases, the form of free vibrations occupies the entire surface of a shell, and in others it is localized in the vicinity of some lines or points. Asymptotic approaches for constructing localized solutions of differential equations were developed in the papers [6,16–18,131] in the study of various problems related to the dynamics of thin shells, wave diffraction and, generally, in various branches of physics. Some of these approaches, their modifications as well as new methods will be used in our book. Let us give a short overview of these methods.

In the paper [6], using the Airy equations method, Alumyae gives the asymptotic solution of equations describing free axisymmetric oscillations of conical shells with turning points. Later, this approach was used and essentially developed in book [64] for the analyses of many other problems on free vibrations of thin elastic shells.

In the monograph [17], Babich and Buldyrev constructed the localized eigenfunctions like the *bouncing ball* and *whispering gallery* for the Helmholtz equations. Quasi-photon solutions of the wave equation oscillating in a small neighborhood of a space-time ray and decreasing exponentially outside this small neighborhood were constructed by Babich, Buldyrev and Molotkov [16], Babich and Ulin [19], Kiselev and Perel [106–108]. In particular, in the paper [108], localized solutions of the linear 3D wave equation were found in the explicit form using functions having Gaussian localization in both the longitudinal and transverse directions.

In the monographs by Maslov [131,132], the complex WKB method for constructing localized solutions of multidimensional nonlinear equations with a small parameter was developed. As examples, non-linear equations of quantum mechanics, equations of a crystal lattice, etc. were considered. The obtained asymptotic solutions are localized in the vicinity of some curves or surfaces. The structure of such solutions is based on the Hamiltonian formalism of narrow-ray mechanics. The modification of Maslov's method will be used below to study the localized dynamics of thin shells.

The analytical construction of modes (including localized ones) of free vibrations of thin shells is, in the general case, an unresolved problem. Even a much simpler problem on transverse vibrations of a membrane (which is reduced to the Dirichlet problem for the two-dimensional Helmholtz equation) in the general case does not have an exact analytical solution. An analytical solution can be only constructed for areas that admit separating variables (among such regions we mention a circle, an ellipse, a rectangle, and some triangles). A number of problems on free vibrations of shells also allow the separation of variables. We point out shells of revolution in which the circumferential coordinate is readily separated, and non-circular cylindrical shells with simply supported curved edges, in which the longitudinal coordinate is also separated. After the separation of variables, the initial 2D problem is reduced to a much simpler 1D problem. This problem contains a natural small parameter (relative shell thickness) at the highest derivative, and asymptotic integration methods can be used to solve it.

We remind that the waveforms are called here localized if they are concentrated in small neighborhoods of some fixed or moving lines or points on the middle surface and decaying exponentially with distance from these lines or points. Let us classify the dynamic problems to be considered in what follows, and which imply the application of a certain asymptotic approach:

- stationary problem on free vibrations localized near an edge;
- stationary problems on free localized vibrations in the vicinity of the *weakest* lines situated far from edges;
- quasi-stationary problems on parametric vibrations concentrated near the *weakest* lines;
- non-stationary problems on travelling WPs localized in the neighborhood of moving lines and points.

The first class of problems is subdivided into (i) the simplest ones on the edge vibrations of thin rectangular plates and (ii) more complicated problems on oscillations of circular plates and shells of revolution near an edge. The edge modes in rectangular plates with free edges are easily represented by the exponential function decaying far from the free edge [118]. In addition, the analysis of free edge vibrations of circular plates and shells of revolution with a large number of waves in the circumferential direction leads to equations containing the turning points [6,64]. Their solutions are found in terms of the Airy function, $\mathrm{Ai}(x)$, which is one of two integrals of the equation

$$y''(x) - xy(x) = 0. \tag{1.5}$$

The last three classes of problems form the main content of the book, and they are united by the use of a single approach based on the modified Maslov-WKB asymptotic method [131, 132]. We outline the idea of this method on an example of a cylindrical shell.

Following Maslov's method, the form of vibrations localized in the neighborhood of the shell generatrix $\varphi = q$ can be approximated by the function

$$U(s, \varphi, t, \varepsilon) = u^* \exp\{\mathrm{i}\left[\varepsilon^{-1} S(\varphi, t, \varepsilon)\right]\}, \tag{1.6}$$

where $\mathrm{i} = \sqrt{-1}$ is the imaginary unit, t is a dimensionless time, s and φ are dimensionless axial coordinate and angle, respectively, ε is a small parameter (the relative thickness which is introduced in the different way for various problems), U is any of required functions (displacement, stress function, etc.), u^* is the amplitude function, and $S(\varphi, t)$ is the complex phase function such that $\Im S(q) = 0$ and $\Im S(\varphi) > 0$ for $\varphi \neq q$. Here and in what follows, the symbol $\Im$ denotes the imaginary part.

For free vibrations localized near the weakest generatrix $\varphi = q$ (when q is the fixed constant), Tovstik [233] proposed the explicit form of the phase function

$$S = \omega t + \varepsilon^{1/2} p\xi + \frac{1}{2}\varepsilon b\xi^2, \quad \xi = \varepsilon^{-1/2}(\varphi - q), \tag{1.7}$$

where ω is the dimensionless frequency, p is a real number (wave parameter) and b is a complex constant such that $\Im b > 0$. The amplitude function $u^*(s, \xi; \varepsilon)$ is sought in the form of a series by powers of $\varepsilon^{1/2}$. Note that, in contrast to the original Maslov method, this representation of the localized mode allows one to predict free vibrations of shells with any boundary conditions on arbitrary edges (which are not necessarily plane curves). The substitution of (1.6), (1.7) into governing equations and boundary conditions permits performing the asymptotic splitting of the original 2D problem into a sequence of 1D boundary-value problems considered on the weakest generatrix $\varphi = q$.

In the quasi-stationary problems on parametric vibrations localized near the weakest line $\varphi = q$, the ansatz remains the same with accuracy up to the amplitude function u^*. In the case of weak excitation (when the periodic component of the load is small with respect to the static one), it is assumed that $u^*(s, \xi, \varepsilon t, \varepsilon^2 t, \ldots)$ depend on *slow* time. Thus, the algorithm for studying localized parametric vibrations is the combination of Tovstik's method [233] and the multiple scale method with respect to time. This approach allows for the prediction of the form of parametric oscillations in the vicinity of the fixed (weakest) generatrix $\varphi = q$ with a slowly growing amplitude function u^* with respect to time in the case of parametric resonance as well as a way to define the main region of instability [115, 151].

The principal idea of the asymptotic method for studying wave packets is to use the Lagrange approach, according to which an "observer" is situated in the center $\varphi = q(t)$ of a travelling WP. Now, $q(t)$ is a function of time and defines the trajectory of a moving WP (for 2D wave packets with the center at a point, a trajectory is a certain curve on the middle surface). The introduction of the local coordinate system with the origin on the moving line $\varphi = q(t)$ leads to differential equations which govern behavior of the nth WP, where n is the number of semi-waves in the axial direction. The displacement components of a shell in the limits of the nth WP are found again by (1.6), but the function S is represented as follows [150, 155]:

$$S(\varphi, t; \varepsilon) = \int_0^{\tau} \omega(\tau) d\tau + \varepsilon^{1/2} p(t)\xi + \frac{1}{2}\varepsilon b(t)\xi^2, \tag{1.8}$$

where ω is an instantaneous frequency, and $p(t)$, $b(t)$ are functions of time, moreover, $\Im b(t) > 0$ for any finite time interval. The amplitude function $u^*(s, t; \varepsilon)$ is sought again as a series

$$u = u_0(s, \xi, t) + \varepsilon^{1/2} u_1(s, \xi, t) + \varepsilon u_2(s, \xi, t) + \ldots, \tag{1.9}$$

where the leading approximation $u_0(s, \xi, t) = z_n(s) P_0(\xi, t)$ is found as the product of the beam function $z_n(s)$ (corresponding to the WP center) and the Hermite polynomial $P_0(\xi, t)$ with coefficients depending on time. The asymptotic algorithm developed generates an equation like the dispersion one, the Hamiltonian system to determine the functions $p(t), q(t)$, the Riccati equation with respect to the complex function $b(t)$, and the amplitude equation for $P_0(\xi, t)$ as well.

As can be seen, the above approach is the generalization of the asymptotic method developed in [233] and intended for investigation of stationary problems. At the same time, it can be considered as a certain modification of the Maslov-WKB method [131, 132] in accordance to which the substitution of ansatz (1.6) into the governing equations yields the canonical system. This system consists of the non-linear Hamilton-Jacobi equation and the transfer equation with respect to the functions $S(\varphi, t, \varepsilon)$ and $u^*(s, \varphi; \varepsilon)$. The last ones are found in the class $O_f(\varepsilon^m)$ of so-called asymptotically equivalent functions.

In our opinion, the advantageous feature of our method is that it allows one to study the unsteady localized dynamics of a shell accounting for its finite dimensions and the effects of boundary conditions at edges of a complex geometry, including oblique and nonplanar edges.

Chapter 2

Equations of the two-dimensional theory of shells

In accordance with the Kirchhoff–Love (KL) or Timoshenko–Reissner (TR) hypotheses, 3D equations of motion of a solid body treated by a thin-walled shell are readily reduced to 2D ones for the middle surface of a shell. The equations describing this motion are composed of the following three groups of relations:

— the geometrical equations coupling the tangential and bending strains of the middle surface and displacements of points on this surface;

— the equilibrium equations of an element of the middle surface;

— and finally, the constitutive equations coupling forces and moments with strains of the middle surface.

The first two groups of relations are independent of the accepted model of the shell: these equations are the same both for the Kirchhoff–Love model and for the Timoshenko-like model.

In this chapter, we formulate the above relations and equations for single layer isotropic shells within the framework of both models. Their derivation and detailed discussions may be found, e.g., in books [9, 44, 45, 50, 65, 67, 120,

173]. In addition, we consider also the equivalent single layer model (ESL) for laminated transversally isotropic shells based on the generalized kinematic hypotheses of Timoshenko [80], and the recently delivered ESL model for geterogeneous in the thickness direction transversally isotropic shells based on the asymptotic expansion of solution of the 3D equations [229]. The error of governing equations based on different models is briefly discussed.

2.1 Geometric relations

Consider the middle surface of a shell and introduce the orthogonal system of curvilinear coordinates α, β (sometimes denoted by α_1, α_2) coinciding with the lines of curvature of this surface. The position of an arbitrary point M on the surface is defined by its radius vector $\mathbf{r} = \mathbf{r}(\alpha, \beta)$. We introduce the local orthogonal coordinate system with the unit vectors $\mathbf{e}_1$, $\mathbf{e}_2$, $\mathbf{n}$ as shown in Fig. 2.1, where

$$\mathbf{e}_1 = \frac{1}{A}\frac{\partial \mathbf{r}}{\partial \alpha}, \quad \mathbf{e}_2 = \frac{1}{B}\frac{\partial \mathbf{r}}{\partial \beta}, \quad \mathbf{n} = \mathbf{e}_1 \times \mathbf{e}_2, \quad A = \left|\frac{\partial \mathbf{r}}{\partial \alpha}\right|, \quad B = \left|\frac{\partial \mathbf{r}}{\partial \beta}\right|. \tag{2.1}$$

The curvilinear coordinates are chosen (if it is possible) in such a way that the unit vector $\mathbf{n}$ is the inner normal to the shell.

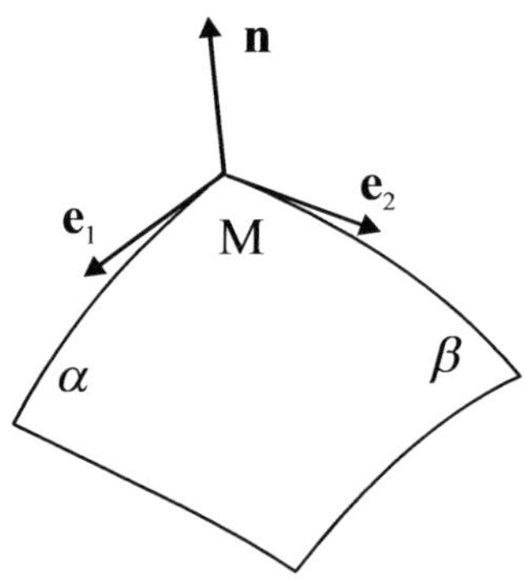

FIGURE 2.1: Curvilinear coordinates.

The first and the second quadratic forms of the surface are as follows:

$$I = ds^2 = A^2 d\alpha^2 + B^2 d\beta^2, \qquad II = \frac{A^2}{R_1} d\alpha^2 + \frac{B^2}{R_2} d\beta^2, \tag{2.2}$$

where ds is the arc length on the surface, R_1 and R_2 are the principal radii of curvature; the principal curvatures are denoted by $k_1 = R_1^{-1}$, $k_2 = R_2^{-1}$.

In the general case, A, B, R_1 and R_2 are functions of α and β and satisfy the Codazzi–Gauss relations

$$\frac{\partial}{\partial \beta}\left(\frac{A}{R_1}\right) = \frac{1}{R_2}\frac{\partial A}{\partial \beta}, \quad \frac{\partial}{\partial \alpha}\left(\frac{B}{R_2}\right) = \frac{1}{R_1}\frac{\partial B}{\partial \alpha},$$
$$\frac{\partial}{\partial \alpha}\left(\frac{1}{A}\frac{\partial B}{\partial \alpha}\right) + \frac{\partial}{\partial \beta}\left(\frac{1}{B}\frac{\partial A}{\partial \beta}\right) = -\frac{AB}{R_1 R_2}. \tag{2.3}$$

In particular, for shells of revolutions it is convenient to take, as curvilinear coordinates, the arc length s of the generator and the angle φ in the circumferential direction. Hence $A = 1$, and the functions B, R_1 and R_2 depend only on s,

$$B = R_2 \sin\theta, \quad \frac{1}{R_1} = \frac{d\theta}{ds}, \quad \frac{d}{ds}\left(\frac{1}{R_2}\right) = \frac{1}{B}\left(\frac{1}{R_1} - \frac{1}{R_2}\right)\frac{dB}{ds}, \tag{2.4}$$

where θ is the angle between the axis of revolution and the normal vector (see Fig. 2.2).

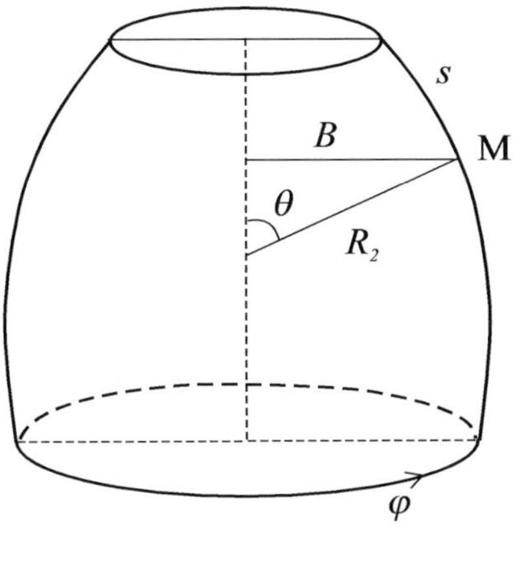

FIGURE 2.2: Shell of revolution.

The formulas for differentiation of the unit vectors read as

$$\frac{\partial \mathbf{e}_1}{\partial \alpha} = -\frac{1}{B}\frac{\partial A}{\partial \beta}\mathbf{e}_2 + \frac{A}{R_1}\mathbf{n}, \quad \frac{\partial \mathbf{e}_2}{\partial \alpha} = \frac{1}{B}\frac{\partial A}{\partial \beta}\mathbf{n}, \quad \frac{\partial \mathbf{n}}{\partial \alpha} = -\frac{A}{R_1}\mathbf{e}_1,$$
$$\frac{\partial \mathbf{e}_2}{\partial \beta} = -\frac{1}{A}\frac{\partial B}{\partial \alpha}\mathbf{e}_1 + \frac{B}{R_2}\mathbf{n}, \quad \frac{\partial \mathbf{e}_1}{\partial \beta} = \frac{1}{A}\frac{\partial B}{\partial \alpha}\mathbf{n}, \quad \frac{\partial \mathbf{n}}{\partial \beta} = -\frac{B}{R_2}\mathbf{e}_2. \tag{2.5}$$

Let us now consider a small deformation of the surface. Assume that after the deformation the point M occupies the position M^* with radius vector $\mathbf{r}^*$. We set

$$\mathbf{r}^* = \mathbf{r} + \mathbf{u}, \quad \mathbf{u} = u_1\mathbf{e}_1 + u_2\mathbf{e}_2 + w\mathbf{n}, \tag{2.6}$$

where $\mathbf{u}$ is the displacement vector and u_1, u_2, w (sometimes denoted by u, v, w or by u_1, u_2, u_3) are its projections onto the unit coordinate vectors before the deformation. In the linear problems considered below, we need not distinguish between the projections to the unit coordinate vectors before and after deformation.

Using the derivation formulas (2.5), one obtains

$$\frac{1}{A}\frac{\partial \mathbf{r}^*}{\partial \alpha} = (1+\varepsilon_1)\mathbf{e}_1 + \omega_1\mathbf{e}_2 - \gamma_1\mathbf{n}, \qquad \frac{1}{B}\frac{\partial \mathbf{r}^*}{\partial \beta} = (1+\varepsilon_2)\mathbf{e}_2 + \omega_2\mathbf{e}_1 - \gamma_2\mathbf{n}, \tag{2.7}$$

where

$$\begin{aligned}
\varepsilon_1 &= \frac{1}{A}\frac{\partial u_1}{\partial \alpha} + \frac{1}{AB}\frac{\partial A}{\partial \beta}u_2 - \frac{w}{R_1}, &\quad \varepsilon_2 &= \frac{1}{B}\frac{\partial u_2}{\partial \beta} + \frac{1}{AB}\frac{\partial B}{\partial \alpha}u_1 - \frac{w}{R_2}, \\
\omega_1 &= \frac{1}{A}\frac{\partial u_2}{\partial \alpha} - \frac{1}{AB}\frac{\partial A}{\partial \beta}u_1, &\quad \omega_2 &= \frac{1}{B}\frac{\partial u_1}{\partial \beta} - \frac{1}{AB}\frac{\partial B}{\partial \alpha}u_2, \\
\gamma_1 &= -\frac{1}{A}\frac{\partial w}{\partial \alpha} - \frac{u_1}{R_1}, &\quad \gamma_2 &= -\frac{1}{B}\frac{\partial w}{\partial \beta} - \frac{u_2}{R_2}.
\end{aligned} \tag{2.8}$$

Here, ε_1 and ε_2 are the tensile strains in the α- and β- directions, respectively, ω_1 and ω_2 are the angles of rotation of the unit vectors $\mathbf{e_1}$ and $\mathbf{e_2}$ in the projection to the tangent plane, and γ_1 and γ_2 are the angles of rotation of the normal vector $\mathbf{n}$. The magnitudes ε_1, ε_2 and $\omega = \omega_1 + \omega_2$ are the tangential strains, besides,

$$\omega = \omega_1 + \omega_2 = \frac{B}{A}\frac{\partial}{\partial \alpha}\left(\frac{u_2}{B}\right) + \frac{A}{B}\frac{\partial}{\partial \beta}\left(\frac{u_1}{A}\right) \tag{2.9}$$

is the shear angle in the tangent plane.

The bending and torsion strains, κ_1, κ_2 and τ, are defined as

$$\begin{aligned}
\kappa_1 &= -\frac{1}{A}\frac{\partial \gamma_1}{\partial \alpha} - \frac{1}{AB}\frac{\partial A}{\partial \beta}\gamma_2, \qquad \kappa_2 = -\frac{1}{B}\frac{\partial \gamma_2}{\partial \beta} - \frac{1}{AB}\frac{\partial B}{\partial \alpha}\gamma_1, \\
\tau &= -\frac{1}{B}\frac{\partial \gamma_1}{\partial \beta} + \frac{1}{AB}\frac{\partial B}{\partial \alpha}\gamma_2 + \frac{\omega_1}{R_2} = -\frac{1}{A}\frac{\partial \gamma_2}{\partial \alpha} + \frac{1}{AB}\frac{\partial A}{\partial \beta}\gamma_1 + \frac{\omega_2}{R_1}.
\end{aligned} \tag{2.10}$$

In view of (2.8), the both expressions for the torsion τ are equivalent, and hence the formula for the torsion can also be written in the symmetric form

$$2\tau = -\frac{B}{A}\frac{\partial}{\partial \alpha}\left(\frac{\gamma_2}{B}\right) - \frac{A}{B}\frac{\partial}{\partial \beta}\left(\frac{\gamma_1}{A}\right) + \frac{\omega_1}{R_2} + \frac{\omega_2}{R_1}. \tag{2.11}$$

2.2 Equations of motion

In the vector form, the motion equations of a shell read as:

$$
\begin{aligned}
&\frac{\partial(B\mathbf{F}_1)}{\partial\alpha} + \frac{\partial(A\mathbf{F}_2)}{\partial\beta} + AB\mathbf{q} = 0,\\
&\frac{\partial(B\mathbf{G}_1)}{\partial\alpha} + \frac{\partial(A\mathbf{G}_2)}{\partial\beta} + AB\,(\mathbf{e}_1 \times \mathbf{F}_1 + \mathbf{e}_2 \times \mathbf{F}_2) + AB\mathbf{m} = 0.
\end{aligned}
\tag{2.12}
$$

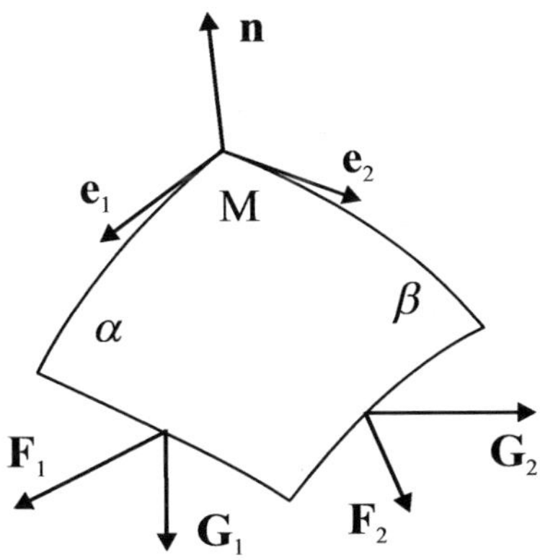

FIGURE 2.3: Forces and moments.

Here, $\mathbf{F}_1$ and $\mathbf{G}_1$ ($\mathbf{F}_2$ and $\mathbf{G}_2$) are the vector stress resultants and stress couples per unit length of the internal stresses acting on the lines α = const (β = const) (see Fig. 2.3), $\mathbf{q}$ and $\mathbf{m}$ are the densities per unit area of the middle surface of the inertia force and moment, respectively. These vectors are expanded into the unit orthogonal vectors as follows:

$$
\begin{aligned}
&\mathbf{F}_1 = T_1\mathbf{e}_1 + S_1\mathbf{e}_2 + Q_1\mathbf{n}, \quad \mathbf{G}_1 = H_1\mathbf{e}_1 - M_1\mathbf{e}_2,\\
&\mathbf{F}_2 = T_2\mathbf{e}_1 + S_2\mathbf{e}_2 + Q_2\mathbf{n}, \quad \mathbf{G}_2 = M_2\mathbf{e}_1 - H_2\mathbf{e}_2,\\
&\mathbf{q} = -\rho h\frac{\partial^2 u_1}{\partial t^2}\mathbf{e}_1 - \rho h\frac{\partial^2 u_2}{\partial t^2}\mathbf{e}_2 - \rho h\frac{\partial^2 w}{\partial t^2}\mathbf{n},\\
&\mathbf{m} = \frac{\rho h^3}{12}\frac{\partial^2\gamma_2}{\partial t^2}\mathbf{e}_1 - \frac{\rho h^3}{12}\frac{\partial^2\gamma_1}{\partial t^2}\mathbf{e}_2,
\end{aligned}
\tag{2.13}
$$

where ρ and h are the material density and shell thickness, respectively.

The form of equations (2.12) is independent of the accepted two-dimensional shell model (the Kirchhoff–Love model or the Timoshenko model accounting for shears). The difference between these models appeared in the elasticity relations (the constitutive equations) to be discussed below. We only note that for the Kirchhoff–Love model one should assume that $\mathbf{m} = 0$ in the second equation of (2.12), because the rotational inertia of the transverse fibers has the same asymptotic order of smallness as the transverse shears, and so it is expedient to consider the effect of the rotational inertia only in the Timoshenko model, which takes into account shears.

Using formulas (2.5), one can find the projections of equations (2.12) onto the unit vectors, $\mathbf{e}_1$, $\mathbf{e}_2$, $\mathbf{n}$:

$$\begin{aligned}
&\frac{\partial(BT_1)}{\partial\alpha} - \frac{\partial B}{\partial\alpha}T_2 + \frac{\partial(AS_2)}{\partial\beta} + \frac{\partial A}{\partial\beta}S_1 - AB\frac{Q_1}{R_1} - AB\rho h\frac{\partial^2 u_1}{\partial t^2} = 0,\\
&\frac{\partial(AT_2)}{\partial\beta} - \frac{\partial A}{\partial\beta}T_1 + \frac{\partial(BS_1)}{\partial\alpha} + \frac{\partial B}{\partial\alpha}S_2 - AB\frac{Q_2}{R_2} - AB\rho h\frac{\partial^2 u_2}{\partial t^2} = 0,\\
&\frac{\partial(BQ_1)}{\partial\alpha} + \frac{\partial(AQ_2)}{\partial\beta} + AB\left(\frac{T_1}{R_1} + \frac{T_2}{R_2}\right) - AB\rho h\frac{\partial^2 w}{\partial t^2} = 0,\\
&\frac{\partial(BH_1)}{\partial\alpha} + \frac{\partial B}{\partial\alpha}H_2 + \frac{\partial(AM_2)}{\partial\beta} - \frac{\partial A}{\partial\beta}M_1 + ABQ_2 - AB\frac{\rho h^3}{12}\frac{\partial^2\gamma_2}{\partial t^2} = 0,\\
&\frac{\partial(AH_2)}{\partial\beta} + \frac{\partial A}{\partial\beta}H_1 + \frac{\partial(BM_1)}{\partial\alpha} - \frac{\partial B}{\partial\alpha}M_2 + ABQ_1 - AB\frac{\rho h^3}{12}\frac{\partial^2\gamma_1}{\partial t^2} = 0,\\
&AB(S_1 - S_2) + AB\left(\frac{H_1}{R_1} - \frac{H_2}{R_2}\right) = 0.
\end{aligned} \tag{2.14}$$

With a correct choice of the constitutive equations, the sixth equilibrium equation from (2.14) should be satisfied identically.

2.3 Elasticity relations and boundary conditions for the Kirchhoff–Love model

Important agreement: in this book, the Kirchhoff–Love (KL) model as well as the Timothenko–Reissner (TR) model are accepted for studying localized dynamics of thin shells. The equivalent single layer (ESL) model for multi-layered shells will be considered also. The KL model is the main one, and if the model is not named, this is the KL model. In all cases the TR or ESL model is named specially.

Assume that a shell is made of a linearly elastic homogeneous isotropic material with the Young modulus E and Poisson's ratio ν. Based on the KL hypothesis on the conservation of the straight normal to the middle surface of a shell under its deformation, V. V. Novozhilov and L. I. Balabukh proposed a simplest variant of the constitutive relations coupling forces and moments with strains of the middle surface of the shell and satisfying the sixth equilibrium equation of (2.14) (see [173]):

$$
\begin{aligned}
&T_1 = K(\varepsilon_1 + \nu\varepsilon_2), && T_2 = K(\varepsilon_2 + \nu\varepsilon_1),\\
&S_1 = \frac{K(1-\nu)}{2}\left(\omega + \frac{h^2\tau}{6R_2}\right), && S_2 = \frac{K(1-\nu)}{2}\left(\omega + \frac{h^2\tau}{6R_1}\right),\\
&M_1 = D(\kappa_1 + \nu\kappa_2), && M_2 = D(\kappa_2 + \nu\kappa_1),\\
&H_1 = H_2 = H = D(1-\nu)\tau, && K = \frac{Eh}{1-\nu^2}, \quad D = \frac{Eh^3}{12(1-\nu^2)}.
\end{aligned}
\tag{2.15}
$$

Formulas for the shear stress resultants S_1 and S_2 can be written in the form

$$
S_1 = S + \frac{H}{R_2}, \quad S_2 = S + \frac{H}{R_1}, \tag{2.16}
$$

where

$$
S = \frac{K(1-\nu)}{2}\omega = Gh\omega, \qquad G = \frac{E}{2(1+\nu)}. \tag{2.17}
$$

Here S is the auxiliary function, and G is the shear modulus. The second terms in equations (2.16) are small, and so in writing the first terms of the asymptotic expansions one may approximately assume that

$$
S_1 \simeq S_2 \simeq S. \tag{2.18}
$$

At the same time, with the use of assumptions (2.18) the boundary-value problem of the shell theory becomes non self-adjoin. Let us consider this problem in more detail. We write the Lagrange function L of the deformed shell (taking into account the KL hypotheses and without consideration of the rotational inertia of the transverse fibers) in the form:

$$
\begin{aligned}
L &= T - \Pi,\\
T &= \frac{1}{2}\int\!\!\int_\Omega \rho h\left(\dot{u}_1^2 + \dot{u}_2^2 + \dot{w}^2\right) AB\, d\alpha\, d\beta,\\
\Pi &= \int\!\!\int_\Omega (\Pi_\varepsilon + \Pi_\kappa)\, AB\, d\alpha\, d\beta,
\end{aligned}
\tag{2.19}
$$

where Ω is the domain of variation of the curvilinear coordinates, T and Π are the kinetic and potential energy, respectively, the dots denote time derivatives, and Π_ε and Π_κ are the densities of the potential energies of stretching and bending defined as

$$
\begin{aligned}
\Pi_\varepsilon &= \frac{1}{2}K\left(\left(\varepsilon_1^2 + 2\nu\varepsilon_1\varepsilon_2 + \varepsilon_2^2\right) + \frac{1-\nu}{2}\omega^2\right),\\
\Pi_\kappa &= \frac{1}{2}D\left(\kappa_1^2 + 2\nu\kappa_1\kappa_2 + \kappa_2^2 + 2(1-\nu)\tau^2\right).
\end{aligned}
\tag{2.20}
$$

In view of (2.8), (2.9) and (2.10) the potential energy (2.19) depends on the displacements u_1, u_2, w and their derivatives. Variation of the functional (2.19) with respect to displacements (after partial integration) leads to the first three equations of (2.14), in which the shearing forces Q_1 and Q_2 are substituted from the fourth and fifth equations. Here, the forces and moments are expressed in terms of the strains of the middle surface by equations (2.15). In the meantime we get the natural boundary conditions.

Let $\alpha = \alpha_0$ be a part of the boundary Γ of the domain Ω. The corresponding terms, as obtained by variation of functional (2.19), can be written in the form

$$T_1 \delta u_1 + S_1^* \delta u_2 + Q_1^* \delta w + M_1 \delta \gamma_1 = 0 \quad \text{for} \quad \alpha = \alpha_0, \tag{2.21}$$

where

$$S_1^* = S_1 + \frac{H}{R_2} = S + \frac{2H}{R_2}, \quad Q_1^* = Q_1 - \frac{1}{B}\frac{\partial H}{\partial \beta}. \tag{2.22}$$

We shall consider only variants of boundary conditions for which the generalized displacements u_1, u_2, w, γ_1 (and hence, their variations) or the corresponding generalized forces T_1, S_1^*, Q_1^*, M_1 vanish on the edge:

$$\begin{array}{llcll} u_1 = 0 & (1) & \text{or} & T_1 = 0 & (0), \\ u_2 = 0 & (1) & \text{or} & S_1^* = 0 & (0), \\ w = 0 & (1) & \text{or} & Q_1^* = 0 & (0), \\ \gamma_1 = 0 & (1) & \text{or} & M_1 = 0 & (0). \end{array} \tag{2.23}$$

The other variants of boundary conditions are not considered. As a result, one obtains 16 variants of boundary conditions, which for brevity will be denoted by four-digit binary numbers with the same order as in (2.23). We put 1 or 0, if the generalized displacement or force, respectively, is equal to zero. For example, the clamped edge $u_1 = u_2 = w = \gamma_1 = 0$ is denoted by 1111, the simply supported edge $T_1 = u_2 = w = M_1 = 0$ is identified by 0110, and the free edge $T_1 = S^* = Q_1^* = M_1 = 0$ has the notation 0000.

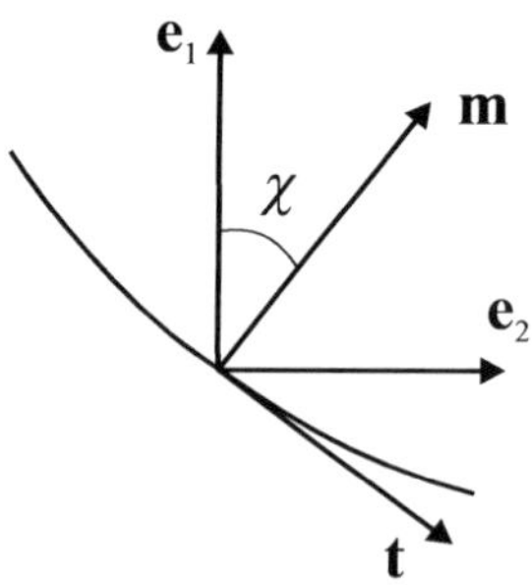

FIGURE 2.4: The slant edge.

Now, we consider the boundary conditions at a shell edge which does not coincide with the coordinate line (see Fig. 2.4). Let **t** be the unit vector tangential to the edge, **m** be the unit normal vector lying in the tangent plane and directed inside of the shell, and χ be the angle between the vectors **m** and $\mathbf{e}_1$. In this setting, the generalized displacements are as follows:

$$u_m = u_1 \cos\chi + u_2 \sin\chi, \quad u_t = -u_1 \sin\chi + u_2 \cos\chi,$$
$$w, \quad \gamma_m = \gamma_1 \cos\chi + \gamma_2 \sin\chi. \tag{2.24}$$

The corresponding generalized stress resultants T_m, S_t, Q_m and stress couples M_m read [29]:

$$\begin{aligned}
&T_m = T_1 \cos^2\chi + (S_1 + S_2)\sin\chi\cos\chi + T_2 \sin^2\chi \\
&-H_{mt}\left(\frac{1}{R_1} - \frac{1}{R_2}\right)\sin\chi\, \cos\chi, \\
&S_t = (T_2 - T_1)\sin\chi\, \cos\chi + S_1\cos^2\chi - S_2 \sin^2\chi \\
&+H_{mt}\left(\frac{\sin^2\chi}{R_1} + \frac{\cos^2\chi}{R_2}\right), \\
&Q_m = \left(Q_1 - \frac{1}{B}\frac{\partial H_{mt}}{\partial\beta}\right)\cos\chi + \left(Q_2 - \frac{1}{A}\frac{\partial H_{mt}}{\partial\alpha}\right)\sin\chi, \\
&M_m = M_1\cos^2\chi + 2H\sin\chi\, \cos\chi + M_2\sin^2\chi,
\end{aligned} \tag{2.25}$$

where

$$H_{mt} = H\left(\cos^2\chi - \sin^2\chi\right) + (M_2 - M_1)\sin\chi\, \cos\chi\,. \tag{2.26}$$

As for the edge $\alpha = \alpha_0$, we shall consider variants of boundary conditions for which the generalized displacements (2.24) or the corresponding generalized forces (2.25) vanish:

$$\begin{aligned}
&u_m = 0 \quad (1) \quad \text{or} \quad T_m = 0 \quad (0), \\
&u_t = 0 \quad (1) \quad \text{or} \quad S_t^* = 0 \quad (0), \\
&w = 0 \quad (1) \quad \text{or} \quad Q_m^* = 0 \quad (0), \\
&\gamma_m = 0 \quad (1) \quad \text{or} \quad M_m = 0 \quad (0).
\end{aligned} \tag{2.27}$$

In (2.23) and (2.27), the boundary conditions in the left-hand columns (marked by 1) impose geometric restrictions on shell displacements. We call all these conditions as *constraints*. In the right-hand columns, there are conditions (marked by 0) containing stress resultants and stress couples.

2.4 Governing equations. Membrane equations

We shall neglect the inertia forces of the rotational motion in the fourth and fifth equations of (2.14). Then system (2.14) can be reduced to the following three differential equations written in terms of displacements:

$$\sum_{j=1}^{3} \left(\mu^4 N_{ij} u_j + L_{ij} u_j\right) - \frac{1}{\omega_0^2} \frac{\partial^2 u_i}{\partial t^2} = 0, \quad i = 1, 2, 3. \tag{2.28}$$

Here, N_{ij} and L_{ij} are linear differential operators with respect to the variables α and β. It is convenient to consider system (2.28) in the dimensionless form, when all the linear dimensions h, $Ad\alpha$, $Bd\beta$, R_1, R_2 are related to the characteristic dimension R of the middle surface, and the forces T_k, S_k, Q_k and moments M_k, H are related to Eh and EhR, respectively. The dimensionless thickness, which is considered as the principal small parameter, is denoted by h_*; for the remaining quantities, we shall retain the same notation after conversion to the dimensionless form. In system (2.28), $\mu > 0$ is a small parameter, $\omega_0 > 0$ is the characteristic frequency of vibrations. Besides,

$$\mu^4 = \frac{h_*^2}{12(1-\nu^2)}, \qquad h_* = \frac{h}{R}, \qquad \omega_0^2 = \frac{E}{R^2 \rho}. \tag{2.29}$$

When examining the spectrum of free vibrations, one can separate variables in system (2.28):

$$\mathbf{U}(\alpha, \beta, t) = \mathbf{U}(\alpha, \beta) e^{i\omega t}, \quad \mathbf{U} = \{u_1, u_2, u_3\}. \tag{2.30}$$

Then we come to the boundary-value problem involving the system of equations

$$\sum_{j=1}^{3} \left(\mu^4 N_{ij} u_j + L_{ij} u_j\right) + \lambda u_i = 0, \quad i = 1, 2, 3, \quad \lambda = \frac{\omega^2}{\omega_0^2} = \frac{R^2 \rho \omega^2}{E} \tag{2.31}$$

and four boundary conditions on each edge of the shell.

For the boundary conditions listed in Section 2.3, the above boundary-value problem is self-adjoint (see [13, 64]). It has real non-negative discrete spectrum with an accumulation point at infinity.

The frequencies and modes of free vibrations can be found from the variational problem

$$\lambda = \min_{u_1, u_2, u_3} \frac{\Pi'_\varepsilon + h_*^2 \Pi'_\kappa}{T'}, \tag{2.32}$$

where

$$\Pi'_\varepsilon = \frac{1}{2}\int\int_\Omega \frac{1}{1-\nu^2}\left((\varepsilon_1^2 + 2\nu\varepsilon_1\varepsilon_2 + \varepsilon_2^2) + \frac{1-\nu}{2}\omega^2\right) AB\,d\alpha d\beta,$$
$$\Pi'_\kappa = \frac{1}{2}\int\int_\Omega \frac{1}{12(1-\nu^2)}\left(\kappa_1^2 + 2\nu\kappa_1\kappa_2 + \kappa_2^2 + 2(1-\nu)\tau^2\right) AB\,d\alpha d\beta,$$
$$T' = \frac{1}{2}\int\int_\Omega \left(u_1^2 + u_2^2 + u_3^2\right) AB\,d\alpha d\beta, \tag{2.33}$$

and the displacements u_1, u_2, u_3 satisfy the geometric boundary conditions; that is, the constraints imposed on the generalized displacements. Note that the force boundary conditions are natural for the functional (2.32).

The eighth-order system (2.31) is singularly perturbed. For $\mu = 0$, this system degenerates into the fourth-order *membrane system*

$$\sum_{j=1}^{3} L_{ij}u_j + \lambda u_i = 0, \quad i = 1, 2, 3. \tag{2.34}$$

Here, instead of four boundary conditions on each edge of the shell, one should specify two (tangential) boundary conditions, which impose constraints on the tangential displacement u_1, u_2 or the membrane stress resultants T_1, T_2, S.

The spectrum of the membrane boundary-value problem (2.34) is much more complicated than that for the full problem (2.31) (see [13,64]). It contains a segment of the continuous spectrum $\lambda \in \Omega$, where Ω is the range of the function

$$\Omega(\theta, \alpha, \beta) = (1-\nu^2)\left(\frac{\sin^2\theta}{R_1^2(\alpha,\beta)} + \frac{\cos^2\theta}{R_2^2(\alpha,\beta)}\right), \qquad 0 \le \theta \le 2\pi, \quad \{\alpha, \beta\} \in G. \tag{2.35}$$

Outside the interval Ω the spectrum of the membrane problem is discrete, the end-points of the interval Ω may be the accumulation points of the discrete spectrum. As for the full problem, $\lambda = \infty$ is an accumulation point for the spectrum.

The explicit form of the differential operators L_{ij} and N_{ij} may be found in [13]. The operators $L_{ij} = (1-\nu^2)L'_{ij}$ and N_{33}, which will be used in what follows, are given by

$$
\begin{aligned}
L'_{11}u_1 &= \frac{1}{A}\frac{\partial}{\partial\alpha}\left(\frac{1}{AB}\frac{\partial(Bu_1)}{\partial\alpha}\right) + \frac{1-\nu}{2B}\frac{\partial}{\partial\beta}\left(\frac{1}{AB}\frac{\partial(Au_1)}{\partial\beta}\right) + \frac{1-\nu}{R_1R_2}u_3,\\
L'_{12}u_2 &= \frac{1}{A}\frac{\partial}{\partial\alpha}\left(\frac{1}{AB}\frac{\partial(Au_2)}{\partial\beta}\right) - \frac{1-\nu}{2B}\frac{\partial}{\partial\beta}\left(\frac{1}{AB}\frac{\partial(Bu_2)}{\partial\alpha}\right),\\
L'_{13}u_3 &= -\frac{1}{A}\frac{\partial}{\partial\alpha}\left(\left(\frac{1}{R_1}+\frac{1}{R_2}\right)u_3\right) + \frac{1-\nu}{AR_2}\frac{\partial u_3}{\partial\alpha},\\
L'_{31}u_1 &= \frac{1}{AB}\left(\frac{1}{R_1}+\frac{1}{R_2}\right)\frac{\partial(Bu_1)}{\partial\alpha} - \frac{1-\nu}{AB}\frac{\partial}{\partial\alpha}\left(\frac{Bu_1}{R_2}\right),\\
L'_{33}u_3 &= -\left(\frac{1}{R_1^2}+\frac{1\nu}{R_1R_2}+\frac{1}{R_2^2}\right)u_3, \quad N_{33} = \Delta\Delta w + N_{33}^{*}w,\\
\Delta w &= \frac{1}{AB}\left(\frac{\partial}{\partial\alpha}\left(\frac{B}{A}\frac{\partial w}{\partial\alpha}\right) + \frac{\partial}{\partial\beta}\left(\frac{A}{B}\frac{\partial w}{\partial\beta}\right)\right).
\end{aligned}
\tag{2.36}
$$

The formulas for the remaining operators L'_{ij} are obtained from equations given in (2.36) by the cyclic permutation (1,2; α,β; A,B). Here, Δ is the Laplace operator in the curvilinear coordinates α,β, N_{33}^{*} is a differential operator of the second order, and the order of operators N_{ij} with $i+j<6$ is at most three. These operators will not be involved in the subsequent analysis.

2.5 Qualitative analysis of the frequency spectrum for free vibrations

In the qualitative analysis of system (2.31), we shall take into account that the shell is relatively thin-walled; this is described by the small parameter $h_* = h/R$. We shall also assume that this system does not involve any other small or large parameters, except the frequency parameter λ. In particular, we shall consider shells of average length; that is, we assume that the shell is not too long or too short. A complete classification of integrals for the differential equations of small free vibrations of thin shells is given in [64,66], which also contain, as a corollary to this classification, a qualitative analysis of the frequency spectrum and modes of vibration.

The subsequent qualitative analysis is based on using the concepts of the index of variation ς introduced in [65]. We assume that

$$
\max\left\{\left|\frac{\partial F}{\partial\alpha}\right|, \left|\frac{\partial F}{\partial\beta}\right|\right\} \sim h_*^{-\varsigma}F, \tag{2.37}
$$

TABLE 2.1: Classification of solutions

(1)	quasi-transverse solutions with small variability	$0 < \varsigma < 1/2,$	$r = 0,$	$w_* \gg u_*,$
(2)	quasi-tangential solutions	$0 < \varsigma < 1,$	$r = \varsigma,$	$w_* \ll u_*,$
(3)	Rayleigh-type solutions	$0 \leq \varsigma < 1/2,$	$r = -1 + 2\varsigma,$	$w_* \gg u_*,$
(4)	quasi-transverse solutions with large variability	$1/2 \leq \varsigma < 1,$	$r = -1 + 2\varsigma,$	$w_* \gg u_*.$

where F is any unknown function which determines the dynamic stress-strain state of the shell. We also introduce the parameter $r < 1$, which determines the asymptotic order of the dimensionless frequency of vibrations as follows

$$\lambda \sim h_*^{-2r}, \qquad \frac{\omega}{\omega_0} \sim h_*^{-r}, \tag{2.38}$$

where ω is the natural frequency of free vibrations. The wave length L on the shell middle surface is coupled with the parameter ς by the relation $L \sim Rh_*^{\varsigma}$. Two-dimensional models of the theory of thin shells can be used only under the condition that $L \gg h$, which is equivalent to the restriction $\varsigma < 1$. For $\varsigma = 1$, the wave length of deformations becomes commensurable with the shell thickness.

In a number of cases it proves convenient to introduce different indexes of variation, ς_1 and ς_2, in the α- and β-directions, respectively:

$$\frac{\partial F}{\partial \alpha} \sim h_*^{-\varsigma_1} F, \qquad \frac{\partial F}{\partial \beta} \sim h_*^{-\varsigma_2} F, \qquad \varsigma = \max\{\varsigma_1, \varsigma_2\}. \tag{2.39}$$

In this case, ς_1 and ς_2 are called the partial indexes of variation, and ς is the general index.

The indexes of variation allow one to compare the orders of separate terms in equations of Sections 2.2–2.4 and to find the consistent values of the parameters ς, r on the assumption that at least two terms in each equation be of maximal order. Possible relations between the parameters ς and r are shown in Fig. 2.5 by lines, which correspond to four principal classes of solutions of system (2.31) (see Table 2.1).

The strong inequalities in Table 2.1 indicate the relations between the normal deflections $u_3 = w$ and the projections u_1, u_2 of displacements onto the tangential directions:

$$w_* = \max\{|w|\}, \qquad u_* = \max\{|u_1|, |u_2|\}. \tag{2.40}$$

These inequalities hold only for $\varsigma > 0$. If $\varsigma = 0$, then for the solutions of classes (1), (2) and (3), the strong inequalities in Table 2.1 should be replaced

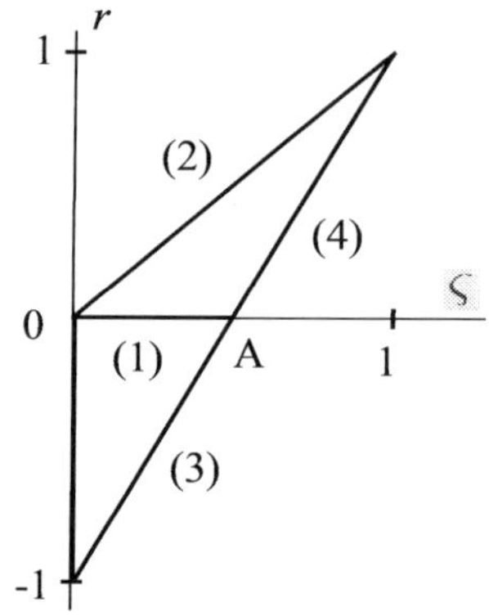

FIGURE 2.5: The asymptotic frequency portrait of free vibrations.

by $w_* \sim u_*$; that is, the normal deflection and the tangential displacement have the same asymptotic order.

To each of these classes there corresponds its own simplified system. The solutions of classes (1) and (2) satisfy the dynamic membrane system of differential equations of the forth order.

The Rayleigh-type solutions (3) obey the dynamic equations of the pure bending stress state [67], the inextensibility conditions $\varepsilon_1 = \varepsilon_2 = \omega = 0$ of the middle surface being satisfied exactly or approximately. These solutions are responsible for extremely low frequencies of vibrations.

The solutions of class (4) (for $\varsigma > 1/2$) satisfy in the principal terms the system of equations for transverse vibrations of a plate.

The cases $\varsigma = 0$ and $\varsigma = 1/2$ deserve special consideration. For $\varsigma = r = 0$, separate construction of the solutions of classes (1) and (2) turns out to be impossible, and for $\varsigma = 1/2$, $r = 0$ the solutions of classes (1), (3) and (4) cannot be separated.

For the analysis of the last case (see point A in Fig. 2.5), one can use the system of equations of the technical shell theory (called also as the theory of shallow shells):

$$D\Delta\Delta w + \rho h \frac{\partial^2 w}{\partial t^2} - \Delta_R \Phi = 0, \qquad \frac{1}{Eh}\Delta\Delta\Phi + \Delta_R w = 0, \tag{2.41}$$

where Δ is the Laplace operator in the curvilinear coordinates (2.36), Δ_R is the second-order operator depending on the shell curvature,

$$\Delta_R w = \frac{1}{AB}\left(\frac{\partial}{\partial \alpha}\left(\frac{B}{AR_2}\frac{\partial w}{\partial \alpha}\right) + \frac{\partial}{\partial \beta}\left(\frac{A}{BR_1}\frac{\partial w}{\partial \beta}\right)\right); \tag{2.42}$$

and the auxiliary force function Φ is coupled with the membrane stress resultants by the relations:

$$T_1 = \frac{1}{B}\frac{\partial}{\partial\beta}\left(\frac{1}{B}\frac{\partial\Phi}{\partial\beta}\right) + \frac{1}{A^2B}\frac{\partial B}{\partial\alpha}\frac{\partial\Phi}{\partial\alpha},$$
$$T_2 = \frac{1}{A}\frac{\partial}{\partial\alpha}\left(\frac{1}{A}\frac{\partial\Phi}{\partial\alpha}\right) + \frac{1}{AB^2}\frac{\partial A}{\partial\beta}\frac{\partial\Phi}{\partial\beta}, \tag{2.43}$$
$$S = -\frac{1}{AB}\left(\frac{\partial^2\Phi}{\partial\alpha\partial\beta} - \frac{1}{A}\frac{\partial A}{\partial\beta}\frac{\partial\Phi}{\partial\alpha} - \frac{1}{B}\frac{\partial B}{\partial\alpha}\frac{\partial\Phi}{\partial\beta}\right).$$

The system of equations (2.41) with a specified (below) error can be used for constructing solutions of classes (1), (3) and (4) not only for $\varsigma = 1/2$, $r = 0$, but also in a wider range.

Each of the four classes of solutions from Table 2.1 generates the corresponding frequency spectrum of free vibrations. This may also result in the appearance of "internal resonances", when frequencies generated by solutions from different classes are equal (or close). In this case, the construction of the vibration mode becomes complicated.

As a rule, the Rayleigh-type solutions with $r < 0$ do not satisfy the given boundary conditions. In this case the additional edge effect solutions decreasing out of the shell edges are used. Near the edge $\alpha = \alpha_0$ these solutions satisfy the equation

$$D\frac{d^4w}{d\alpha^4} + \frac{Eh}{R_2^2}w = 0 \tag{2.44}$$

and have the index of variation $\varsigma = 1/2$.

2.6 Dynamic equations for pre-stressed shells

In subsequent chapters, we shall consider free vibrations, parametric dynamic response and wave motions of pre-stressed shells. Governing equations for these shells are readily obtained from non-linear equations by means of the linearization in the neighborhood of the initial stress state (which may be non-stationary). It should be noted that the procedure of deducing these equations is identical to that used to derive the buckling equations (e.g., see in [222, 226]) and is not given here.

Let the edge forces and surface load $\mathbf{q}^\circ = q_1^\circ\mathbf{e}_1 + q_2^\circ\mathbf{e}_2 + q_n^\circ\mathbf{n}$ be such that the following conditions are valid:

- if the external forces are not stationary, then the characteristic time of their variation is much greater than the minimum period of free vibrations;

- the displacements $u_1^\circ, u_2^\circ, w^\circ$ and the membrane stress resultants $T_1^\circ, T_2^\circ, S^\circ$ describing the pre-stressed state are weakly varying functions of α, β;

- the deformation of the neutral (middle) surface is far from bending;

- the maximum value of the non-linear membrane strains satisfies the inequality

$$\max\left\{|\varepsilon_{11}^\circ|, |\varepsilon_{22}^\circ|, |\varepsilon_{12}^\circ|\right\} \lesssim h_*, \tag{2.45}$$

  where

$$\begin{aligned} \varepsilon_{ii}^\circ &= \varepsilon_i^\circ + \tfrac{1}{2}\left(\varepsilon_i^{\circ 2} + \omega_i^{\circ 2} + \gamma_i^{\circ 2}\right), \quad i = 1, 2, \\ \varepsilon_{12}^\circ &= \omega^\circ + \varepsilon_1^\circ \omega_2^\circ + \varepsilon_2^\circ \omega_1^\circ + \gamma_1^\circ \gamma_2^\circ, \end{aligned} \tag{2.46}$$

  and all the magnitudes with superscript $^\circ$ are determined by Eqs. (2.8), (2.9), where u_1, u_2, w have to be substituted by $u_1^\circ, u_2^\circ, w^\circ$, respectively.

The first condition presupposes that the load may be considered as quasi-static, and the remained ones mean that "the shell is stressed, but not strained" [3]. Then the impact of the quasi-static variable load $\mathbf{q}^\circ(\alpha_1, \alpha_2, t)$ on the shell dynamics can be taken into account by incorporating the additional term

$$\Delta_T w = \frac{1}{AB}\left[\frac{\partial}{\partial \alpha}\left(\frac{BT_1^\circ}{A}\frac{\partial w}{\partial \alpha} + S^\circ \frac{\partial w}{\partial \beta}\right) + \frac{\partial}{\partial \beta}\left(\frac{AT_2^\circ}{B}\frac{\partial w}{\partial \beta} + S^\circ \frac{\partial w}{\partial \alpha}\right)\right] \tag{2.47}$$

into the third equation of system (2.28). The neglect of additional terms related to the initial stresses in the first two equations results in an error comparable with that of the initial equations.

After introducing the initial stresses, equations of the technical shell theory to be frequently used in what follows will take the form

$$D\Delta\Delta w - \Delta_T w - \Delta_R \Phi + \rho h \frac{\partial^2 w}{\partial t^2} = 0, \qquad \frac{1}{Eh}\Delta\Delta\Phi + \Delta_R w = 0. \tag{2.48}$$

The initial stress resultants $T_1^\circ, T_2^\circ, S^\circ$ are found from the coupled equations of the membrane shell theory

$$\begin{aligned} &\frac{1}{AB}\left[\frac{\partial(BT_1^\circ)}{\partial \alpha} - \frac{\partial B}{\partial \alpha}T_2^\circ + \frac{\partial(A^2 S^\circ)}{A\partial \beta}\right] + q_1^\circ = 0, \\ &\frac{1}{AB}\left[\frac{\partial(AT_2^\circ)}{\partial \beta} - \frac{\partial A}{\partial \beta}T_1^\circ + \frac{\partial(B^2 S^\circ)}{B\partial \alpha}\right] + q_2^\circ = 0, \\ &\frac{T_1^\circ}{R_1} + \frac{T_2^\circ}{R_2} - q_n^\circ = 0. \end{aligned} \tag{2.49}$$

The problem on the existence of the membrane stress state in a thin shell was discussed in papers [67, 173].

In some chapters, the dynamics of shells resting on an elastic isotropic foundation will be studied. The simplest model describing the response of an elastic foundation is the Winkler model. According to this model, the foundation response is proportional to the normal displacement,

$$q_f = -c_f w, \tag{2.50}$$

where c_f is the spring constant depending only on elastic properties of a surrounding medium. To take this response into account, force (2.50) should be introduced into the third equation of system (2.28). If the shell motion is predicted in the framework of the technical shell theory, then Eqs. (2.41) or (2.48) are to be also supplemented by term (2.50). In particular, for an initially pre-stressed thin shells, one obtains the following equations:

$$\begin{aligned} &D\Delta\Delta w - \Delta_T w + c_f w - \Delta_R \Phi + \rho h \frac{\partial^2 w}{\partial t^2} = 0, \\ &\frac{1}{Eh}\Delta\Delta\Phi + \Delta_R w = 0. \end{aligned} \tag{2.51}$$

The detailed analysis of the effect of an elastic foundation on buckling and vibrations of thin plates and shells was given in book [89] and articles [223, 224]. These studies showed that the spring constant c_f is influenced by an expected number of waves on the shell surface. Let the transversal deflection be the doubly-periodic function of coordinates, $w = w_0 \sin k_1\alpha \sin k_2\beta$. Then, assuming the full contact between the shell and foundation, one has

$$c_f = \frac{2E_f(1-\nu_f)k}{(1+\nu_f)(3-4\nu_f)}, \quad k = \sqrt{k_1^2 + k_2^2}, \tag{2.52}$$

where E_f, ν_f are the Young's modulus and Poisson's ratio, respectively, of a surrounding elastic medium. It should be noted that formula (2.52) was derived for an infinite plate resting on an elastic half-space. Therefore, the range of applicability of Eq. (2.51) with c_f defined by (2.52) is restricted by the following conditions: 1) they are valid far from the plate (shell) edges; 2) a foundation has to be sufficiently deep; 3) forces of inertia of a foundation are not taking into account.

Equations (2.48), (2.51) are suitable for describing free vibrations accompanied by formation of a large number of short waves.

2.7 The Timoshenko–Reissner model for a transversely isotropic shell

Assume that a shell of constant thickness h is made of a transversely isotropic linearly elastic homogeneous material. The classic TR model [203,

219] is based on the assumption that after a deformation a normal fiber remains straight, but it is inclined to the neutral surface normal. In addition to projections, u_1, u_2, w, of displacement, two shear angles, δ_1, δ_2, are introduced. The shear angles are coupled with the angles φ_1, φ_2 of a normal fiber rotation as

$$\delta_1 = \varphi_1 - \gamma_1, \qquad \delta_2 = \varphi_2 - \gamma_2, \tag{2.53}$$

where the angles γ_1, γ_2 are given in Eqs. (2.8).

For the TR model the geometric relations and the equilibrium equations, as described in Sections 2.1 and 2.2, remain the same as for the KL model. The elasticity relations (2.15) for the tangential stress-resultants T_1, T_2, S_1, S_2 also remain the same. The remaining elasticity relations read as

$$\begin{gathered}
M_1 = D(\kappa_1 + \nu\kappa_2), \quad M_2 = D(\kappa_2 + \nu\kappa_1), \quad H_1 = H_2 = H = D(1-\nu)\tau, \\
\kappa_1 = -\frac{1}{A}\frac{\partial \varphi_1}{\partial \alpha} - \frac{1}{AB}\frac{\partial A}{\partial \beta}\varphi_2, \quad \kappa_2 = -\frac{1}{B}\frac{\partial \varphi_2}{\partial \beta} - \frac{1}{AB}\frac{\partial B}{\partial \alpha}\varphi_1, \\
2\tau = -\frac{B}{A}\frac{\partial}{\partial \alpha}\left(\frac{\varphi_2}{B}\right) - \frac{A}{B}\frac{\partial}{\partial \beta}\left(\frac{\varphi_1}{A}\right), \quad Q_1 = \Gamma\delta_1, \quad Q_2 = \Gamma\delta_2, \\
K = \frac{Eh}{1-\nu^2}, \quad D = \frac{Eh^3}{12(1-\nu^2)}, \quad \Gamma = G_{13}h.
\end{gathered} \tag{2.54}$$

Here Γ is the transversal shear stiffness, G_{13} is the transversal shear elastic modulus for a transversely isotropic material. The modulus G_{13} is independent of the moduli E, ν. In particular, the ratio G_{13}/E can be small. For an isotropic material $G_{13} = G_{12} = E/[2(1+\nu)]$.

As a rule, the face shell surfaces are free, and the classic TR hypotheses do not satisfy this condition. For this reason the correcting factor $k = 5/6$ in the elasticity relation $\Gamma = kG_{13}h$ was introduced [218, 229].

The equilibrium equations (2.14) with the elasticity relations (2.15), (2.54) give a system of 10th order differential equations with respect to the unknown functions $u_1, u_2, w, \varphi_1, \varphi_2$. Instead of Eqs. (2.23) for the KL model, the solution satisfies five boundary conditions. We will consider some of the following 32 variants at the edge $\alpha = \alpha_0$

$$\begin{array}{llcll}
u_1 = 0 & (1) & \text{or} & T_1 = 0 & (0), \\
u_2 = 0 & (1) & \text{or} & S_1 = 0 & (0), \\
w = 0 & (1) & \text{or} & Q_1 = 0 & (0), \\
\varphi_1 = 0 & (1) & \text{or} & M_1 = 0 & (0), \\
\varphi_2 = 0 & (1) & \text{or} & H = 0 & (0).
\end{array} \tag{2.55}$$

The natural frequencies can be found from the variational problem similar to (2.32)

$$\lambda = \frac{\rho R^2 \omega^2}{E} = \min_{u_1, u_2, u_3, \varphi_1, \varphi_2} \frac{\Pi'_\varepsilon + h_*^2 \Pi'_\kappa + \Pi'_\delta}{T'}, \tag{2.56}$$

where the generalized displacements, $u_1, u_2, u_3, \varphi_1, \varphi_2$, satisfy the conditions on the left of Eqs.(2.55) (if they are imposed). The values Π_ε and Π'_κ are the same as in (2.32) and

$$\begin{aligned} \Pi'_\delta &= \frac{\Gamma}{2Eh}\int\int_\Omega (\delta_1^2+\delta_2^2)AB\,d\alpha d\beta, \\ T' &= \frac{1}{2}\int\int_\Omega \left(u_1^2+u_2^2+u_3^2+J(\varphi_1^2+\varphi_2^2)\right)AB\,d\alpha d\beta, \quad J=\frac{h_*^2}{12}. \end{aligned} \tag{2.57}$$

For simplicity, consider the TR model for an transversally isotropic shell of revolution. In Eqs. (2.14) we take φ_1, φ_2 instead of γ_1, γ_2 and separate the variables

$$\begin{aligned} &Z_s(s,\varphi,t) = Z_s(s)\sin m\varphi e^{i\omega t} \quad \text{or} \quad Z_c(s,\varphi,t) = Z_c(s)\cos m\varphi e^{i\omega t}, \\ &Z_s = \{u, w, \varphi_1, T_1, T_2, M_1, M_2, Q_1, \delta_1, \gamma_1\}, \quad Z_c = \{v, \varphi_2, S, H, Q_2, \delta_2, \gamma_2\}, \end{aligned} \tag{2.58}$$

where the curvilinear co-ordinates s, φ are introduced in Eqs. (2.4), m is the number of waves in the circumferential direction. Now Eqs. (2.14) lead to a system of ordinary differential equations (ODE) of 10th order. For $m \gg 1$, $m \sim h_*^{-\varsigma_2}$ we seek for asymptotic solutions with positive indexes of variation, $\varsigma > 0$, of this system in the form

$$Z(s,h_*,m) = Z_0(s)(1+O(h_*^\varsigma))e^{\int p(s)ds}, \quad p(s)\sim h_*^{-\varsigma}. \tag{2.59}$$

Substituting solutions (2.59) into Eqs. (2.14) we get a characteristic equation for $p(s)$ in the dimensionless form

$$(\Delta_0 - g_1)\left[D_*\Delta_0^4 + (1-g\Delta_0))\left(\left(\frac{p^2}{R_2^2}-\frac{r^2}{R_1^2}\right)^2 - \lambda\Delta_0^2\right)\right] = 0, \tag{2.60}$$

with

$$\begin{aligned} &\Delta_0 = p^2 - r^2, \quad r = \frac{m}{B} \sim h_*^{-\varsigma_2}, \quad D_* = \frac{h_*^2}{12(1-\nu^2)} \sim h_*^2, \\ &\eta = \frac{Eh}{\Gamma} \sim h_*^{-\varsigma_3}, \quad g = D_*\eta \sim h_*^{2-\varsigma_3}, \quad g_1 = \frac{2}{(1-\nu)g} \sim h_*^{\varsigma_3-2}. \end{aligned} \tag{2.61}$$

Here η is the ratio between the tangential stiffness Eh and the transversal shear stiffness Γ, and we put $\eta \sim h_*^{-\varsigma_3}$. Here the values s, B, R_1, R_2 are related to R.

Equation (2.60) has two roots

$$p = \pm\sqrt{g_1 + r^2} \sim h_*^{-\varsigma}, \quad \varsigma = \max\{1-\varsigma_3/2, \varsigma_2\} \tag{2.62}$$

which correspond to a boundary layer. The boundary layer is used to satisfy the additional boundary condition $\varphi_2 = 0$. The rest of the 8 roots of Eq. (2.60) are close to the corresponding roots of the KL model if $g \ll 1$.

For an isotropic homogeneous material the following estimates have place: $\eta \sim 1$, $g \sim h_*^2$, $g_1 \sim h_*^{-2}$. From these estimates and Eq. (2.62) it follows that the boundary layer has the index of variation $\varsigma = 1$. It is in contradiction with the main principle of the 2D shell theories, in accordance to which the wave length in the tangential directions should be much larger than the shell thickness [225].

Let the normal deflection be $w = 1$. It is easy to verify that $\{\delta_1, \delta_2\} \sim h_*^2$. If we ignore the boundary layer, then the relative difference between the internal solutions obtained by the KL and the TR models has the order of h_*^2. Therefore, for isotropic homogeneous shell the TR model is contradictory and more complex, and is not more exact than the KL model.

In contrast to an isotropic material, the TR model for a transversely isotropic material with small shear stiffness Γ and $\eta \sim h_*^{-\varsigma_3}$, $\varsigma_3 > 0$ gives the results, which are sometimes much better than those obtained by the KL model. In particular, the boundary layer has the index of variation $\varsigma < 1$ (see Eq. (2.62)).

For the following analysis of the TR model, it is convenient to use the (approximate) technical shell theory [53, 251], similar to Eqs. (2.41) for the KL model. Instead of the angles φ_1 and φ_2 of a normal fibers rotation we introduce the new unknown functions Ψ, Θ defined by

$$\varphi_1 = -\frac{1}{A}\frac{\partial \Psi}{\partial \alpha} + \frac{1}{B}\frac{\partial \Theta}{\partial \beta}, \qquad \varphi_2 = -\frac{1}{B}\frac{\partial \Psi}{\partial \beta} - \frac{1}{A}\frac{\partial \Theta}{\partial \alpha}. \tag{2.63}$$

Then, after simplifying Eqs. (2.14), the technical shell theory equations for the TR model in the dimensionless form read as

$$\Delta\Theta - g_1\Theta = 0, \tag{2.64}$$

$$\begin{aligned} &\Delta\Delta\Phi_* + \Delta_R w_* = 0, \\ &D_*\Delta\Delta\Psi - \Delta_R\Phi_* - \lambda w_* = 0, \\ &w_* = \Psi - g\Delta\Psi, \end{aligned} \tag{2.65}$$

with

$$\begin{aligned} &g = D\eta, \quad \eta = \frac{K_0}{\Gamma}, \quad g_1 = \frac{2}{g(1-\nu)}, \quad K_0 = Eh, \\ &D = \frac{h_*^2}{12(1-\nu^2)}, \quad \Phi_* = \frac{\Phi}{K_0}, \quad w_* = \frac{w}{R}, \quad \lambda = \frac{\rho R^2\omega^2}{E}. \end{aligned} \tag{2.66}$$

Here the operators Δ and Δ_R are the same as in Eqs. (2.41). Equation (2.64) describes the boundary layer, and the system (2.65) describes the internal stress-strain state (SSS). The characteristic equation of Eqs. (2.64), (2.65) coincides with Eq. (2.60). Equations (2.65) (likewise Eqs. (2.41)) are acceptable only to describe the SSS with the large variability ($\varsigma > 0$).

Consider a circular cylindrical shell of (dimensionless) length L and radius $R = 1$ made of a transversely isotropic material. Let the shell edges $s = 0$ and $s = L$ be simply supported. Then the transversely vibration modes are

$$w_{nm}(s, \varphi) = w_0 \sin p_n s \sin m\varphi, \qquad n = 1, 2, \dots, \qquad m = 2, 3, \dots \tag{2.67}$$

and Eqs. (2.65) give the corresponding eigenvalues

$$\lambda_{nm} = \frac{\rho R^2 \omega_{nm}^2}{E} = \frac{D_* \Delta^2}{1 + g\Delta} + \frac{\Delta_R^2}{\Delta^2} = \frac{D_*(p_n^2 + m^2)^2}{1 + g(p_n^2 + m^2)} + \frac{p_n^4}{(p_n^2 + m^2)^2} \tag{2.68}$$

where

$$\Delta = p_n^2 + m^2, \quad \Delta_R = \frac{p_n^2}{R_2^2} + \frac{m^2}{R_1^2} = p_n^2, \quad p_n = \frac{n\pi}{L}. \tag{2.69}$$

For small values, $g\Delta = q(p_n^2 + m^2)$, we get $\Psi \approx w$, and Eqs. (2.65) coincide with Eqs. (2.41). In this case the transversal shear may be neglected. The transversal shear effect (described in Eq. (2.68) by the term $1 + g\Delta$ compared with 1) is comparatively small if

$$\varsigma_3 < \min\{1, 2 - 2\varsigma_2\} \quad \text{or in the original notation} \quad \frac{K_0}{\Gamma} \gg \max\{h_*, m^2 h_*^2\}. \tag{2.70}$$

In the opposite case the transversal shear influence is prevailing.

It is easy to generalize Eqs. (2.65) to the case of the initial pre-stressed state. Let the initial stress resultants T_1^0, T_2^0, S^0 be given. Then the second Eq. (2.65) should be replaced by

$$D_* \Delta\Delta\Psi - \Delta_R \Phi_* - \lambda w_* - \Delta_T w_* = 0, \quad \Delta_T w_* = t_1 \kappa_1 + 2t_{12}\tau + t_2 \kappa_2, \tag{2.71}$$

with $\{T_1^0, T_2^0, S^0\} = Eh\{t_1, t_2, t_{12}\}$.

In particular, for the studied cylindrical shell pre-stressed by the initial stresses T_1^0, T_2^0 Eq. (2.68) reads as

$$\frac{\rho R^2 \omega_{nm}^2}{E} = \frac{D_*(p_n^2 + m^2)^2}{1 + g(p_n^2 + m^2)} + \frac{p_n^4}{(p_n^2 + m^2)^2} + t_1 p_n^2 + t_2 m^2. \tag{2.72}$$

For negative values, t_1 and/or t_2, the right-hand side of Eq. (2.72) can be negative. In this case a shell buckling takes place.

2.8 Timoshenko–Reissner shell inhomogeneous in the thickness direction

In addition to the assumptions accepted in the previous section, we shall assume that the elastic moduli of a transversely isotropic shell depend on the transversal co-ordinate z, $0 \le z \le h$. For a multi-layer shell the moduli are piecewise functions of z. We assume that there is a full contact between layers.

We seek a homogeneous shell that is equivalent with respect to elastic properties of an inhomogeneous shell under consideration. This problem has various approximate solutions. One of them is based on the asymptotic expansions of solutions of 3D equations of the theory of elasticity [166,167,230,231] and it is presented here. The other is described in the next Section. For a transversely isotropic inhomogeneous in the thickness direction plate (in particular, for a multilayer plate), the parameters of an equivalent single layered plate read as follows:

$$K_0 = \int_0^h E(z)dz, \quad D = \int_0^h E_*(z)(z-a)^2 dz, \quad K = \int_0^h E_*(z)dz,$$
$$E_*(z) = \frac{E(z)}{1-\nu^2(z)}, \quad \frac{1}{\Gamma} = \frac{1}{D^2}\int_0^h \frac{\phi^2(z)dz}{G_{13}(z)}, \quad \phi(z) = \int_0^z E_*(\zeta)(\zeta - a)d\zeta,$$
$$a = \frac{1}{K}\int_0^h E_*(z)z\,dz, \quad g = \frac{D}{R^2\Gamma}, \quad \rho_0 = \int_0^h \rho(z)dz. \tag{2.73}$$

Here $E(z)$, $\nu(z)$, $G_{13}(z)$ the are given elastic moduli of a transversely isotropic inhomogeneous material, $\rho(z)$ is the mass density. Equations (2.73) give, respectively, the equivalent elastic stiffness K_0, D and Γ, for a plate, and the mass density per unit areas ρ_0 . We accept approximately that the same values have stiffness for a shell in Eqs. (2.65). The value $z = a$ is the co-ordinate of a neutral surface that for a inhomogeneous shell may not coincide with the middle surface, $z = h/2$.

In [228], the error of the approximate solution was estimated for a multi-layer plate; besides, it was established that the error is small enough for large limits of heterogeneity. Here we study a test example for which the exact numerical solution was found.

Example 2.1. Consider a three-layer circular cylindrical shell of dimensionless radius $R = 1$, length $L = 4$, thickness $h = 0.01$. The thicknesses of layers are $h_1 = 0.12h$, $h_2 = 0.8h$, $h_3 = 0.08h$ $(h = h_1 + h_2 + h_3)$. All the layers are made of a homogeneous isotropic material. The parameters of the outer (hard) layers are $E_1 = E_3 = 1$, $\nu_1 = \nu_3 = 0.3$. Consider four variants of parameters of the intermediate (soft) layer

$$E_2 = 10^{-i}, \quad \nu_2 = 0.3 + 0.05i, \quad \rho_2 = 1 - 0.3i, \quad i = 0, 1, 2, 3. \tag{2.74}$$

With $i = 0$ this material is homogeneous.

TABLE 2.2: Dependence of the approximate values ω_{nm}^a on the wave numbers n and m and on the level of heterogeneity E_1/E_2.

		$E_2 = 1$		$E_2 = 0.1$		$E_2 = 0.01$		$E_2 = 0.001$	
n	m	δ_{nm}	g_*	δ_{nm}	g_*	δ_{nm}	g_*	δ_{nm}	g_*
	2	−12.8	1.0	−12.9	1.0	−12.9	1.0	−12.9	1.1
1	5	−5.1	1.0	−5.2	1.0	−5.2	1.0	−5.3	1.3
	10	−0.8	1.0	−0.8	1.0	−0.8	1.1	−0.7	2.2
	20	0.2	1.0	0.2	1.1	0.3	1.5	1.3	5.9
	2	−2.8	1.0	−2.8	1.0	−2.8	1.0	−2.7	1.2
5	5	−2.8	1.0	−3.0	1.0	−2.9	1.1	−2.8	1.5
	10	−1.3	1.0	−1.4	1.0	−1.3	1.1	−1.2	2.1
	20	−0.5	1.0	−0.7	1.1	−0.5	1.3	−0.2	4.3
	2	−0.4	1.0	−0.6	1.1	−0.5	1.3	−0.1	4.1
20	5	−0.5	1.0	−0.7	1.0	−0.5	1.3	−0.2	4.3
	10	−0.6	1.0	−0.7	1.0	−0.5	1.4	0.0	5.3
	20	−0.2	1.0	−0.2	1.1	0.1	1.8	1.5	9.0

Let the shell edges, $s = 0$ and $s = L$, be simply supported. Then the transversely vibration modes are

$$w_{nm}(s, \varphi) = w_0 \sin p_n s \sin m\varphi, \qquad n = 1, 2, \ldots, \qquad m = 2, 3, \ldots \tag{2.75}$$

Equation (2.68) for the approximate frequency $(\omega_{nm}^a)^2$ yields

$$\frac{\rho_0(\omega_{nm}^a)^2}{K_0} = \frac{(D/K_0)(p_n^2 + m^2)^2}{1 + g(p_n^2 + m^2)} + \frac{p_n^4}{(p_n^2 + m^2)^2}, \quad n = 1, 2, \ldots, \quad m = 2, 3, \ldots, \tag{2.76}$$

where the parameters D, Γ, K_0, ρ_0 are given in Eqs. (2.73).

The relative error δ_{nm} of the of the approximate frequencies ω_{nm}^a

$$\delta_{nm} = \frac{\omega_{nm}^a - \omega_{nm}}{\omega_{nm}} \cdot 100\% \tag{2.77}$$

is given in Table 2.2 versus the exact ones ω_{nm}.

Firstly, Table 2.2 shows that the error δ_{nm} does not change essentially when the level of heterogeneity $E_1/E_2 = 1/\eta$ increases, instead of the shear parameter $g_* = 1 + g(p_n^2 + m^2)$ is not close to 1. Secondly, the error δ_{nm} decreases with increasing variability $\sqrt{p_n^2 + m^2}$. The error δ_{12} of the natural frequency is large. Mainly, the large error in the cases of small variability is connected with the fact that in the technical shear models (2.41) and (2.65) the tangential inertia forces are excluded from consideration.

The studied model describes only the bending part of spectrum. The longitudinal modes and high-frequency thickness modes are not studied.

2.9 The equivalent single layer model for laminated transversally isotropic cylindrical shells

In this section, we consider the equivalent single layer (ESL) model, developed by Grigoliuk and Kulikov [80], for laminated cylindrical shells.

Let a thin laminated cylindrical shell consist of N transversely isotropic layers characterized by thickness h_n, density ρ_n, Young's modulus E_n, shear modulus G_n, and Poisson's ratio ν_n, where $n = 1, 2, \ldots, N$. The middle surface of any fixed layer is taken as the reference surface. We introduce the local orthogonal coordinate system with origin at the point O and unit vectors $\boldsymbol{e}_1, \boldsymbol{e}_2$, $\boldsymbol{n} = \boldsymbol{e}_1 \times \boldsymbol{e}_2$ as shown in Fig. 2.1. Let α and β be the axial and circumferential coordinates, respectively, and z be the normal coordinate. The radius of curvature of the reference surface is R_2. The shell is bounded by two not necessarily plane edges

$$L_1(\beta) \leq \alpha \leq L_2(\beta) \tag{2.78}$$

and may be not closed in the β- direction (the case of a non-circular cylindrical panel).

We introduce the additional notations: $z = z_n$ is the coordinate of the upper bound of the nth layer, $z = z_0$ is the coordinate of the inner surface, u_1, u_2 and w are the in-plane (tangential) and transversal (normal) displacements of points on the reference surface of the shell, respectively, $h = \sum_{n=1}^{N} h_n$ is the total thickness of the laminate, $u_1^{(n)}$, $u_2^{(n)}$ are the tangential displacements of points of the nth layer, and σ_{n3} are the transverse shear stresses in the nth layer.

The following hypotheses firstly stated in [80] are assumed here:

1. The distribution law of the transverse tangent stresses across the thickness of the nth layer is assumed to be of the form

$$\sigma_{i3} = f_0(z)\mu_i^{(0)}(\alpha, \beta, t) + f_n(z)\mu_i^{(n)}(\alpha, \beta, t) \quad i = 1, 2, \tag{2.79}$$

where t is time, $f_0(z), f_n(z)$ are defined continuous functions satisfying the following conditions:

$$\begin{aligned} &f_0(z_0) = f_0(z_N) = 0, \quad f_n(z_{n-1}) = f_n(z_n) = 0, \\ &f_n(z) = 0 \quad \text{if} \quad z \notin [z_{n-1}, z_n]. \end{aligned} \tag{2.80}$$

2. The normal stresses σ_{33} are negligible with respect to other components of the stress tensor.

3. The transversal displacement $w(\alpha, \beta, t)$ does not depend on z.

4. The in-plane displacements are distributed across thickness according to the generalized kinematic hypothesis of Timoshenko:

$$u_i^{(n)}(\alpha, \beta_2, z, t) = u_i(\alpha, \beta, t) + z\gamma_i(\alpha, \beta, t) + \mathcal{G}(z)\psi_i(\alpha, \beta, t), \qquad (2.81)$$

where $\mathcal{G}(z) = \int\limits_0^z f_0(x)\mathrm{d}x$.

As seen from Eqs. (2.79), (2.80), the transverse stresses are continuous functions everywhere in the shell, including the interface surfaces $z = z_n$, and satisfy the traction-free boundary conditions on the face surfaces. In Eq. (2.81), ψ_i are additional unknown parameters characterizing the transverse shear in the shell. Hypothesis (2.81) allows predicting the non-linear distribution of the tangential displacements in the $z-$ direction. If $\mathcal{G} \equiv 0$, it turns into the linear Timoshenko hypothesis, which coincides with the classical KL hypothesis if the angles γ_i are defined by (2.8). The functions $\mu_i^{(0)}, \mu_i^{(n)}$ are coupled with the vector $\bar{\Psi} = (\psi_1, \psi_2)^{\mathrm{T}}$ and depend on elements of a matrix characterizing the shear deformability of the nth layer (see in [80]).

2.9.1 Governing equations in terms of displacements

In what follows, we assume that $\gamma_1 = -\partial w/\partial\alpha$, $\gamma_2 = -\partial w/\partial\beta$. Hence in the framework of ESL model based on the stated above hypotheses, the five components $w, u_i, \psi_i (i = 1, 2)$ can be considered as independent functions describing dynamic behavior of the laminated shell taking into account the shear effect.

Assuming

$$\begin{aligned}
f_0(z) &= \frac{1}{h^2}(z - z_0)(z_N - z) && \text{for} \quad z \in [z_0, z_N], \\
f_n(z) &= \frac{1}{h_n^2}(z - z_{n-1})(z_n - z) && \text{for} \quad z \in [z_{n-1}, z_n], \\
f_n(z) &= 0 && \text{for} \quad z \notin [z_{n-1}, z_n]
\end{aligned} \qquad (2.82)$$

and using the mixed variational principle, Grigoliuk and Kulikov [80] derived governing equations in terms of stress resultants and generalized stress couples.

In [158], these equations were reduced to the following five coupled non-linear differential equations with respect to the five unknowns $w, \hat{u}_i, \psi_i (i = 1, 2)$:

$$\frac{\partial^2 \hat{u}_1}{\partial \alpha^2} + \frac{1-\nu}{2}\frac{\partial^2 \hat{u}_1}{\partial \beta^2} + \frac{1+\nu}{2}\frac{\partial^2 \hat{u}_2}{\partial \alpha \partial \beta} + \frac{\nu}{R_2}\frac{\partial w}{\partial \alpha} + \frac{\partial w}{\partial \alpha}\frac{\partial^2 w}{\partial \alpha^2}$$
$$+\nu\frac{\partial w}{\partial \beta}\frac{\partial^2 w}{\partial \alpha \partial \beta} + \frac{1-\nu}{2}\left(\frac{\partial w}{\partial \alpha}\frac{\partial^2 w}{\partial \beta^2} + \frac{\partial w}{\partial \beta}\frac{\partial^2 w}{\partial \alpha \partial \beta}\right) - \frac{\rho(1-\nu^2)}{E}\frac{\partial^2 \hat{u}_1}{\partial t^2} = 0,$$
$$\frac{1+\nu}{2}\frac{\partial^2 \hat{u}_1}{\partial \alpha \partial \beta} + \frac{1-\nu}{2}\frac{\partial^2 \hat{u}_2}{\partial \alpha^2} + \frac{\partial^2 \hat{u}_2}{\partial \beta^2} + \frac{\partial}{\partial \beta}\left(\frac{w}{R_2}\right) + \frac{\partial w}{\partial \beta}\frac{\partial^2 w}{\partial \alpha^2} + \nu\frac{\partial w}{\partial \alpha}\frac{\partial^2 w}{\partial \alpha \partial \beta}$$
$$+\frac{1-\nu}{2}\left(\frac{\partial w}{\partial \beta}\frac{\partial^2 w}{\partial \alpha^2} + \frac{\partial w}{\partial \alpha}\frac{\partial^2 w}{\partial \alpha \partial \beta}\right) - \frac{\rho(1-\nu^2)}{E}\frac{\partial^2 \hat{u}_2}{\partial t^2} = 0,$$
$$\eta_2\frac{\partial \Delta w}{\partial \alpha} - \eta_1\left(\frac{\partial^2 \psi_1}{\partial \alpha^2} + \frac{1+\nu}{2}\frac{\partial^2 \psi_2}{\partial \alpha \partial \beta} - \frac{1-\nu}{2}\frac{\partial^2 \psi_1}{\partial \beta^2}\right) + \frac{12(1-\nu^2)\Gamma}{Eh^3}\psi_1 = 0,$$
$$\eta_2\frac{\partial \Delta w}{\partial \beta} - \eta_1\left(\frac{\partial^2 \psi_2}{\partial \beta^2} + \frac{1+\nu}{2}\frac{\partial^2 \psi_1}{\partial \alpha \partial \beta} - \frac{1-\nu}{2}\frac{\partial^2 \psi_2}{\partial \alpha^2}\right) + \frac{12(1-\nu^2)\Gamma}{Eh^3}\psi_2 = 0,$$
$$\frac{h^2}{12(1-\nu^2)}\Delta\left[\eta_3\Delta w - \eta_2\left(\frac{\partial \psi_1}{\partial \alpha} + \frac{\partial \psi_2}{\partial \beta}\right)\right] + \frac{1}{R_2(1-\nu^2)}\left(\nu\frac{\partial \hat{u}_1}{\partial \alpha} + \frac{\partial \hat{u}_2}{\partial \beta}\right)$$
$$-\frac{1}{1-\nu^2}\left\{\frac{\partial^2 w}{\partial \alpha^2}\left[\frac{\hat{u}_1}{\partial \alpha} + \nu\left(\frac{\partial \hat{u}_2}{\partial \beta} + \frac{w}{R_2}\right) + \frac{1}{2}\left(\left(\frac{\partial w}{\partial \alpha}\right)^2 + \nu\left(\frac{\partial w}{\partial \beta}\right)^2\right)\right]\right.$$
$$+\frac{\partial^2 w}{\partial \beta^2}\left[\nu\frac{\hat{u}_1}{\partial \alpha} + \frac{\partial \hat{u}_2}{\partial \beta} + \frac{w}{R_2} + \frac{1}{2}\left(\left(\frac{\partial w}{\partial \beta}\right)^2 + \nu\left(\frac{\partial w}{\partial \alpha}\right)^2\right)\right]$$
$$(1-\nu)\frac{\partial^2 w}{\partial \alpha \partial \beta}\left(\frac{\partial \hat{u}_1}{\partial \beta} + \frac{\partial \hat{u}_2}{\partial \alpha} + \frac{\partial w}{\partial \alpha}\frac{\partial w}{\partial \beta}\right)$$
$$\left.-\frac{1}{2R_2}\left(\left(\frac{\partial w}{\partial \beta}\right)^2 + \nu\left(\frac{\partial w}{\partial \alpha}\right)^2\right)\right\} + \frac{\rho}{E}\frac{\partial^2 w}{\partial t^2} = 0, \tag{2.83}$$

where E, ν, Γ, ρ are the reduced Young modulus, Poisson ratio, shear stiffness and density, respectively, introduced by the formulas

$$E = \frac{1-\nu^2}{h}\sum_{n=1}^{N}\frac{E_n h_n}{1-\nu_n^2}, \quad \nu = \sum_{n=1}^{N}\frac{E_n h_n \nu_n}{1-\nu_n^2}\left(\sum_{n=1}^{N}\frac{E_n h_n}{1-\nu_n^2}\right)^{-1},$$
$$\Gamma = \frac{\left[\sum\limits_{n=1}^{N}\left(\lambda_n - \frac{\lambda_{n0}^2}{\lambda_{nn}}\right)\right]^2}{\sum\limits_{n=1}^{N}\left(\lambda_n - \frac{\lambda_{n0}^2}{\lambda_{nn}}\right)G_n^{-1}} + \sum_{n=1}^{N}\frac{\lambda_{n0}^2}{\lambda_{nn}}G_n, \quad \rho = \frac{1}{h}\sum_{n=1}^{N}\rho_n h_n, \tag{2.84}$$
$$\lambda_n = \int\limits_{z_{n-1}}^{z_n} f_0^2(z)\mathrm{d}z, \qquad \lambda_{nj} = \int\limits_{z_{n-1}}^{z_n} f_n(z)f_j(z)\mathrm{d}z, \qquad j = 0, n,$$

$\hat{u}_1, \hat{u}_2$ are the generalized in-plane displacements (axial and circumferential, respectively) coupled with u_1, u_2 by the equations

$$u_i = \hat{u}_i + \frac{1}{2} h c_{13} w_{,i} - \frac{1}{2} h c_{12} \psi_i, \tag{2.85}$$

and parameters $\eta_1, \eta_2, \eta_3, c_{12}, c_{13}, q_{44}$ are determined as follows:

$$\begin{aligned}
&c_{12} = \sum_{n=1}^{N} \xi_n^{-1} \pi_{3n} K_n, \quad c_{13} = \sum_{n=1}^{N} (\zeta_{n-1} + \zeta_n) K_n, \\
&K_n = \frac{E_n h_n}{1-\nu_n^2} \left(\sum_{n=1}^{N} \frac{E_n h_n}{1-\nu_n^2} \right)^{-1}, \quad \frac{1}{12} h^3 \pi_{1k} = \int_{\delta_{k-1}}^{\delta_k} \mathcal{G}^2(z) \mathrm{d}z, \\
&\frac{1}{12} h^3 \pi_{2n} = \int_{z_{n-1}}^{z_n} z\, \mathcal{G}(z) \mathrm{d}z, \quad \frac{1}{2} h^2 \pi_{3n} = \int_{z_{n-1}}^{z_n} \mathcal{G}(z) \mathrm{d}z, \\
&\eta_1 = \sum_{n=1}^{N} \xi_n^{-1} \pi_{1n} K_n - 3c_{12}^2, \quad \eta_2 = \sum_{n=1}^{N} \xi_n^{-1} \pi_{2n} K_n - 3c_{12} c_{13}, \\
&\eta_3 = 4 \sum_{n=1}^{N} \left(\xi_n^2 + 3\zeta_{n-1}\zeta_n \right) K_n - 3c_{13}^2, \\
&h\xi_n = h_n, \quad h\zeta_j = z_j \, (j = 0, n),
\end{aligned} \tag{2.86}$$

Equations (2.83) can be used to predict localized waves running in the axial direction in very long laminated cylindrical shells.

2.9.2 Technical theory equations

Equations (2.83) can be simplified for the analysis of some class of vibrations. Let us consider here the stress state characterized by the index of variation $\varsigma = 1/2$ and the following estimates:

$$w \sim h_* R, \quad u_i \ll w. \tag{2.87}$$

It is also obvious that $\hat{u}_i \ll w$. Let

$$\max\{\hat{u}_i\} \sim h_*^{\zeta_u} R, \quad \max\{\psi_i\} \sim h_*^{\zeta_\psi}, \quad G \sim h_*^{\zeta_G} E, \tag{2.88}$$

where ζ_u, ζ_ψ are the indexes of intensity of $\hat{u}_i, \psi_i$, respectively, and $h_*^{\zeta_G}$ is the order of the reduced shear modulus G with regard to the reduced Young's modulus E. Now analyzing the orders of all terms in Eqs. (2.83), one obtains

$$\zeta_u = 3/2, \quad \zeta_\psi = 1/2, \quad \zeta_G = 1. \tag{2.89}$$

The stress-strain state characterized by the index of variation $\varsigma = 1/2$ and the indexes of intensity (2.88) is called the nonlinear *combined* stress state [222, 226]. For this state all the terms in Eqs. (2.83), including non-linear ones, have the same order. If $w \ll h_* R$, then the non-linear terms in the governing equations may be omitted.

Let $q_i = 0$ and the inertia forces in the tangential directions are very small. Then the first two equations from Eqs. (2.83) become homogeneous. Their counterparts written in terms of the stress resultants T_1, T_2 may be identically satisfied by introducing the force function Φ, see Eqs. (2.43). Instead of these two equations, we write down the non-linear differential equation coupling the stress function Φ and the normal displacement w. This equation is readily derived from the strain compatibility condition and has the following form:

$$\triangle^2 \Phi - Eh \left[\frac{1}{R_2} \frac{\partial^2 w}{\partial \alpha^2} + \left(\frac{\partial^2 w}{\partial \alpha \partial \beta} \right)^2 - \frac{\partial^2 w}{\partial \alpha^2} \frac{\partial^2 w}{\partial \beta^2} \right] = 0. \tag{2.90}$$

Let us introduce new functions Ψ and Θ such that

$$\psi_1 = \frac{\partial \Psi}{\partial \alpha} + \frac{\partial \Theta}{\partial \beta}, \quad \psi_2 = \frac{\partial \Psi}{\partial \beta} - \frac{\partial \Theta}{\partial \alpha}. \tag{2.91}$$

The substitution of (2.90) into the third and fourth equations from (2.83) results in the following equations

$$\begin{aligned} \frac{Eh^3}{12(1-\nu^2)} \Delta \frac{\partial}{\partial \alpha} (\eta_1 \Psi - \eta_2 w) + \frac{Eh^3}{24(1+\nu^2)} \eta_1 \Delta \frac{\partial \Theta}{\partial \beta} &= \Gamma \left(\frac{\partial \Psi}{\partial \alpha} + \frac{\partial \Theta}{\partial \beta} \right), \\ \frac{Eh^3}{12(1-\nu^2)} \Delta \frac{\partial}{\partial \beta} (\eta_1 \Psi - \eta_2 w) - \frac{Eh^3}{24(1+\nu^2)} \eta_1 \Delta \frac{\partial \Theta}{\partial \alpha} &= \Gamma \left(\frac{\partial \Psi}{\partial \beta} - \frac{\partial \Theta}{\partial \alpha} \right). \end{aligned} \tag{2.92}$$

The above equations are identically satisfied if

$$\frac{Eh^3}{12(1-\nu^2)} \Delta (\eta_1 \Psi - \eta_2 w) = \Gamma \Psi, \tag{2.93}$$

$$\frac{Eh^3}{24(1+\nu)} \eta_1 \Delta \Theta = \Gamma \Theta. \tag{2.94}$$

Following to [80], we introduce the displacement function χ so that

$$w = \left(1 - \frac{h^2}{\varpi} \triangle \right) \chi, \tag{2.95}$$

$$\Psi = -\frac{\eta_2}{\eta_1} \frac{h^2}{\varpi} \triangle \chi \tag{2.96}$$

and substitute them into Eq. (2.93). It can be seen that this equation is identically satisfied if and only if

$$\varpi = \frac{12(1-\nu^2)\Gamma}{Eh\eta_1}. \tag{2.97}$$

Hence Eq. (2.94) can be rewritten as

$$\frac{1-\nu}{2}\frac{h^2}{\varpi}\triangle\Theta = \Theta. \tag{2.98}$$

Consider the last equation of (2.83) and take into account the following relations coupling the stress resultants and displacements [80]:

$$\begin{aligned}
T_1 &= \frac{Eh}{1-\nu^2}\left\{\frac{\partial \hat{u}_1}{\partial \alpha} + \nu\frac{\partial \hat{u}_2}{\partial \beta} + \frac{w}{R_2} + \frac{1}{2}\left[\left(\frac{\partial w}{\partial \alpha}\right)^2 + \nu\left(\frac{\partial w}{\partial \beta}\right)^2\right]\right\},\\
T_2 &= \frac{Eh}{1-\nu^2}\left\{\frac{\partial \hat{u}_2}{\partial \beta} + \nu\left(\frac{\partial \hat{u}_1}{\partial \alpha} + \frac{w}{R_2}\right) + \frac{1}{2}\left[\left(\frac{\partial w}{\partial \beta}\right)^2 + \nu\left(\frac{\partial w}{\partial \alpha}\right)^2\right]\right\},\\
S &= \frac{Eh}{2(1+\nu)}\left(\frac{\partial \hat{u}_1}{\partial \beta} + \frac{\partial \hat{u}_2}{\partial \beta} + \frac{\partial w}{\partial \alpha}\frac{\partial w}{\partial \beta}\right).
\end{aligned} \tag{2.99}$$

Now it may be rewritten as

$$\begin{aligned}
&\frac{Eh^3}{12(1-\nu^2)}\triangle\left[\eta_3\triangle w - \eta_2\left(\frac{\partial \psi_1}{\partial \alpha} + \frac{\partial \psi_2}{\partial \beta}\right)\right]\\
&-T_1\frac{\partial^2 w}{\partial \alpha^2} - 2S\frac{\partial^2 w}{\partial \alpha \partial \beta} - T_2\frac{\partial^2 w}{\partial \beta^2} + \frac{T_2}{R_2} + h\rho\frac{\partial^2 w}{\partial t^2} = 0.
\end{aligned} \tag{2.100}$$

Substituting of Eqs. (2.43), (2.91), (2.95) and (2.96) into (2.100) we get the following equation

$$\begin{aligned}
&D_3\left(1 - \frac{\theta h^2}{\varpi}\triangle\right)\triangle^2\chi + \frac{1}{R_2}\frac{\partial^2 \Phi}{\partial \alpha_1^2} + h\rho\frac{\partial^2 w}{\partial t^2}\\
&-\frac{\partial^2 \Phi}{\partial \beta^2}\frac{\partial^2 w}{\partial \alpha^2} - 2S\frac{\partial^2 w}{\partial \alpha \partial \beta} - \frac{\partial^2 \Phi}{\partial \beta^2}\frac{\partial^2 w}{\partial \beta^2} = 0,
\end{aligned} \tag{2.101}$$

where

$$D_3 = \frac{Eh^3}{12(1-\nu^2)}\eta_3 \tag{2.102}$$

is the reduced bending stiffness of the laminated cylindrical shell, and

$$\theta = 1 - \frac{\eta_2^2}{\eta_1\eta_3}. \tag{2.103}$$

Here, θ is a small parameter. So, the calculations performed for a single layer isotropic shell [80] give $\theta = 1/85$.

We consider here only two groups of boundary conditions in terms of the displacement and stress functions [80], the group of simple support and that of clamping. The group of simple support is subdivided into two variants:

- the edge $\alpha = \alpha^*$ is provided with a diaphragm preventing shear along the edge plane, then

$$\chi = \Delta\chi = \Delta^2\chi = \frac{\partial\Theta}{\partial\alpha_1} = 0 \quad \text{for} \quad \alpha = \alpha^*; \tag{2.104}$$

- the edge diaphragm is absent, then

$$\begin{gathered}\left(1 - \frac{h^2}{\varpi}\Delta\right)\chi = 0, \quad \frac{\partial^2}{\partial\alpha_1^2}\left(1 - \frac{h^2}{\varpi}\Delta\right)\chi = 0, \\ \left(\frac{\partial^2}{\partial\alpha_1^2} + \nu\frac{\partial^2}{\partial\alpha_2^2}\right)\chi - (1-\nu)\frac{\partial^2\Theta}{\partial\alpha_1\alpha_2} = 0, \\ 2\frac{\partial^2\chi}{\partial\alpha_1\partial\alpha_2} + \frac{\partial^2\Theta}{\partial\alpha_1^2} - \frac{\partial^2\Theta}{\partial\alpha_2^2} = 0\end{gathered} \tag{2.105}$$

for $\alpha = \alpha^*$.

The second group for a clamped edge also contains two variants:

- for the edge $\alpha = \alpha^*$ with the diaphragm

$$\begin{gathered}\left(1 - \frac{h^2}{\varpi}\Delta\right)\chi = 0, \quad \frac{\partial}{\partial\alpha_1}\left(1 - \frac{h^2}{\varpi}\Delta\right)\chi = 0, \\ \frac{\partial\chi}{\partial\alpha_1} - \frac{\partial\Theta}{\partial\alpha_2} = 0, \quad \frac{\partial\chi}{\partial\alpha_2} + \frac{\partial\Theta}{\partial\alpha_1} = 0.\end{gathered} \tag{2.106}$$

- and for the edge $\alpha = \alpha^*$ without a diaphragm

$$\left(1 - \frac{h^2}{\varpi}\Delta\right)\chi = 0, \quad \frac{\partial\chi}{\partial\alpha_1} = \frac{\partial}{\partial\alpha_1}(\Delta\chi) = \Theta = 0. \tag{2.107}$$

It is seen that each variant of boundary conditions is incomplete because it does not contain conditions for the generalized in-plane displacements $\hat{u}_i$ or stress resultants T_{i1}. For example, the conditions of free support results in the additional conditions for the stress function [80]

$$\Phi = \Delta\Phi = 0 \quad \text{at} \quad \alpha = \alpha^*. \tag{2.108}$$

The system of five differential equations (2.90), (2.95), (2.96), (2.98) and (2.101) with some variant of boundary conditions describes non-linear dynamics of a laminated cylindrical shell. If $G \to \infty$ ($\varpi \to \infty$), then $\Psi, \Theta \to 0$, $\chi \to w$ and this system degenerates into the well-known system of two non-linear equations with respect to w, Φ which are similar to von Karman's equations for plates.

Linearizing Eqs. (2.90), (2.101) and taking into account (2.95), we arrive at the following linear governing equations

$$D_3\left(1-\frac{\theta h^2}{\varpi}\Delta\right)\Delta^2\chi-\frac{1}{R_2}\frac{\partial^2\Phi}{\partial\alpha^2}+h\rho\frac{\partial^2}{\partial t^2}\left(1-\frac{h^2}{\varpi}\Delta\right)\chi=0,$$
$$\Delta^2\Phi-\frac{Eh}{R_2}\left(1-\frac{h^2}{\varpi}\Delta\right)\frac{\partial^2\chi}{\partial\alpha^2}=0. \tag{2.109}$$

If a laminated shell is pre-stressed by the membrane stress resultants $T_1^\circ, T_2^\circ, S^\circ$, then the first equation in system (2.109) should be supplemented by the additional term (2.47). When assuming $\theta = 0$, one obtains equations similar to Eqs. (2.65) based on the TR model.

In what follows, Eqs. (2.109) will be used to study localized free and parametric vibrations of thin laminated cylindrical shells taking into account the shear effect.

2.10 On the error of 2D shell theories

In [172] it was established that the error Δ of the KL hypotheses in static problems is of the order of the relative shell thickness:

$$\Delta \sim h_*. \tag{2.110}$$

In the subsequent investigations [65], [110], this estimate was confirmed for the stress states with the index of variation $\varsigma < 1/2$ (see Es. (2.37)). The more general error estimate, acceptable at $\varsigma < 1$, reads as

$$\Delta \sim \max\{h_*,\, h_*^{2-2\varsigma}\}. \tag{2.111}$$

At $\varsigma \leq 1/2$, estimates (2.110) and (2.111) coincide, and for $\varsigma \geq 1$ the 2D shell theory is inacceptable.

Consider a problem about the additional error which appears when the technical (or shallow) shell theory (2.41) is used. We note, first, that Eqs. (2.41) are acceptable only for the analysis of quasi-transversal vibrations for which

$$u_* \ll w_*, \tag{2.112}$$

(see Eqs. (2.40)), because in Eqs. (2.41) the tangential inertia forces are not contained.

When deriving Eqs. (2.41), in the first two Eqs. (2.14) the shear stress resultants Q_1 and Q_2 have not been taken into account that leads to the relative error of the order $h_*^{2-2\varsigma}$.

Next, the tangential inertia forces were also omitted, and the expressions for curvatures, κ_1, κ_2, and torsion, τ, were simplified. All these assumptions resulted in the relative error of order $h_*^{2\varsigma}$.

As a result, the error Δ_1 of the technical shell theory is of the order

$$\Delta_1 \sim \max\{h_*^{2\varsigma},\, h_*^{2-2\varsigma}\}. \tag{2.113}$$

Therefore, for the vibration modes with small variability, $0 < \varsigma < 1/2$, the error of Eqs. (2.41) of the technical theory is larger than the error of the full system of Eqs. (2.14), and for $\varsigma = 0$ Eqs. (2.41) become unacceptable.

Functional (2.32) gives the natural frequencies in the assumption that all the shell mass is concentrated in the middle surface of a shell. In the framework of the KL hypotheses, we consider an influence of a rotation motion of normal fibers on the frequencies of quasi-transversal vibrations. The dimensionless kinetic energy including a rotation motion reads as

$$T' = \frac{1}{2}\int\int_{\Omega}\left(u_1^2 + u_2^2 + u_3^2 + \frac{h_*^2}{12}\left(\gamma_1^2 + \gamma_2^2\right)\right) AB\, d\alpha d\beta. \tag{2.114}$$

Taking into account the estimates $\{\gamma_1,\, \gamma_2\} \sim h_*^{-\varsigma} w$, we conclude that the relative influence of rotation motion is of the order $\Delta_2 = h_*^{2-2\varsigma}$. Hence, the error Δ_2 does not exceed the error of Eqs. (2.14) (see (Eq. (2.111)), and the inertia of rotation motion may be excluded of Eqs. (2.14).

Now we discuss the TR model accounting for a transversal shear. The differential system is of the 10th order, and when compared with the KL model, the additional solutions (named as the boundary layer solutions) appear.

Consider, firstly, a shell made of an isotropic material. In this case the longitudinal stiffness, Eh, and the transversal shear stiffness, Γ, are of the identical asymptotical orders, and the boundary layer solutions have the index of variation $\varsigma = 1$ (see Eq. (2.61). Such approach to the 2D shell theory is incorrect from the asymptotic point of view [225, 227], because for the correct theory the typical length of deformations in tangential directions is to be much larger than the shell thickness.

At the same time, the TR model is often used in the non-stationary problems of waves propagation. The TR model leads to a hyperbolic system of equations in partial derivatives, and the KL model leads to a parabolic system, which gives an infinite velocity of waves.

The TR model is used for transversely isotropic shells and for multi-layered shells for which the transverse shear stiffness is essentially smaller than the stretch stiffness, $\eta = Eh/\Gamma = O(h_*^{-\varsigma_3})$, $\varsigma_3 > 0$. In this case for the boundary layer the index of variation $\varsigma = 1 - \varsigma_3/2 < 1$ (see Eq. (2.62)), and the 2D shell theory becomes asymptotically correct and gives the more exact results than the KL model. Moreover, for the very small η the KL model is not applicable. [225, 230, 231].

For the TR model the technical shell theory, described by Eqs. (2.63)–(2.65), also may be introduced. This theory is acceptable for any values of the shear parameter η (or of often used shear parameter $g \sim \eta h_*^2$).

In Sections 4.2 and 4.3, for simplicity a semi-momentless SSS [173] is used to describe localized vibration modes of shells with the zero Gaussian curvature. The full SSS may be presented as a sum of the semi-momentless state and of the edge effect. In the TR model they are augmented by the boundary layer. In the last case the transversal shear stiffness cannot be too small

$$\eta \gg h_*^{3/2} \ (\varsigma_3 < 3/2), \quad \text{or} \quad g \gg h_*^{1/2}. \tag{2.115}$$

In the opposite case only the full Eqs. (2.14), (2.54) may be used.

Chapter 3

Localized vibration modes of plates and shells of revolution

In this chapter, we give a definition of localized vibrations modes of shells to be studied below and present their short classification. The problems on free vibrations admiting separation of variables are considered. At first, the localized modes of a plate with free rectilinear edge and of a circular plate are constructed.

For shells of revolution, a separation of variables is performed. In the case of a large number of waves in the circumferential direction, the governing equations in the framework of the Kirchhoff-Love model is reduced to the eighth order linear system of odinary differential equations with variable coefficients and a small parameter at derivatives. The asymptotic solution of this system leads to a characteristic equation of eighth order whose roots describe the behavior of partial solutions. The purely imaginary roots correspond to oscillating solutions. The concept of the *weakest parallel*, near which the purely imaginary roots firstly appear, is introduced.

It is shown that between the shell area for which some roots are imaginary, and the area for which all roots have the non-zero real parts, there is a point called the *turning point*. The constructions of asymptotic expansions proposed by R.E.Langer [117] and V.P.Maslov [131] are used. The localization of vibration modes near both an edge and strictly within the shell are considered. Solutions of some problems on localized vibrations of such shells as an ellipsoid, a pre-stressed cylindrical shell, and a long cylinder resting on an elastic foundation are demonstrated. The effect of transverse shear on the non-uniform vibration modes near a weakly supported edge is investigated in of presence of the initial stresses. A comparison of solutions for the Kirchhoff-Love model and the Timoshenko like models is performed for various boundary conditions and for the shear parameter varying in a wide range.

3.1 On the modes of localized vibrations of shells

An analytic description of modes of free vibrations of shells with arbitrary form of a middle surface is a very difficult problem, and we will not promise to solve it. Even the essentially more simple problem about vibrations a membrane with an arbitrary boundary Γ, which is described by the two-dimensional Helmholtz equation

$$\Delta w + \lambda w = 0, \qquad w(x, y) = 0 \quad \text{at} \quad (x, y) \in \Gamma, \tag{3.1}$$

has no analytic solution. This suggests studying the narrow class of vibration modes, namely the localized vibration modes.

Definition 3.1. The vibration modes are called *localized modes* near a line or a point on the middle surface if the deflection $\mathbf{U}$ exponentially decreases away from this line or point.

Further we shall be mostly concerned with the localization near the lines that lies within a shell or coincides with one of its edges. The lines near which the localization takes place are called the *weakest lines*. So we are confined with the cases in which the decrease of the deflection away from a weakest line is so quick that the index of variation is positive ($\varsigma > 0$), see (2.37). In this case under the asymptotic expansions of vibration modes it is possible to neglect the influence of the boundary conditions on the edges that does not coincide with the weakest line.

In this chapter, shells of revolution that are bounded by two parallels or have a form of a dome are considered. The homogeneous boundary conditions

(2.23) are given on the parallels $s = s_1$ and $s = s_2$, so this problem admits the separation of variables

$$\begin{aligned} w(s,\varphi) &= w_m(s)\cos m\varphi, \quad \{u_1,\, w, \varepsilon_i,\, \gamma_1,\, \kappa_i,\, T_i,\, M_i,\, Q_1\}, \\ u_2(s,\varphi) &= u_{2m}(s)\sin m\varphi, \quad \{u_2,\, \omega,\, \omega_i,\, \gamma_2,\, \tau,\, S_i,\, H,\, Q_2\}, \end{aligned} \qquad i = 1, 2, \quad (3.2)$$

where in the brackets we have the functions that are transformed by the same formulas, s is the length of the generatrix ($s_1 \le s \le s_2$), φ and $m = 0, 1, \ldots$ are the angle and the number of waves in the circumferential direction. After the separation of variables we get a sequence of one-dimensional boundary value problems.

Under a fixed m, the problem is reduced to the system of ODE of the 8th order,

$$\frac{d\mathbf{X_0}}{ds} = \mathbf{A^0}(s, h_*, m)\mathbf{X_0}. \qquad (3.3)$$

For convenience, when analyzing the main unknowns, we take the functions contained in the boundary conditions (2.23)

$$\mathbf{X_0} \equiv \{u_1,\, u_2,\, w,\, \gamma_1,\, T_1,\, S_1^*,\, Q_1^*,\, M_1\}. \qquad (3.4)$$

Then Eqs. (3.3) read as:

$$\begin{aligned} & u_1' = k_1 w + \varepsilon_1, \quad u_2' = \omega_1, \quad w' = -\gamma_1 - k_1 u_1, \quad \gamma_1' = -\kappa_1, \\ & T_1' = b_1(T_2 - T_1) - q(S_1^* + 2k_2 H) + k_1 Q_1^* - \lambda u_1, \\ & (S_1^*)' = q(T_2 + k_2 M_2) - 2b_1 S_1^* - \lambda u_2, \\ & (Q_1^*)' = q(2b_1 H - qM_2) - b_1 Q_1^* - k_1 T_1 - k_2 T_2 - \lambda w, \\ & M_1' = b_1(M_2 - M_1) - 2qH - Q_1^*, \end{aligned} \qquad (3.5)$$

where, the derivative with respect to s is denoted by a stroke. The remaining unknowns are coupled to the main ones as

$$\begin{aligned} & \varepsilon_2 = b_1 u_1 - q u_2 - k_2 w, \quad \varepsilon_1 = (1-\nu^2)T_1 - \nu\varepsilon_2, \quad b_1 = \frac{B'}{B}, \quad q = \frac{m}{B}, \\ & T_2 = (\varepsilon_2 + \nu\varepsilon_1)/(1-\nu^2), \quad \gamma_2 = qw - k_2 u_2, \\ & \kappa_2 = -b_1\gamma_1 - q\gamma_2, \quad \kappa_1 = \mu^{-4} M_1 - \nu\kappa_2, \quad k_1 = 1/R_1,\ k_2 = 1/R_2, \\ & M_2 = \mu^4(\kappa_2 + \nu\kappa_1), \quad \omega_2 = -q u_1 - b_1 u_2, \\ & H = \mu^4(1-\nu)(q\gamma_1 + b_1\gamma_2 + k_2\omega_1), \\ & \omega_1 = \frac{S_1^* - 2\mu^4(1-\nu)k_2(q\gamma_1 + b_1\gamma_2) - G\omega_2}{G + 2(1-\nu)k_2^2\mu^4}, \quad G = \frac{1}{2(1+\nu)}. \end{aligned} \qquad (3.6)$$

Equations (3.3) or (3.5) contain two main parameters: the small thickness parameter, μ, and the number of waves in the circumferential direction, m.

Assume that $m \gg 1$, because only in this case all solutions of Eqs.(3.3) have the positive index of variation ($\varsigma > 0$). We will put $m = \mu^{-1}\rho$. For small $m \sim 1$, Eqs. (3.3) have four linearly independent momentless solutions with the index of variation $\varsigma = 0$ that cannot be used in a construction of localized vibration modes.

Equations (3.5) are convenient for a numerical solution, but for an asymptotic analysis it is better to present it in the standard form of a system with the small parameter at derivatives:

$$\mu \frac{d\mathbf{X}}{ds} = \mathbf{A}(s, \mu)\mathbf{X}. \tag{3.7}$$

Here, the vector

$$\mathbf{X} \equiv \{\mu^{-1}u_1,\, \mu^{-1}u_2,\, w,\, \mu\gamma_1,\, T_1,\, S_1^*,\, \mu^{-3}Q_1^*,\, \mu^{-2}M_1\}, \tag{3.8}$$

contains the main unknowns.

Under some assumptions (of which the principal one is that $m \sim \mu^{-1}$ or $\rho = \mu m \sim 1$) the main unknowns have the same asymptotic orders, the matrix $\mathbf{A}$ can be regularly expanded in powers of the small parameter μ,

$$\mathbf{A}(s, \mu) = \sum_{n=0}^{\infty} \mu^n \mathbf{A}_n(s). \tag{3.9}$$

Under these assumptions we seek all solutions of Eqs. (3.7) in the form of formal asymptotic series,

$$\begin{aligned} &\boldsymbol{X}_j = \boldsymbol{X}_j^0(s,\mu)\exp\left\{\mu^{-1}\textstyle\int p_j(s)\,ds\right\}, \quad p_j(s) = a_j(s) + \mathrm{i}b_j(s),\ \mathrm{i} = \sqrt{-1}, \\ &\boldsymbol{X}_j^0(s,\mu) = \sum_{k=0}^{\infty} \mu^k \boldsymbol{X}_{jk}^0(s), \quad j = 1, 2, \ldots, 8, \end{aligned} \tag{3.10}$$

where $p_j(s)$ are the roots of the characteristic equation

$$\det(\mathbf{A}_0(s) - p\mathbf{E}) = 0. \tag{3.11}$$

Expansion (3.10) is unacceptable near the points $s = s_*$ at which two or more roots $p_j(s)$ coincide to each other, because at these points the coefficients of series (3.10), $\boldsymbol{X}_{jk}(s_*) = \infty$. Such points are called *turning points.*

The exactness of the initial Eqs. (3.5) allows us to find only the first two terms of these series (with $k = 0$ and $k = 1$).

If the real part a_j of the root p_j is not zero, then the corresponding solution (3.10) is an exponentially decreasing or increasing (with an oscillation at $b_j \neq 0$) function of s. With $a_j = 0$, $b_j \neq 0$, Eq. (3.10) describes an oscillating function. If among the solutions (3.10) there are oscillating functions, then some points of the spectrum lie near the studied value λ (see [64]).

By $\lambda_{0,m}$ with a fixed value of m we denote the minimum value of the parameter λ such that at $\lambda+\varepsilon$, $\varepsilon>0$, there are purely imaginary roots of Eq. (3.11) at least with some $s\in[s_1, s_2]$.

By $s_{0,m}$ we denote the *weakest parallel* in the neighborhood of which the purely imaginary roots appear for the first time (with the growth of λ, starting from $\lambda=0$) .

The value $\lambda_{0,m}$ exists, because at $\lambda=0$ Eq. (3.11) has no purely imaginary roots, and at the large enough values λ this equation has always purely imaginary roots.

In the cases when the roots p_j are independent of s, the weakest parallel, $s_{0,m}$, is not defined (for example, for a cylindrical shell).

Let the weakest parallel $s_{0,m}$ exist. The asymptotic of natural frequencies and eigenfunctions depends substantially on the position of this parallel. If the parallel $s_{0,m}$ coincides with one of the boundary parallels (for definiteness, let $s_{0,m}=s_1$), then for the asymptotic expansions the standard Airy functions are used, the boundary conditions with $s=s_1$ should be satisfied (see Section 3.6). If the weakest parallel lies within the shell ($s_1<s_{0,m}<s_2$), then the influence of the boundary conditions may be neglected, and the asymptotic expansions use the standard functions of parabolic cylinder and Hermit polynomials (see Section 3.7). We shall call such localized vibration modes *oscillating modes.*

There exist other types of localized modes. A free or a weakly supported edge may be a cause of localized modes. In this case the eigen-value λ satisfies the inequality $\lambda<\lambda_{0,m}$, the localized mode being a linear combination of four decreasing solutions (3.10) with $a_j<0$ (see Section 3.6). We call such localized modes *non-uniform localized modes.*

Localization modes of free vibrations are also possible for plate vibrations near a free edge. The problem for plates is more simple than that for shells. In the next section, we shall begin with the problem for a plate.

3.2 Localized vibration modes of a plate with free rectilinear edge

The plate vibrations are divided into the bending vibrations and the in-plane vibrations. The equations and boundary conditions of plate vibrations can be obtained from equations of Chapter 2 with $R_1^{-1}=R_2^{-1}=0$.

3.2.1 The bending vibrations

Consider, at first, the *bending vibrations* of a plate lying in the semi-plane $0\le x<\infty$, $-\infty<y<\infty$, with the free edge $x=0$. These vibrations are described by the equation

$$D\Delta\Delta w - \rho h\omega^2 w = 0, \qquad D = \frac{Eh^3}{12(1-\nu^2)} \tag{3.12}$$

and the boundary conditions at $x = 0$:

$$M_x = D\left(\frac{\partial^2 w}{\partial x^2} + \nu\frac{\partial^2 w}{\partial y^2}\right) = 0, \quad Q_x^* = -D\left(\frac{\partial^3 w}{\partial x^3} + (2-\nu)\frac{\partial^3 w}{\partial x \partial y^2}\right) = 0. \tag{3.13}$$

We shall seek a solution of the boundary-value problem (3.12), (3.13) decreasing away from the edge $x = 0$ in the form

$$w(x, y) = w_q(x)\sin qy, \qquad w_q(x) \to 0 \quad at \quad x \to \infty. \tag{3.14}$$

After the separation of variables we get

$$w_q(x) = C_1 e^{-qr_1 x} + C_2 e^{-qr_2 x}, \qquad r_{1,2} = \sqrt{1 \pm \xi}, \tag{3.15}$$

where C_1 and C_2 are arbitrary constants, and r_1 and r_2 are the roots of the equation

$$(r^2-1)^2 - \xi^2 = 0, \qquad \omega = \omega_0\xi, \quad \omega_0 = \sqrt{\frac{Eh^2q^4}{12(1-\nu^2)\rho}}, \tag{3.16}$$

where ξ is the dimensionless frequency.

From the boundary conditions (3.13) we get the following equation for ξ:

$$r_2(r_1^2 - \nu)^2 - r_1(r_2^2 - \nu)^2 = 0. \tag{3.17}$$

We are interested only in the roots $\xi < 1$, because for $\xi > 1$ the function (3.15) is not localized at $x = 0$. Equation (3.17) has a single root $\xi < 1$ depending on the Poisson ratio ν and its approximate value is as follows (see [222]):

$$\xi^2 = 1 - \frac{4\nu^4}{(2-\nu)^4}. \tag{3.18}$$

The frequency parameter $\xi(\nu)$ exists only when $0 < \nu \leq 0.5$. It is close to 1 and changes from 1 at $\nu = 0$ to 0.975 with $\nu = 0.5$. The frequency ω is close to ω_0. The coefficient $r_2 = \sqrt{2}\nu^2/(2-\nu)^2 \ll 1$ is small and so the decreasing of a localized vibration mode is slow.

The similar localization was first described in [92]. In this work, the buckling problem of a long compressed rectilinear plate with a free edge was studied. The buckling problem is reduced to the same boundary-value problem (3.12), (3.13), and the value $\xi = 0.9987$ with $\nu = 0.3$ is obtained.

The bending localized vibration modes are connected with the Rayleigh waves. The function $w(x, y, t) = w_q(x)\sin(qy - \omega t)$ describes the Rayleigh wave moving along the plate edge $x = 0$.

3.2.2 The in-plane vibrations

The *in-plane vibrations*, namely, the vibrations with displacements in the plane of a plate, are described by the system of equations

$$
\begin{aligned}
&K\left(\frac{\partial^2 u}{\partial x^2}+\frac{1-\nu}{2}\frac{\partial^2 u}{\partial y^2}+\frac{1+\nu}{2}\frac{\partial^2 v}{\partial x\partial y}\right)+\rho h\omega^2\, u=0,\\
&K\left(\frac{1+\nu}{2}\frac{\partial^2 u}{\partial x\partial y}+\frac{1-\nu}{2}\frac{\partial^2 v}{\partial x^2}+\frac{\partial^2 v}{\partial y^2}\right)+\rho h\omega^2\, v=0.
\end{aligned}
\tag{3.19}
$$

where $K = Eh/(1-\nu^2)$.

Let the edge $x = 0$ be free, then

$$
T_1 = K\left(\frac{\partial u}{\partial x}+\nu\frac{\partial v}{\partial y}\right)=0,\quad S = K\frac{1-\nu}{2}\left(\frac{\partial u}{\partial y}+\frac{\partial v}{\partial x}\right)=0. \tag{3.20}
$$

We seek a solution localized near the free edge $x = 0$ in the form

$$
\begin{aligned}
&u(x,y)=u_q(x)\cos qy,\quad v(x,y)=v_q(x)\sin qy,\\
&u_q(x), v_q(x)\to 0 \quad \text{as} \quad x\to\infty.
\end{aligned}
\tag{3.21}
$$

Equations (3.19) have the solution

$$
u_q(x)=C_1 r_1 e^{-qr_1x}-C_2e^{-qr_2x},\quad v_q(x)=-C_1e^{-qr_1x}+C_2r_2e^{-qr_2x}, \tag{3.22}
$$

with arbitrary constants, C_1 and C_2, and

$$
r_1=\sqrt{1-(1-\nu)\xi_1^2/2},\quad r_2=\sqrt{1-\xi_1^2},\quad \omega=\omega_1\xi_1,\quad \omega_1=\sqrt{\frac{Eq^2}{2(1+\nu)\rho}}. \tag{3.23}
$$

The parameter $\xi_1 = \xi_1(\nu)$ can be found on substituting Eqs. (3.22) into the boundary conditions (3.20); it is given in Table 3.1.

TABLE 3.1: Dimensionless frequencies of the in-plane vibrations.

ν	0	0.1	0.2	0.3	0.4	0.5
ξ_1	0.874	0.892	0.905	0.916	0.925	0.933

Compare the bending and the in-plane localized vibrations. We note the inequality $\xi_1 < \xi$ for the dimensionless frequencies.

But for the fixed wave number q the opposite inequality for the dimension frequencies $\omega_1 \gg \omega_0$ holds. Indeed,

$$
qh=\frac{2\pi h}{L}\ll 1,\qquad \frac{\omega_0}{\omega_1}=\frac{qh}{\sqrt{6(1-\nu)}}\ll 1, \tag{3.24}
$$

where L is the length of wave, which in the two-dimensional plate and shell theories should be much larger than the thickness h.

For short waves with $qh > 1$, the two-dimensional theory is unacceptable, while for very short waves, $qh \gg 1$ and the three-dimensional theory leads to the vibration modes localized near the free selvages of a plate (see [71], [102]). We shall not be concerned here with such vibrations.

The bending vibrations of plates are similar to the quasi-transversal vibrations of shells and the in-plane vibrations of plates are similar to the quasi-tangential vibrations of shells. Unlike to the plates, the problem for shells is not split exactly into the quasi-transversal and the quasi-tangential problems. That is why the quasi-transversal localized vibrations of shells are possible, and the quasi-tangential localized vibrations are impossible, because they are accompanied with quasi-transversal vibrations with small amplitudes (due to inequality (3.24)) .

3.3 Localized vibration modes of a circular plate

The bending vibrations of a circular plate of a radius R with a free edge are described by Eq. (3.12). We separate the variables $w(s, \varphi) = w_m(s) \cos m\varphi$ in the polar coordinates s, φ. Now the solution bounded in the plate center reads as

$$w_m(s) = C_1 w_m^{(1)}(s) + C_2 w_m^{(2)}(s), \quad w_m^{(1)}(s) = J_m(x), \quad w_m^{(2)}(s) = I_m(x), \tag{3.25}$$

where C_1 and C_2 are arbitrary constants. Here $J_m(x)$ and $I_m(x)$ are the Bessel functions of real and imaginary arguments, which vanish at $x = 0$. Besides,

$$x = \frac{\alpha s}{R}, \qquad \alpha^4 = \frac{\rho h \omega^2 R^4}{D}, \quad \lambda = \frac{\rho \omega^2 R^2}{E} = \lambda_0 \xi^2, \quad \lambda_0 = \frac{h^2 m^4}{12(1-\nu^2)R^2}. \tag{3.26}$$

The boundary conditions at the free edge $s = R$ are as follows:

$$M_1 = D\left(\frac{d^2 w}{ds^2} + \nu\left(\frac{1}{s}\frac{dw}{ds} - \frac{m^2 w}{s^2}\right)\right) = 0,$$

$$Q_1^* = D\left(-\frac{d^3 w}{ds^3} - \frac{1}{s}\frac{d^2 w}{ds^2} + \frac{(2-\nu)m^2 + 1}{s^2}\frac{dw}{ds} - \frac{(3-\nu)m^2 w}{s^3}\right) = 0. \tag{3.27}$$

We find the parameter x and the ratio of constants C_1 and C_2 from these boundary conditions. The equation for x reads as

$$f_m(x) = M_1^{(1)}(x) Q_1^{*(2)}(x) - M_1^{(2)}(x) Q_1^{*(1)}(x) = 0. \tag{3.28}$$

For a fixed number m of the nodal diameters, Eq. (3.28) has a countable number of roots that define the position of nodal (immovable) circles. For $m \gg 1$, the first vibration modes are localized near the plate edge $s = R$.

We shall perform an asymptotic simplification of Eq. (3.28) for $m \gg 1$ and use asymptotic formulas for the Bessel functions $J_m(x)$ and $I_m(x)$ as $m \to \infty$, $x \to \infty$, $m \approx x$. It is possible to use the well known results (see [1]), but we deliver these formulas, because the same method will be used below for the construction of the localized modes for shells of revolution.

The function $w_m^{(2)}(s)$ satisfies the equation

$$\Delta_m w - \frac{\alpha^2}{R^2} w = 0, \qquad \Delta_m w = \frac{1}{s}\frac{d}{ds}\left(s\frac{dw}{ds}\right) - \frac{m^2}{s^2} w. \tag{3.29}$$

Putting

$$w_m^{(2)}(s) = \frac{v(y)}{\sqrt{y}}, \qquad s = Ry, \quad 0 \le y \le 1, \tag{3.30}$$

one can see that the function $v(y)$ obeys the equation

$$\frac{1}{m^2}\frac{d^2 v}{dy^2} - \left(p_2(y) - \frac{1}{4m^2}\right) v = 0, \qquad p_2 = \xi + \frac{1}{y^2}, \quad \xi = \frac{\alpha^2}{m^2} = \frac{\omega}{\omega_0} \tag{3.31}$$

in which the asymptotic WKB method can be used. The asymptotic expansion as $m \to \infty$ of the solution, decreasing away from the edge $y = 1$, is as follows:

$$v(y, m) \simeq \sum_{k=0}^{\infty} m^{-k} v_k(y) \exp\{m \int_1^y \sqrt{p_2(y)}\, dy\}, \qquad v_0(y) = (p_2(y))^{-1/4}. \tag{3.32}$$

The remaining functions $v_k(y)$, $k > 0$, can be expressed as integrals of the previous ones, but we use only the first approximation.

The asymptotic behavior of the function $w_m^{(1)}(s)$ as $m \to \infty$ is fairly complicated. As in Eq. (3.30), we put $w_m^{(1)}(s) = v(y)/\sqrt{y}$ and obtain the equation

$$\frac{1}{m^2}\frac{d^2 v}{dy^2} - \left(p_1(y) - \frac{1}{4m^2}\right) v = 0, \qquad p_1 = -\xi + \frac{1}{y^2}. \tag{3.33}$$

This equation with $y = y_* = \xi^{-1/2}$ has a turning point. For $y > y_*$, all solutions of this equation oscillate, while for $y < y_*$, one of the solutions decreases exponentially with decreasing y. This solution is bounded in the center of a round plate.

The uniform asymptotic expansion of the solution to Eq. (3.33) was first presented in [117], and in our designation it reads as

$$v(y, m) = a(y, m)\mathrm{Ai}(\eta) + m^{-4/3} b(y, m)\mathrm{Ai}'(\eta), \qquad \mathrm{Ai}''(\eta) - \eta \mathrm{Ai}(\eta) = 0, \tag{3.34}$$

where Ai(η) is the Airy function , and the derivatives with respect to η are denoted by strokes. Besides,

$$a(y,m) = \sum_{k=0}^{\infty} m^{-k} a_k(y), \quad \omega_0(y) = \left(\frac{d\zeta}{dy}\right)^{-1}, \qquad b(y,m) = \sum_{k=0}^{\infty} m^{-k} b_k(y),$$

$$\eta = m^{2/3}\zeta(y), \qquad \zeta(y)\left(\frac{d\zeta}{dy}\right)^2 = p_1(y), \quad \zeta(y) = \left(\frac{3}{2}\int_{y_*}^{y} \sqrt{p_1(y)}\,dy\right)^{2/3}, \tag{3.35}$$

and a_k, b_k, ζ are holomorphic functions of y. For the function $p_1(y)$ in Eq. (3.33), the function $\zeta(y)$ reads as [23–25]

$$\zeta(y) = \begin{cases} \left(\dfrac{3}{2y_*}\ln\dfrac{y_* + \sqrt{y_*^2 - y^2}}{y} - \dfrac{3\sqrt{y_*^2 - y^2}}{2y_*^2}\right)^{2/3}, & y \le y_*, \\ -\left(\dfrac{3\sqrt{y^2 - y_*^2}}{2y_*^2} - \dfrac{3}{2y_*}\arccos\dfrac{y_*}{y}\right)^{2/3}, & y \ge y_*. \end{cases} \tag{3.36}$$

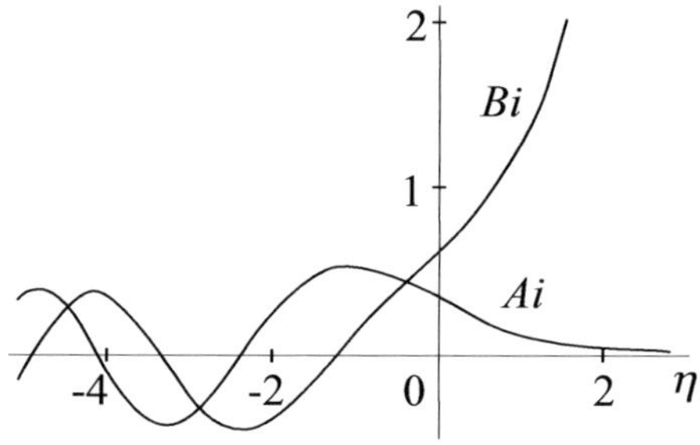

FIGURE 3.1: Functions $Ai(\eta)$ and $Bi(\eta)$.

The plots of the standard solutions $Ai(\eta)$ and $Bi(\eta)$ of the Airy equation are shown in the Fig. 3.1.

The boundary conditions (3.27) give the equation for the frequency parameter ξ,

$$\sqrt{p_2(1)}\mathrm{Ai}(\eta_1)(1-\xi-\nu)^2 - m^{-1/3}\zeta'(1)\mathrm{Ai}'(\eta_1)(1+\xi-\nu)^2 = 0 \tag{3.37}$$

with $\eta_1 = \eta(1)$.

This equation is the asymptotic approximation as $m \to \infty$ of the exact equation (3.28) with $x = m\sqrt{\xi}$. The relative error of Eq. (3.37) is of the order m^{-1}.

The asymptotic expansions of the Airy function at $\eta \to \pm\infty$ are as follows (see also in [1])

$$Ai(\eta) \simeq \frac{1}{2\sqrt{\pi}}\eta^{-1/4}e^{-z}\sum_{k=0}^{\infty}\frac{(-1)^k c_k}{z^k},$$
$$Ai(-\eta) \simeq \frac{1}{\sqrt{\pi}}\eta^{-1/4}\left[\sum_{k=0}^{\infty}\frac{(-1)^k c_{2k}}{z^{2k}}\sin z_1 - \sum_{k=0}^{\infty}\frac{(-1)^k c_{2k+1}}{z^{2k+1}}\cos z_1\right], \tag{3.38}$$

with

$$z = \frac{2}{3}\eta^{3/2}, \quad z_1 = z + \frac{\pi}{4}, \qquad c_k = \frac{\Gamma(3k+1/2)}{54^k\, k!\,\Gamma(k+1/2)}. \tag{3.39}$$

Making $\eta_1 \to \infty$, we have $Ai'(\eta_1) \sim -\sqrt{\eta_1}Ai(\eta_1)$. Then for a plate with free edge, Eq. (3.37) transforms into Eq. (3.17) which with $\nu = 0.3$ has the solution $\xi_0 = 0.9987$. But to obtain the value ξ_0 from Eq. (3.37), it is necessary to take a very large value m (in this case $m > 3566$). Therefore, the semi-plate is not an acceptable approximation to calculate, for a given m, the first frequency of a round plate. Moreover, Eq. (3.37) has a countable set of roots, while Eq. (3.17) has only one root.

Example 3.1. As an example, we give the first three dimensionless frequencies of a round plate with $m = 10$, $\nu = 0.3$. Columns 1 and 2 of Table 3.2 contain the values ξ (see (3.31)) for a plate with the free edge, as found from the exact equation (3.28) and the approximate equation (3.37), respectively. It is worth noting that the first frequency for a round plate differs substantially from the value $\xi = 0.9987$ for a semi-plate.

TABLE 3.2: Dimensionless frequencies of a round plate.

1 (2.28)	2 (2.37)	3 (2.31)	4 (2.31)
1.246	1.212	2.088	2.316
2.509	2.398	3.391	3.609
3.825	3.802	4.854	5.204

In Table 3.2, columns 3 and 4 contain the values of ξ for a round plate with the simply supported and clamped edges, respectively.

In Fig. 3.2, the first three vibration modes with $m = 10$ for a plate with a free edge are shown.

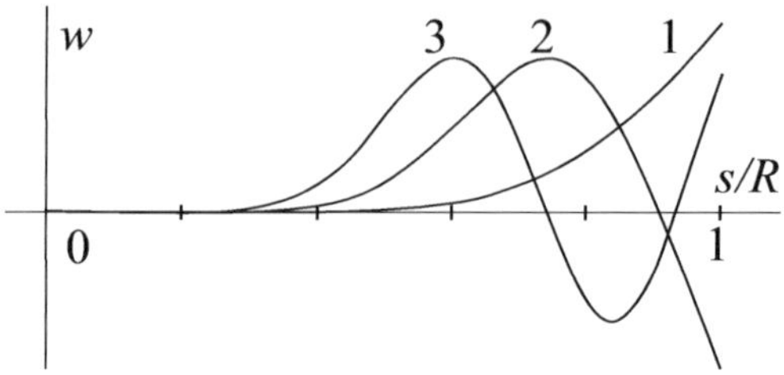

FIGURE 3.2: Three first vibration modes of a round plate with a free edge.

3.4 Vibrations of a shell of revolution

Consider Eqs. (3.3) or Eqs. (3.5). We shall seek a solution of these systems satisfying the boundary conditions with $s = s_1$ and $s = s_2$ and localized near one of the parallels.

We shall assume that $m \gg 1$, because only in this case the system has localized solutions. For $m \sim 1$ the system has 4 momentless solutions, which are not localized. The case $m \sim 1$ is discussed in details in [64]. For $m \gg 1$ all the solutions have the positive index of variation $\varsigma > 0$; they can be presented (except near the turning points) in the form (3.10)

$$\begin{aligned} u_i(s,\varphi) &= u_i^0(s)\exp\left\{\textstyle\int p_0(s)\,ds\right\}\cos m\varphi, \quad i=1,3, \quad u_3 = w, \\ u_2(s,\varphi) &= u_2^0(s)\exp\left\{\textstyle\int p_0(s)\,ds\right\}\sin m\varphi. \end{aligned} \tag{3.40}$$

with $p_0(s) = \mu^{-1}p(s)$.

The behavior of the solution depends substantially on the function $p_0(s)$. This function satisfies an algebraic equation of eighth order, which can be obtained by "freezing" the coefficients in Eqs. (3.5) and by substituting Eqs. (3.40). The homogeneous system for the coefficients u_i^0 reads as

$$\begin{aligned} &(p_0^2 - \nu_1 q_0^2 + \lambda_1)\,u_1^0 + \nu_2 p_0 q_0\,u_2^0 - (k_1 + \nu k_2)p_0\,u_3^0 = 0, \qquad \nu_1 = (1-\nu)/2, \\ &-\nu_2 p_0 q_0\,u_1^0 + (\nu_1 p_0^2 - q_0^2 + \lambda_1)\,u_2^0 - (k_2 + \nu k_1)q_0\,u_3^0 = 0, \qquad \nu_2 = (1+\nu)/2, \\ &(k_1 + \nu k_2)p_0\,u_1^0 + (k_2 + \nu k_1)q_0\,u_2^0 \\ &\qquad +(\lambda_1 - \mu_1^4(p_0^2 - q_0^2)^2 - k_1^2 - 2\nu k_1 k_2 - k_2^2)\,u_3^0 = 0, \end{aligned} \tag{3.41}$$

where all the linear values are related to R, the following notation being used

$$q_0 = \frac{m}{B}, \quad \lambda = \frac{\nu_3 R^2 \rho \omega^2}{E}, \quad \lambda_1 = (1-\nu^2)\lambda, \quad \mu_1^4 = (1-\nu^2)\mu^4, \quad \mu^4 = \frac{h_*^2}{12(1-\nu^2)}. \tag{3.42}$$

Some small terms in Eqs. (3.41) are omitted.

The determinant of Eqs. (3.41) gives the equation for $p_0(s)$,

$$\begin{aligned} F(p_0, s) &= (\lambda_1 + \Delta)(\lambda_1 + \nu_1\Delta)(\lambda_1 - \mu_1^4\Delta^2 - k_1^2 - 2\nu k_1 k_2 - k_2^2) \\ &+ (k_1 + \nu k_2)^2 p_0^2(\lambda_1 + \nu_1 p_0^2 - q_0^2) + (1+\nu)(k_1 + \nu k_2)(k_2 + \nu k_1) p_0^2 q_0^2 \\ &- (k_2 + \nu k_1)^2 q_0^2(\lambda_1 + p_0^2 - \nu_1 q_0^2) = 0, \qquad \Delta = p_0^2 - q_0^2. \end{aligned} \tag{3.43}$$

Assume that in Eq. (3.43) $\{k_1, k_2\} = O(1)$, $\mu \sim h_*^{1/2}$ is a small parameter, q_0 is a large parameter, $1 << q_0 << h_*^{-1}$, the order of the frequency parameter λ can change in a wide range, $0 < \lambda << h_*^{-2}$. We put

$$\lambda \sim h_*^{-2r}, \quad p_0 \sim h_*^{-\varsigma_1}, \quad q_0 \sim h_*^{-\varsigma_2}, \tag{3.44}$$

where the values r and ς_2 are given, and the parameter ς_1 is unknown. The parameter ς_1 may assume various values, because Eq. (3.43) has 8 roots. We find the approximate equations for the roots p_0 and give the ratio between the amplitudes u_i^0.

To find the orders ς_1 of the roots of Eq. (3.43) we shall use the Newton diagram (see, for example, [22]). The domain of the examined values of the parameters $r < 1$, $0 < \varsigma_2 < 1$, is divided into 6 (sub)domains (see Fig. 3.3), in which the orders ς_1 of the roots p of Eq. (3.43) are calculated by different formulas.

With the exception of the domain 4, the Newton diagram has two segments, each of which approximately defines 4 roots of Eq. (3.43). In this case 4 roots have the order ς_1' (it is the first group of roots), and the remaining 4 roots are of the order ς_1'' (the second group) ($\varsigma_1'' \geq \varsigma_1'$) (see Table 3.3). For cylindrical and conic shells ($k_1 = 0$), the domain 1 is divided into additional two parts, in which the orders ς_1' are calculated by different formulas. In the last line of Table 3.3 the orders corresponding to the point A in Fig. 3.3 are given. In this point with the relative error of the order h_* the equations of the technical shell theory (2.41) can be used.

TABLE 3.3: The order of the roots of Eq. (3.43)

N				ς_1'	ς_1''
1	$\varsigma_2 \leq 1/2$	$r \leq 0$	$k_1 \neq 0$	ς_2	$1/2$
1′	$\varsigma_2 \leq 1/2$	$2\varsigma_2 - 1 \leq r \leq 0$	$k_1 = 0$	$\varsigma_2 + r/2$	$1/2$
1″	$\varsigma_2 \leq 1/2$	$r \leq 2\varsigma_2 - 1$	$k_1 = 0$	$2\varsigma_2 - 1/2$	$1/2$
2	$\varsigma_2 \leq 1/2$	$0 \leq r \leq \varsigma_2$		ς_2	$(1+r)/2$
3	$\varsigma_2 \leq 1/2$	$r \geq \varsigma_2$		r	$(1+r)/2$
4	$\varsigma_2 \geq 1/2$	$r \leq 2\varsigma_2 - 1$		ς_2	ς_2
5	$\varsigma_2 \geq 1/2$	$\varsigma_2 \leq r \leq 2\varsigma_2 - 1$		$(1+r)/2$	ς_2
6	$\varsigma_2 \geq 1/2$	$r \geq \varsigma_2$		$(1+r)/2$	r
A	$\varsigma_2 = 1/2$	$r = 0$		$1/2$	$1/2$

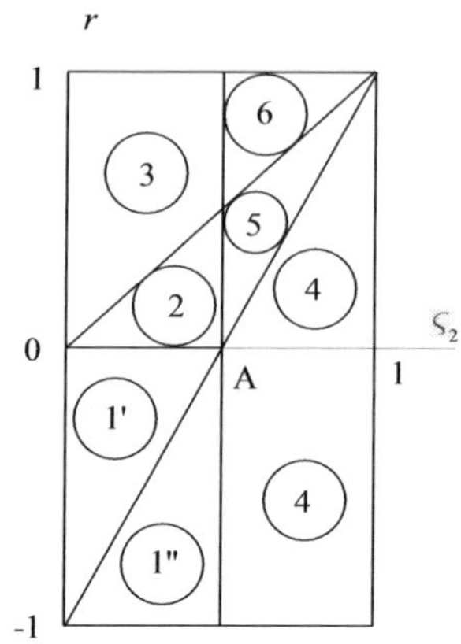

FIGURE 3.3: Division of the domain of parameters.

The following approximate equations can be used to calculate the roots both of the first and the second groups strictly within the domains 1–6:

Domain 1. The roots of the second group satisfy the equation $\mu^4 p^4 + k_2^2 = 0$, which gives the simple edge effect solutions. The roots of the first group with $k_1 \neq 0$ satisfy the equation $(k_2 p_0^2 - k_1 q^2)^2 = 0$, and with $k_1 = 0$, satisfy the equation $k_2^2 p_0^4 - \lambda_1 q_0^4 + \mu^4 q_0^8 = 0$, in which in the sub-domains $1'$ and $1''$ the second and the third summands are the main, respectively.

Domains 2 and 3. In both domains, the roots of the second group satisfy the equation $\mu^4 p_0^4 - \lambda_1 = 0$. The roots of the first group in domain 2 satisfy the equation $(p_0^2 - q_0^2)^2 = 0$, and in domain 3, satisfy the equation $(p_0^2 + \lambda_1)(\nu_1 p_0^2 + \lambda_1) = 0$.

Domain 4. In this domain, the Newton diagram has only one segment and all 8 roots of Eq. (3.43) approximately satisfy the equation $(p_0^2 - q_0^2)^4 = 0$.

Domains 5 and 6. In these domains, the roots of the first group satisfy the equation $\mu^4 p_0^4 - \lambda_1 = 0$. The roots of the second group in domain 5 satisfy the equation $(p_0^2 - q_0^2)^2 = 0$, and in domain 6, satisfy the equation $(p_0^2 + \lambda_1)(\nu_1 p_0^2 + \lambda_1) = 0$.

If the point (ς_2, r) in Fig. 3.3 approaches to one of the boundaries of the domains, then the exactness of these approximate equations decreases. If the point (ς_2, r) lies on one of the boundaries, then these approximate equations are unacceptable. At the point A ($\varsigma_2 = 1/2, r = 0$), the approximate equation for the both groups read as

$$\mu^4 (p_0^2 - q_0^2)^4 - \lambda (p_0^2 - q_0^2)^2 + (k_2 p_0^2 - k_1 q_0^2)^2 = 0, \qquad (3.45)$$

This equation is a characteristic equation of Eqs. (3.41) of the technical shell theory. After the separation of variables, (3.40), this system assumes the form

$$\mu^4 \Delta_q \Delta_q w - \lambda w + \mu^2 \Delta_k \Phi = 0, \qquad \mu^4 \Delta_q \Delta_q \Phi - \mu^2 \Delta_k w = 0, \qquad (3.46)$$

where

$$\Delta_q w = \frac{1}{B}\frac{d}{ds}\left(B\frac{dw}{ds}\right) - q_0^2 w, \quad \Delta_k w = \frac{1}{B}\frac{d}{ds}\left(Bk_2\frac{dw}{ds}\right) - q_0^2 w, \qquad q_0 = \frac{m}{B}. \tag{3.47}$$

Equations (3.46) describe the quasi-transversal solutions, because for them the inequalities $\{|u_1|, |u_2|\} \ll |w|$ are valid. Equations (3.46) can be used not only at the point A, but also in domains 1, 2, 4 and 5. The approximate equations acceptable in these domains can be obtained from Eqs. (3.46) by omitting the small terms.

Let the initial stress resultants $T_1^0(s)$ and $T_2^0(s)$ be given and let a shell rest on an elastic foundation. The effect of an initial torsion S^0 is not studied, because in this case the separation of variables (3.40) is impossible.

Now Eqs. (3.46) are generalized as

$$\begin{aligned} &\mu^4\Delta_q\Delta_q w + 2\mu^2\lambda_t\Delta_t w - \lambda w + c_1 w + \mu^2\Delta_k\Phi = 0, \\ &\mu^4\Delta_q\Delta_q\Phi - \mu^2\Delta_k w = 0, \end{aligned} \tag{3.48}$$

with

$$\Delta_t w = \frac{1}{B}\frac{d}{ds}\left(Bt_1\frac{dw}{ds}\right) - q_0^2 t_2 w, \quad T_i^0 = -2Eh\lambda_1 t_i, \; i = 1, 2, \quad c_1 = \frac{c}{Eh}. \tag{3.49}$$

Here, $\lambda_1 \geq 0$ is the load parameter, t_i are the dimensionless initial stress-resultants that are positive at compression, c_1 is a dimensionless stiffness of foundation.

Equation (3.45) also changes its form

$$\mu^4(p_0^2-q_0^2)^4+(p_0^2-q_0^2)^2(2\lambda_t(t_1p_0^2-t_2q_0^2)+c_1-\lambda)+(k_2p_0^2-k_1q_0^2)^2 = 0. \tag{3.50}$$

Here, for the Winkler model of a foundation, the coefficient c_1 is constant, and for the Il'gamov model (2.52) for the purely imaginary roots p_0 we get $c_1 = c_2\sqrt{q_0^2 - p_0^2}$.

3.5 Non-uniform localized modes of shell vibrations

We recall that non-uniform localized modes are linear combinations of decreasing solutions; they do not contain oscillating solutions. In domains 3 and 6 (see Fig. 3.3), the non-uniform modes do not exist, because in these domains some solutions are oscillating. In the remaining domains, Eq. (3.45) can be used to find the approximate values of the roots of the characteristic equation (3.43). We rewrite Eq. (3.45) as

$$(p^2 - q^2)^4 - \lambda(p^2 - q^2)^2 + (k_2p^2 - k_1q^2)^2 = 0, \qquad p = \mu p_0, \quad q = \mu q_0. \tag{3.51}$$

We seek non-uniform localized modes near the edge $s = s_2 > s_1$ in the form

$$w(s, \lambda) = \sum_{k=1}^{4} C_k w_k e^{\mu^{-1} p_k s}, \tag{3.52}$$

where p_k are the roots of Eq. (3.51) satisfying the condition

$$\mathrm{Re}(p_k) > 0, \tag{3.53}$$

and C_k are arbitrary constants. We seek the rest of the unknowns u_1, u_2, γ_1, T_1, S_1^*, Q_1^*, M_1 contained in the boundary conditions (2.27) by the same form (3.52). Instead of w_k in Eq. (3.52) we take

$$\begin{gathered}
u_{1k} = \frac{k_1 - (q^2 + \nu p_k^2)\Phi_k}{p_k}, \quad u_{2k} = \frac{k_2 + (p_k^2 + \nu q^2)\Phi_k}{q}, \quad \Phi_k = -\frac{k_2 p_k^2 - k_1 q^2}{(p_k^2 - q^2)^2}, \\
w_k = 1, \quad \gamma_{1k} = -\mu^{-1} p_k, \quad T_{1k} = -Ehr^2\Phi_k, \quad S_{1k}^* = -Ehp_k q\Phi_k, \\
Q_{1k}^* = -Eh\mu p_k(p_k^2 - (2-\nu)q^2), \quad M_{1k} = Eh\mu^2(p_k^2 - \nu q^2).
\end{gathered} \tag{3.54}$$

All the values in Eqs. (3.54) are calculated for $s = s_2$. Of course, Eqs. (3.52) and (3.54) are approximate. Their exactness decreases away from the edge.

Equation (3.51) contains only even powers p, and hence the presence of 4 roots satisfying condition (3.53) is equivalent to the absence of (pure) imaginary roots. The frequency equation has the form of a determinant of fourth order

$$\Delta(\lambda) = 0, \tag{3.55}$$

whose rows contain the values (3.54).

For example, for a free edge ($T_1 = S_1^* = Q_1^* = M_1 = 0$), the frequency equation reads as

$$\Delta(\lambda) = \begin{vmatrix} T_{11} & T_{12} & T_{13} & T_{14} \\ S_{11}^* & S_{12}^* & S_{13}^* & S_{14}^* \\ Q_{11}^* & Q_{12}^* & Q_{13}^* & Q_{14}^* \\ M_{11} & M_{12} & M_{13} & M_{14} \end{vmatrix} = 0. \tag{3.56}$$

For $\lambda = 0$, Eq. (3.51) has no imaginary roots. We denote by $\lambda = \lambda_0$ such a value for which at $\lambda < \lambda_0$ Eq. (3.51) has no imaginary roots and at $\lambda > \lambda_0$ some roots of this equation are imaginary. Rewriting Eq. (3.51), we have

$$\lambda = f(z), \qquad f(z) = (z + q^2)^2 + k_2^2 \frac{(z + \kappa q^2)^2}{(z + q^2)^2}, \tag{3.57}$$

with

$$z = -p^2, \qquad \kappa = \frac{k_1}{k_2}. \tag{3.58}$$

Therefore,

$$\lambda_0 = \min_{z \geq 0} f(z) = \begin{cases} f(0), & q^4 + k_1(k_2 - k_1) \geq 0, \\ f(z_0), & q^4 + k_1(k_2 - k_1) < 0, \end{cases} \tag{3.59}$$

where z_0 is the positive root of the equation

$$f'(z) = 0 \quad \text{or} \quad (z + q^2)^4 + (k_2 - k_1)(k_2 z + k_1 q^2) = 0. \tag{3.60}$$

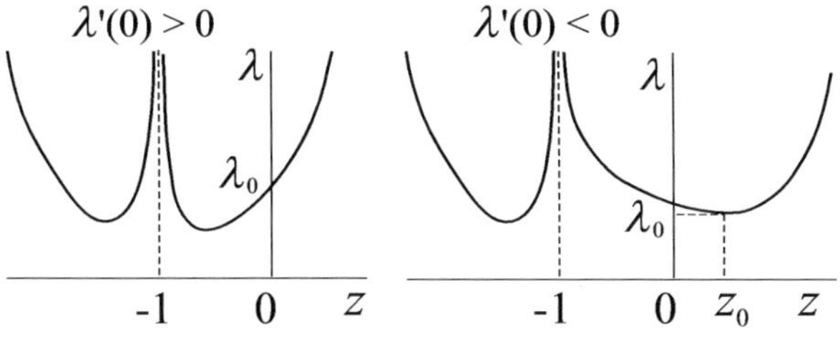

FIGURE 3.4: Functions $\lambda(z)$ at $\lambda'(0) > 0$ and at $\lambda'(0) < 0$.

The plots of the function $\lambda(z)$ is shown Fig. 3.4. For $0 < \lambda < \lambda_0$, Eq. (3.55) has roots only for some variants of boundary conditions. We call a variant of the boundary conditions for which Eq. (3.55) has roots a *weak support.* There are non-uniform localized vibration modes for weak supports.

Calculations show that Eq. (3.55) may have a root only for four variants of boundary conditions: 0000, 0001, 0010 and 0011, containing an edge free or non-tangential constraints $w = 0$ and/or $\gamma_1 = 0$. The presence of tangential constraints $u_1 = 0$ and/or $u_2 = 0$ prevents an appearance of non-uniform vibration modes

Figure 3.5 shows some solutions of Eq. (3.55) for a free edge (0000) and for boundary conditions $T_1 = S = Q_1^* = \gamma_1 = 0$ (0001) with $\nu = 0.3$ in the dependence of two parameters: the wave parameter r and the curvature parameter κ. Besides,

$$\lambda_* = \frac{\lambda}{\lambda_0}, \qquad r = \frac{\mu m}{B(s_2)}, \qquad \kappa = \frac{k_1}{k_2}. \tag{3.61}$$

Without loss of generality, we shall assume that $k_2 = 1$, which can be attained by the choice of the scale R to which are related all linear dimensions of a shell.

For small values of r, the curves $\lambda(r, \kappa)$ for shells with positive ($\kappa > 0$) and negative ($\kappa < 0$) Gaussian curvature are essentially different, and for $r \gg 1$, these curves are close to each other if the values $|\kappa|$ are the same (see Fig. 3.5).

For cylindrical and conical shells ($\kappa_1 = 0$) non-uniform localized forms do not exist.

A rate of decreasing the non-uniform form can be estimated by the value

$$\zeta = \min_k \{\exp(-\mu^{-1}\Re(p_k))\}, \qquad k = 1, 2, 3, 4. \tag{3.62}$$

For a precise localization the condition

$$\zeta \ll 1 \tag{3.63}$$

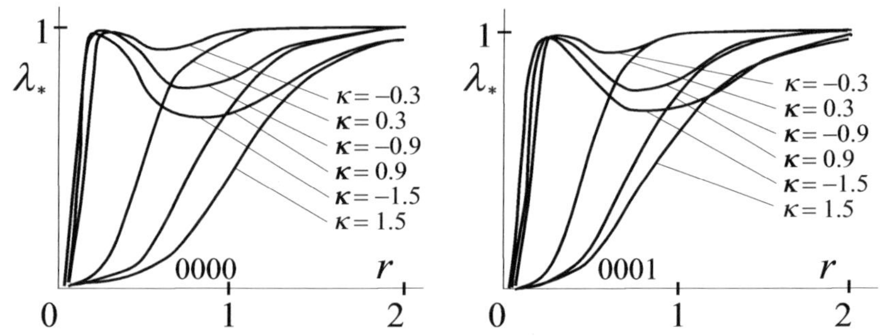

FIGURE 3.5: Relative frequencies of non-uniform localized vibrations.

is to be fulfilled. In the opposite case, the mode does not decrease enough until the opposite edge of a shell or it is incorrect to assume that the values like $w_k(s)$ in the solution (3.52) are constant.

For $\lambda = \lambda_0$ (or $\lambda_* = 1$), Eq. (3.51) has multiple roots; for $\lambda_* < 1$, purely imaginary roots are absent, and for $\lambda_* > 1$, there are one or two pairs of purely imaginary roots. Due to the continuous dependence of roots $p(\lambda)$ on λ with $\lambda_* \simeq 1$, the real parts of some roots of Eq. (3.51) are small, and the condition (3.63) is not fulfilled. Also this condition is not fulfilled if $r \ll 1$, because in this case the moduli of some roots of Eq. (3.51) are small. Therefore, for the values of parameters corresponding to the upper and lower parts of the graphs in Fig. 3.5, the localization is not precise.

For the boundary conditions 0010 and 0011, including the constraint $w = 0$, the non-uniform localized modes are possible in the comparatively narrow domain of parameters. For $\kappa \leq 0.3$ the non-uniform modes are absent, for $\kappa = 0.9$, such forms are present with $r < 0.6$, and for $\kappa = 1.5$ they are present with $r < 1$.

The difference of shells with positive and negative Gaussian curvature for the boundary conditions 0000 and 0001 was already discussed (see Fig. 3.5). For the boundary conditions 0010 and 0011, this difference consists in the absence of precise localized modes for shells with negative Gaussian curvature, for which the system of dimensionless equations is a hyperbolic one. Excitations in hyperbolic systems are diffused on the asymptotic lines without decreasing (see [68], [69]), and for small q, the vibration modes are not localized near the edge.

To compare the asymptotic and numerical results, we consider the truncated paraboloid of revolution with the parameters: $B = z^2/2$, $z_0 = 0.3 \leq z \leq z_1 = 1$, $m = 10$, $h = 0.01$, $\nu = 0.3$ (see Fig. 3.6). The wide edge $z = 1$ is free (0000) or a constraint $\gamma_1 = 0$ is imposed (0001). The boundary conditions at the narrow edge $z = 0.3$ may be arbitrary.

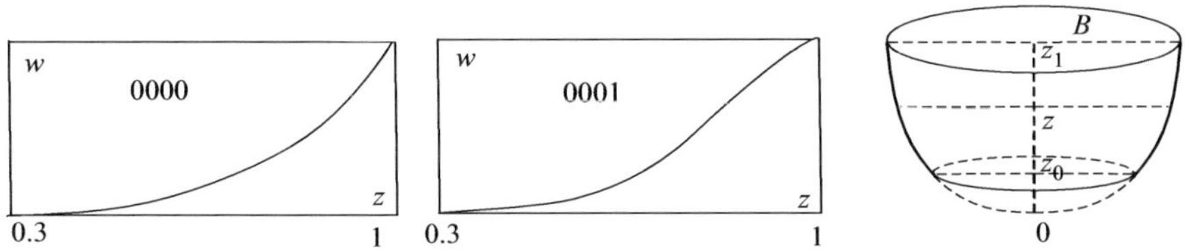

FIGURE 3.6: Non-uniform vibration modes of a paraboloid of revolution.

In Fig. 3.6, the vibration modes localized near the free (0000) or near the weakly supported (0001) edge are shown. Table 3.4 gives the frequency parameters λ obtained by the asymptotic approach from Eq. (3.55) (λ_{as}) and by a direct numerical solution of Eqs. (3.5) (λ_{num}). The numerical solution was performed by the method described in [237]. The asymptotic value λ_{as} does not depend on the boundary conditions at the narrow edge $z = z_0$. The numerical solution shows that the dependence of the frequency parameter λ on the boundary conditions at the edge $z = z_0$ is very small. In Table 3.4, the limits for λ_{num} are given. The difference between λ_{as} and λ_{num} is not small. To explain this difference, we note that the value λ_{as} is found for the shell parameters at the edge $z = 1$, and within the shell the curvature (and hence, the stiffness) is larger, that explains the inequality $\lambda_{as} < \lambda_{num}$.

TABLE 3.4: Comparison of the asymptotic and numerical results.

	λ_{as}	λ_{num}
0000	0.02872	0.03344 — 0.03346
0001	0.03419	0.04325 — 0.04326

3.5.1 Effect of initial momentless stresses on the non-uniform vibration modes

Consider an initial axisymmetric stress state described by stress-resultants $T_1^{(0)}$, $T_2^{(0)}$. The problem on the non-uniform localized vibrations for the pre-stressed shell is similar to that of the existence of localized near the edge non-uniform buckling modes of a pre-stressed shell. In contrast to the vibrational problem, the non-uniform buckling modes are possible in 8 variants of weak boundary conditions (see [222, 226]). These variants are shown in legends for Fig. 3.7.

As a result, the non-uniform vibration modes of a pre-stressed shell are possible with the same 8 variants of boundary conditions. To investigate the pre-stressed non-uniform vibration modes, we use Eqs. (3.51), (3.55) with $c_1 = 0$. We assume that the loading parameter λ_t is less than its critical value λ_t^* for which the buckling takes place.

In the presence of initial stresses, the boundary condution $Q_{1k}^* = 0$ in Eqs. (3.54) is to be replaced by

$$Q_{1k}^* = -Eh\mu p_k(p_k^2 - (2-\nu)q^2 + \lambda_t)w_k = 0. \tag{3.64}$$

As an example, we consider non-uniform vibrations of a cylindrical shell pre-stressed by an axial loading. Note at first that such a shell without initial stresses has no localized vibration modes.

In Eq. (3.50), we put without loss of generality

$$k_1 = 1, \quad k_2 = 0, \quad t_1 = 1, \quad t_2 = 0. \tag{3.65}$$

At $\lambda_t^*(q)$ we denote the dimensionless critical axial compression for the given $q = \mu m$. For the well supported shell

$$\min_q \lambda_t^*(q) = 1, \tag{3.66}$$

that corresponds to the classic critical Lorentz–Timoshenko load [122], [218], and a buckling mode occupies all shell surface. For a weakly supported edge the localization of a buckling mode near this edge takes place, and the critical load decreases. There are 8 variants of the weak support, the corresponding functions $\lambda_t^*(q)$ are shown in Fig. 3.7.

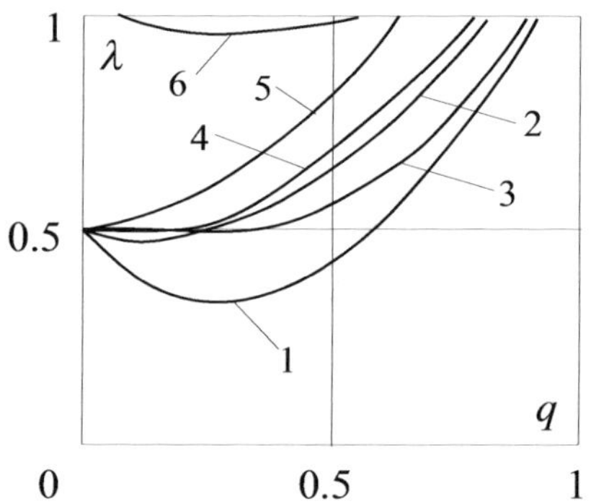

FIGURE 3.7: Graphics of functions $\lambda = \lambda_t^*(q)$ for various boundary conditions (*1* — 0000, *2* — 0100, *3* — 1000, *4* — 1100, *5* — 0010, 1010, 0001, *6* — 0101).

A construction of the non-uniform vibration modes can be performed by solving Eq. (3.55) in the domain

$$\lambda < \lambda_0, \qquad \lambda_t < \lambda_t^*(q). \tag{3.67}$$

Calculations show that the non-uniform vibration modes appear in the intervals $\lambda_t^- < \lambda_t < \lambda_t^*(q)$ of the axial loads for all 8 variants of boundary conditions, see Fig. 3.7.

Example 3.2. Consider the boundary conditions 1100 or $u_1 = u_2 = Q_1^* = M_1 = 0$. Three values of the wave parameter q are studied.

TABLE 3.5: Frequencies $\lambda(\lambda_t, q)$ of non-uniform vibrations.

q	λ_t	$\lambda(\lambda_t, q)$	q	λ_t	$\lambda(\lambda_t, q)$	q	λ_t	$\lambda(\lambda_t, q)$
0.4	0.557	0.000	0.5	0.618	0.000	0.6	0.710	0.000
0.4	0.554	0.005	0.5	0.611	0.012	0.6	0.698	0.026
0.4	0.550	0.010	0.5	0.603	0.025	0.6	0.686	0.052
0.4	0.546	0.015	0.5	0.594	0.038	0.6	0.671	0.078
0.4	0.541	0.020	0.5	0.582	0.050	0.6	0.649	0.104

In the first line of Table 3.5, the values $\lambda_t = \lambda_t^*$ are given (see curve 4 in Fig. 3.7). In the rest lines, the parameter $\lambda_t < \lambda_t^*$ is taken, and the corresponding value of the frequency parameter, $\lambda(\lambda_t, q)$, of the localized vibration mode is given. The parameter, $\lambda(\lambda_t, q)$, grows with decreasing the parameter, λ_t.

Remark 3.1. For the boundary conditions shown in Table 3.5, $\lambda_t^{**} = \min_q \lambda_t^*(q) = 1/2$, therefore for the values λ_t containing in this table a shell loses its stability for the other values q and the results become doubtful. These results are correct for a compressed cylindrical panel with simply supported rectilinear edges for which it is possible to separate the variables, $w(s, \varphi) = w(s) \sin m\varphi,\ \ 0 \le \varphi \le \varphi_0$, with the appropriate value φ_0.

The analysis of non-uniform vibration modes for the TR model is given in Section 3.10.

3.6 Oscillating vibration modes localized near the edge

Consider vibrations of a shell of revolution with large number m of waves in the circumferential direction. We write again Eqs. (3.7) in the standard form with a small parameter at the derivatives

$$\mu \frac{d\mathbf{X}}{ds} = \mathbf{A}(s, \mu)\mathbf{X}, \tag{3.68}$$

and the characteristic equation in the forms (3.51) and (3.57)

$$(p^2 - q^2)^4 - \lambda(p^2 - q^2)^2 + (k_2 p^2 - k_1 q^2)^2 = 0, \tag{3.69}$$

$$\lambda = f(z, s), \qquad f(z, s) = (z + q^2)^2 + \frac{(k_2 z + k_1 q^2)^2}{(z + q^2)^2}, \quad z = -p^2. \tag{3.70}$$

Consider the case when Eq. (3.69) has for all s no more than one pair of purely imaginary roots. This case takes place if for all s the inequality

$$f_z'(0, s) = q^4 + k_1(k_2 - k_1) > 0. \tag{3.71}$$

is fulfilled (see the left part of Fig 3.4). This inequality is fulfilled for a conic shell, for an elongated paraboloid, and also for an arbitrary shell of revolution with the sufficiently large m.

Let the function $g(s) = f(0, s) = q^4 + k_1^2$ be monotonically decreasing. We set

$$\lambda_0 = \min_s g(s) = g(s_2), \quad \lambda_+ = \max_s g(s) = g(s_1). \tag{3.72}$$

Then for $\lambda < \lambda_0$ Eq. (3.69) does not have purely imaginary roots and only non-uniform vibration modes are possible. For $\lambda_0 \leq \lambda < \lambda_+$, there exists a turning point s_*, $s_1 < s_* \leq s_2$, such that for $s = s_*$ Eq. (3.69) has a twofold zero root. At $s < s_*$, Eq. (3.69) does not have purely imaginary roots, and for $s > s_*$ this equation has one pair of purely imaginary roots.

An asymptotic analysis of localized oscillating modes is described in [64]. If the root $p_k(s)$ of Eq. (3.69) is simple, then the asymptotic solution of Eq. (3.68) read as

$$\begin{aligned} &w_k(s, \mu) = \textstyle\sum_{n=0}^{\infty} \mu^n w_k^{(n)}(s) \exp\left\{\mu^{-1} \int p(s) ds\right\}, \\ &w_k^{(0)}(s) = (p^2 - q^2)\left(B\, f_p'\right)^{-1/2}. \end{aligned} \tag{3.73}$$

The remaining unknowns of Eqs. (3.68) have the similar expansions; the main coefficients of series (3.73) (instead $w_k = w_k^{(0)}$) are given in Eqs. (3.54). Here and below, it is supposed that the coefficients in Eqs. (3.68) are holomorphic functions in s.

For the twofold with $s = s^*$ roots $p_1(s)$ and $p_2(s)$ of Eq. (3.69) such that $p_{1,2} \sim \pm\sqrt{s_* - s}$, the main terms of the asymptotic expansions read as [64], [23, 25]

$$w_{1,2}(s, \mu) = \frac{w_1^{(0)}(s) p_1(s)}{\sqrt{\zeta'(s)}} v(\eta), \tag{3.74}$$

where $v(\eta)$ is the solution of the Airy equation, $w_1^{(0)}(s)$ is given in Eq. (3.73) with $k = 1$,

$$\eta = \mu^{-2/3} \zeta(s), \qquad \zeta(s) = \left(\frac{3}{2} \int_{s_*}^{s} \sqrt{p_1(s)} ds\right)^{2/3}. \tag{3.75}$$

We take the branch of the function $\zeta(s)$, for which $\zeta'(s) > 0$. Then for the real $v(\eta)$ the functions (3.74) are real and holomorphic.

Formulas similar to Eq. (3.74) take place for the functions u_2, T_1, M_1 with even powers of p in Eqs. (3.54). For example,

$$u_2(s, \mu) = u_{21}(s) w_{1,2}(s, \mu), \quad \{u_2, T_1, M_1\}. \tag{3.76}$$

where u_{21} is given in Eqs. (3.54).

For the functions u_1, γ_1, S, Q_1^* with the odd powers of p, we get

$$u_1(s,\mu) = \mu^{1/3}\, u_{11}(s)\, w_1^{(0)}(s)\sqrt{\zeta'(s)}\,\frac{dv}{d\eta}, \qquad \{u_1,\, \gamma_1,\, S,\, Q_1^*\}. \tag{3.77}$$

To construct a vibration mode, localized near the edge s_2, we take solutions (3.74), (3.76) and (3.77) with the Airy function $v(\eta) = Ai(\eta)$, decreasing as $\eta \to \infty$ (see Fig. 3.1) and three solutions (3.73) with $\Re(p_k) > 0$ (we take $k = 3, 4, 5$). We ignore the boundary conditions at the edge, $s = s_1$. Then the boundary conditions (2.27) with $s = s_2$ lead to an equation for a frequency parameter λ (as a determinant of 4th order). After a simplification, this equation reads as [64]

$$\begin{gathered} Ai(-\eta_2) + \mu^{1/3}\zeta'(s_2)\, d(s_2)\, Ai'(-\eta_2) = 0, \\ \eta_2 = \left(\frac{3}{2\mu}\int_{s_*}^{s_2} |p_1(s)|\, ds\right)^{2/3} > 0, \end{gathered} \tag{3.78}$$

where $d(s_2)$ depends on the selected variant of the boundary conditions with $s = s_2$. For example, for a free edge (0000) we get

$$\det\begin{vmatrix} d\cdot T_{11} & T_{13} & T_{14} & T_{15} \\ p_1^{-1}S_1 & S_3 & S_4 & S_5 \\ p_1^{-1}Q_{11}^* & Q_{13}^* & Q_{14}^* & Q_{15}^* \\ d\cdot M_{11} & M_{13} & M_{14} & M_{15} \end{vmatrix} = 0, \tag{3.79}$$

where the elements of this determinant are given in Eqs. (3.54) with $s = s_2$. In the first column, the function in Eqs. (3.77) has the multiplier p_1^{-1} and the functions in Eqs. (3.73), (3.76) have the multiplier d.

For a simply supported edge (0110), in the first column all the functions have the multiplier d; therefore, $d = 0$, and Eq. (3.78) becomes simpler,

$$Ai(-\eta_2) = 0. \tag{3.80}$$

Equation (3.78) has roots for any boundary conditions (see Fig. 3.1). The asymptotic expansions of the first roots of Eq. (3.78) read as

$$\lambda^{(k)} = \lambda_0 + \mu^{2/3}a_k + \mu b_k + O\left(\mu^{4/3}\right), \quad a_k = \eta^{(k)}(f_z'(0,s_2))^{1/3}(-g'(s_2))^{2/3}, \tag{3.81}$$

where $\eta^{(k)}$ $(k = 1, 2, ...)$ is the kth root of Eq. (3.80) $(\eta^{(1)} = 2.338,\ \eta^{(2)} = 4.088, \ldots)$. The coefficient a_k in Eq. (3.81) does not depend on the boundary conditions, and the coefficient b_k depends on $d(s_2)$.

If the value $s_2 - s_*$ is not small, then $\eta_2 \gg 1$, and by using the asymptotic equations (3.38), Eq. (3.78) is simplified to read

$$\mathrm{tg}\left(\mu^{-1}\int_{s_*}^{s_2} |p_1(s)|\, ds + \frac{\pi}{4}\right) = d\,|p_1(s_2)|. \tag{3.82}$$

From Eqs. (3.81) and (3.72) it follows that for a fixed number m of waves in the circumferential direction with $\lambda > \lambda_0$ the spectrum becomes compact. The order of the distance between the neighbouring points of spectrum at first is $O(\mu^{2/3})$, and then it is $O(\mu)$.

This asymptotic analysis is useful for qualitative estimates, while in order to construct the exact natural frequencies and vibration modes one should solve Eqs. (3.5) numerically.

Example 3.3. Consider vibrations of an truncated circular cone (see Fig. 3.8 with parameters $m = 15$, $s_1 = 0.5 \le s \le s_2 = 1$, $\nu = 0.3$, $\alpha = 60°$, $h = 0.01$, where 2α is the angle at the cone vertex, s is the distance from the vertex. For $\lambda_0 = 0.0129 < \lambda < \lambda_+ = 0.206$ within the studied interval $S = (s_1, s_2)$, there is a turning point. Depending on the boundary conditions, the interval $[\lambda_0, \lambda_+]$ contains 3 or 4 eigenvalues. Fig. 3.8 shows the first four vibration modes, $w(s)$ if the narrow edge s_1 is clamped and the wide edge s_2 is free.

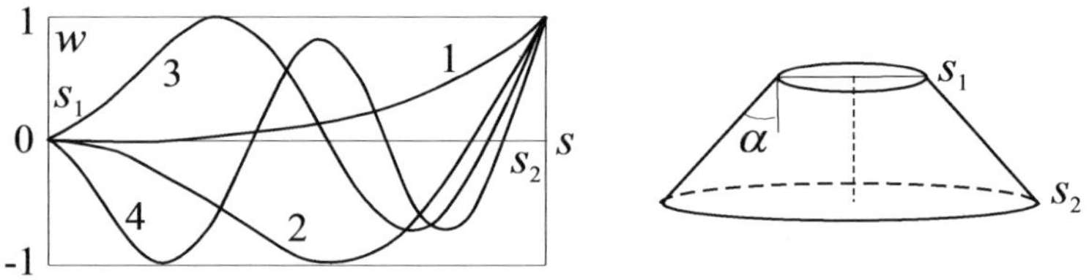

FIGURE 3.8: Vibration modes of a conic shell.

In Table 3.6 the first four eigenvalues $\lambda^{(k)}$ are given for three variants of boundary conditions at the wide edge: free edge (0000), simply supported edge (0110), and clamped edge (1111). In all cases the narrow end is clamped.

TABLE 3.6: Frequency parameters of a truncated cone.

Boundary conditions at $s = s_2$	$\lambda^{(1)}$	$\lambda^{(2)}$	$\lambda^{(3)}$	$\lambda^{(4)}$
0000	0.01825	0.05298	0.10792	0.19067
0110	0.03999	0.08735	0.16044	0.27620
1111	0.04635	0.10637	0.17508	0.30760

Figure 3.8 shows that only the first vibration mode is localized near the edge, the remaining modes occupying the entire shell surface. The manifestation of the localization is that the first eigenvalues $\lambda^{(k)}$ practically do not depend on the boundary conditions on the narrow edge. Calculations show that, depending on the boundary conditions, only the third or fourth digit changes. For a thinner shell the localization effect is more clear.

3.7 Construction algorithm of localized solutions

In Sections 3.5 and 3.6, we constructed the vibration modes localized near the edge. In Section 3.8, we shall study vibration modes for a shell of revolution, localized near the *weakest parallel* lying away from the shell edges. Chapter 4 will be concerned with the modes localized near the *weakest generatrix* for a cylindrical and conic shells. Here, to test ODE of order $2n$ with small parameter, the general algorithm of localized modes construction is given (see [222], [22], [23]). A similar algorithm for partial differential equations was proposed in [131]. In the next chapters this algorithm will be generalized to describe the motion of wave packages.

Consider the self-adjoined ODE of order $2n$,

$$\sum_{k=0}^{n}(-\mathrm{i}\mu)^{2k}\frac{d^k}{dx^k}\left(a_k(x)\frac{d^k y}{dx^k}\right)=0, \quad x_1\le x\le x_2, \tag{3.83}$$

where $\mu>0$ is a small parameter, the coefficients a_k are real holomorphic and linearly depending on the unknown parameter $\lambda>0$,

$$a_k=a_{1k}-\lambda a_{2k}, \quad \lambda>0, \quad a_{1k},a_{2k}\in C^{\infty}. \tag{3.84}$$

We search for eigenvalues λ, for which near the point x_0 $(x_1<x_0<x_2)$ there exist localized solutions of Eq. (3.83), which exponentially decreases as $|x-x_0|$ grows. The point x_0 is referred as a *weakest point.*

Some shell vibration problems can be reduced to this problem after separation of variables. As a rule, we get a system of equations of type (3.83) with coefficients regularly depending on the parameter μ. The algorithm, which will be described below, can be used in this case without major changes.

We seek, at first, the solution of Eq. (3.83) as a formal series:

$$y(x,\mu)=\sum_{n=0}^{\infty}\mu^n y_n(x)\exp\left\{\frac{\mathrm{i}}{\mu}\int p(x)\,dx\right\}. \tag{3.85}$$

Here, $p(x)$ satisfies the algebraic equation

$$\sum_{k=0}^{n}a_k(x)p^{2k}=0, \tag{3.86}$$

which can be obtained by freezing the coefficients in Eq. (3.83). It is worth noting that the solution (3.85) is holomorphic only at x for which the root $p(x)$ is simple.

The function (3.85) satisfies the above decay condition if there exist two points $x_*^{(1)}, x_*^{(2)}$ such that

$$\begin{aligned} \Im(p(x)) &> 0 \qquad \text{for} \quad . > x_*^{(2)}, \\ \Im(p(x)) &= 0 \qquad \text{for} \quad x_*^{(1)} \le x \le x_*^{(2)}, \\ \Im(p(x)) &< 0 \qquad \text{for} \quad x < x_*^{(1)}. \end{aligned} \tag{3.87}$$

Conditions (3.87) are fulfilled only if the points $x_*^{(1)}$ and $x_*^{(2)}$ are turning points. Assume that these point are close to each other and approach the point x_0 as $\mu \to 0$. In this case, the asymptotic expansion based on the standard Airy functions is unacceptable, and so we shall describe another algorithm.

We solve Eq. (3.86) with respect to the parameter λ:

$$\lambda = f(p, x) \equiv \sum_{k=0}^{n} a_{1k}(x) p^{2k} \left(\sum_{k=0}^{n} a_{2k}(x) p^{2k} \right)^{-1}. \tag{3.88}$$

Let the minimum

$$\lambda_0 = \min_{p,x} \{ f(p, x) \} = f(p_0, x_0) \tag{3.89}$$

exist for real $p \ge 0$, and $x_1 \le x_0 \le x_2$, and let the function $f(p, x)$ have the strict minimum such that the second differential of the function f at the point (p_0, x_0) is a positively definite quadratic form

$$d^2 f = f_{pp}^0 dp^2 + 2 f_{px}^0 dp dx + f_{xx}^0 dx^2 > 0, \qquad f_{px}^0 = \frac{\partial^2 f(p_0, x_0)}{\partial p_0 \partial x_0}, \dots \tag{3.90}$$

The derivatives f_{pp}^0, f_{ps}^0, and f_{ss}^0 are calculated at the point (p_0, x_0).

For $\lambda < \lambda_0$, Eq. (3.86) does not have real roots. At $\lambda = \lambda_0$, $x = x_0$, this equation has a twofold root $p = p_0$. At $\lambda = \lambda_0 + d\lambda$ ($d\lambda > 0$, $d\lambda = O(\mu)$), there exists the interval $x_*^{(1)} < x < x_*^{(2)}$ in which Eq. (3.86) has real roots and does not have them outside this interval.

Under these assumptions, the asymptotic solution of Eq. (3.83), as localized near the point x_0, reads as

$$y(x, \mu) = \sum_{k=0}^{\infty} \mu^{k/2} y_k(\xi) \exp\left[\mathrm{i}\mu^{-1/2} p_0 \xi - (1/2) a \xi^2) \right], \tag{3.91}$$

where

$$\xi = \mu^{-1/2}(x - x_0), \qquad \lambda = \lambda_0 + \mu\lambda_1 + \dots, \tag{3.92}$$

and $y_k(\xi)$ are the polynomials in ξ, the values p_0, x_0, λ_k are real, $\Re(a) > 0$. The last inequality shows that the function (3.91) has a decreasing amplitude as $|x - x_0|$ grows.

Inserting the solution (3.91) into Eq. (3.83), we get a sequence of equations for the polynomials $y_k(\xi)$. Simultaneously we find the values a and λ_k. In the zero approximation, we get

$$y_0^{(m)}(\xi) = \mathcal{H}_m(z), \quad z = c^{1/2}\xi, \qquad m = 0, 1, 2, \dots,$$

$$c = \frac{b}{f_{pp}^0}, \qquad b = \sqrt{f_{pp}^0 f_{xx}^0 - (f_{px}^0)^2} > 0, \qquad a = \frac{b + \mathrm{i} f_{px}^0}{f_{pp}^0},$$

$$\lambda^{(m)} = \lambda_0 + \mu\lambda_1^{(m)} + O(\mu^2), \qquad \lambda_1^{(m)} = b(m + 1/2), \qquad m = 0, 1, 2, \dots, \tag{3.93}$$

where $\mathcal{H}_m(z)$ are the Hermite polynomials of degree m (in particular, $\mathcal{H}_0 = 1$, $\mathcal{H}_1 = z$, $\mathcal{H}_2 = z^2 - 1/2$). Equations (3.93) give the series of eigenvalues $\lambda^{(m)}$, whose minimum is $\lambda^{(0)}$.

There are two cases to consider

$$\begin{aligned} &(\text{A}) \qquad p_0 = 0, \\ &(\text{B}) \qquad p_0 > 0, \end{aligned} \tag{3.94}$$

in which the vibration modes are different.

In the case $p_0 = 0$, the vibration modes are as follows:

$$y^{(m)}(x, \mu) = (\mathcal{H}_m(z) + O(\mu)^{1/2})e^{-\Re a z^2/2}, \qquad z = \sqrt{c/\mu}(x - x_0), \tag{3.95}$$

the corresponding eigenvalues $\lambda^{(m)}$ being simple. An illustrative example is given in Section 3.8.

In the case $p_0 > 0$, the function (3.91) is complex-valued. The coefficients a_k of Eq. (3.83) are real, therefore both the real and the pure imaginary parts of the function (3.91) are the solutions of Eq. (3.83). But it is not correct to declare that the eigenvalues $\lambda^{(m)}$ are double and the arbitrary linear combination of the real and the pure imaginary parts of the function (3.91) is also an eigenfunction.

In reality, the series (3.91) is not converging, but asymptotic; to the parameter $\lambda^{(m)}$, there correspond two real eigenfunctions

$$y^{(m,j)}(x, \mu) = \left[\mathcal{H}_m(z)\cos\Psi_j + O(\mu^{1/2})\right]\exp\left(-\Re(a)z^2/2\right), \tag{3.96}$$

with

$$\Psi_j = \frac{p_0 x}{\mu} - \frac{\Im(a)z^2}{2} + \Theta_j, \quad j = 1, 2, \quad z = \left(\frac{c}{\mu}\right)^{1/2}(x - x_0), \tag{3.97}$$

and the phases Θ_j are fixed ($0 \le \Theta_1, \Theta_2 < \pi$). The corresponding exact eigenvalues $\lambda^{(m,1)}$ and $\lambda^{(m,2)}$ are different, but very close to each other,

$$\Delta\lambda = \lambda^{(m,2)} - \lambda^{(m,1)} = O(\mu^N), \tag{3.98}$$

for an arbitrary N. Such eigenvalues are known as *asymptotically double*. These eigenvalues are discussed in more detail in Chapter 4.

3.8 Vibration modes of a shell of revolution localized near a parallel lying strictly within the shell

If we change p by ip, then Eq. (3.88) gives

$$\lambda = f(p,s) \equiv (p^2+q^2)^2 + \frac{(k_2(s)p^2 + k_1(s)q^2)^2}{(p^2+q^2)^2}, \quad q = q(s) = \frac{\rho}{B(s)} \tag{3.99}$$

with $\rho = \mu m$.

Let the inequality (3.71) be fulfilled. Then the minimum of the function $f(p,s)$ in p is attained at $p = p_0 = 0$ and so

$$\lambda_0 = f(0,s) = g(s) = q^4(s) + k_1^2(s). \tag{3.100}$$

In contrast with Section 3.6, we shall assume that the minimum of the function $g(s)$ is attained at a point s_0 lying within the shell $s_1 < s_0 < s_2$ and $g''(s_0) > 0$. Then to construct the first forms of localized vibrations we may use Eqs. (3.93)–(3.95):

$$w^{(n)}(s,\mu) = (\mathcal{H}_n(z) + O(\mu)^{1/2})e^{-cz^2/2}, \qquad z = \sqrt{c/\mu}(s-s_0), \tag{3.101}$$

$$\lambda^{(n)} = g(s_0) + \mu b(n+1/2) + O(\mu^2), \qquad n = 0,1,2,\ldots, \tag{3.102}$$

with

$$c = \sqrt{\frac{g''(s_0)}{f_{pp}^0}}, \qquad b = \sqrt{g''(s_0)f_{pp}^0}, \quad f_{pp}^0 = 4q^2 + 4k_1(k_2-k_1)q^{-2}, \tag{3.103}$$

and all functions in Eq. (3.103) are calculated for $s = s_0$.

The eigenvalues (3.102) are simple.

Example 3.4. Consider vibrations with m waves in the circumferential direction of an ellipsoid of revolution with the thickness h and the semi-axes a_0 and b_0. As a linear scale we take $R = a_0$. Then the parameters of the shell surface read as

$$k_2 = \sqrt{\sin^2\theta + \delta^2\cos^2\theta}, \quad k_1 = \frac{d\theta}{ds} = \frac{k_2^3}{\delta^2}, \quad B = \frac{\sin\theta}{k_2}, \quad \delta = \frac{b_0}{a_0}, \tag{3.104}$$

where θ $(0 \le \theta \le \pi)$ is the angle between the normal to the shell surface and the axis of rotation (see Fig. 3.9).

Consider the function

$$g(s) = q^4 + k_1^2 = \frac{\rho^4 k_2^4}{\sin^4\theta} + \frac{k_2^6}{\delta^4}, \quad \rho = \mu m, \tag{3.105}$$

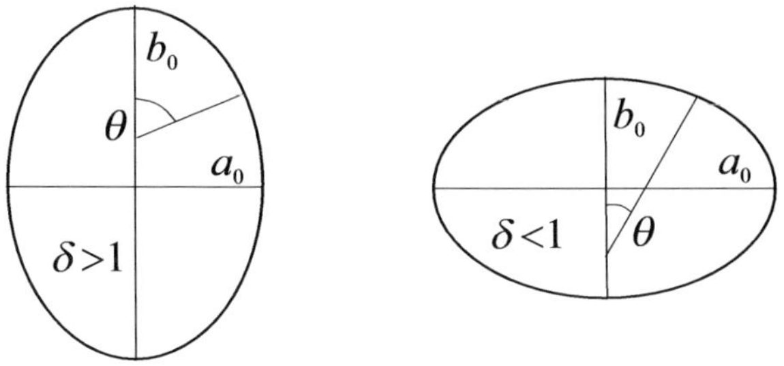

FIGURE 3.9: A prolate ($\delta > 1$) and an oblate ($\delta < 1$) ellipsoid.

which has the minimum in the weakest parallel. We set

$$a = \frac{2\rho^4\delta^6}{3(1-\delta^2)}. \tag{3.106}$$

If $a > 1$, the weakest parallel is the equator, $\theta = \pi/2$. The inequality $a > 1$ is fulfilled for all prolate ellipsoids, for a sphere, and for some oblate ellipsoids (for example, with $\rho = 1$ for $\delta > 0.87$).

If $a = 1$, the weakest parallel is again the equator, but $g''(s_0) = 0$, and Eqs. (3.101) and (3.102) being unacceptable.

If $a < 1$, then there are two weakest parallels in the shell surface: $\theta = \theta_0 \neq \pi/2$ and $\theta = \pi - \theta_0$ (see Fig. 3.9), which satisfy the equation $g'(s) = 0$ or

$$\sin^6\theta(\sin^2\theta + \delta^2\cos^2\theta) = \frac{2\rho^4\delta^6}{3(1-\delta^2)}. \tag{3.107}$$

In Eqs. (3.103),

$$\begin{aligned} g''(s_0) &= \frac{4k_2^6\rho^4}{\sin^6\theta}(k_2^2\sin^2\theta + 5\delta^2\cos^2\theta) \\ &+ \frac{3(1-\delta^2)k_2^8}{2\delta^8}(4k_2^2\cos 2\theta + 7(1-\delta^2)\sin^2 2\theta). \end{aligned} \tag{3.108}$$

We take the parameters: the dimensionless thickness $h/a_0 = 0.005$, the Poisson ratio $\nu = 0.3$, the number of waves in the circumferential direction $m = 15$. The ratio $\delta = b_0/a_0$ of semi-axes is varied in the range $0.6 \leq \delta \leq 1.5$. We calculate the first two eigenvalues using the asymptotic equation (3.102) and by numerically solving Eq. (3.5).

The results are given in Table 3.7 and in Fig. 3.11. In Table 3.7, the exact eigenvalues corresponding to the even and odd with respect to a diameter modes are denoted by $\lambda^{(even)}$ and $\lambda^{(odd)}$, and the asymptotic eigenvalues, as obtained from Eq. (3.101), are denoted by $\lambda_a^{(even)}$ and $\lambda_a^{(odd)}$, respectively. In the 6th column, the zero approximation $\lambda_0 = g(s_0)$ is given, and the 7th column contains the weakest parallel co-ordinate, θ_0.

TABLE 3.7: Frequency parameters of a paraboloid of revolution.

1	2	3	4	5	6	7
δ	$\lambda^{(even)}$	$\lambda_a^{(even)}$	$\lambda^{(odd)}$	$\lambda_a^{(odd)}$	$g(s_0)$	θ_0
0.60	1.71720	1.7204	1.71718	1.7204	1.5693	0.4926
0.70	1.75013	1.7596	1.75011	1.7596	1.6396	0.5823
0.80	1.68689	1.7018	1.68700	1.7018	1.6203	0.6996
0.86	1.59124	1.6085	1.59090	1.6085	1.5529	0.8014
0.90	1.49492	1.5144	1.49595	1.5144	1.4768	0.9019
0.94	1.36411	1.3846	1.36285	1.3846	1.3658	1.0796
0.96	1.27611	1.2988	1.28974	1.2988	1.2906	1.2824
0.98	1.19287	1.2082	1.22016	1.2244	1.2001	1.5708
1.00	1.11656	1.1314	1.15493	1.1623	1.1159	1.5708
1.10	0.82111	0.8324	0.88932	0.8993	0.7989	1.5708
1.20	0.62671	0.6351	0.70196	0.7090	0.5982	1.5708
1.30	0.49493	0.5013	0.56795	0.5718	0.4660	1.5708
1.40	0.40320	0.4081	0.47051	0.4720	0.3762	1.5708
1.50	0.33782	0.3418	0.39853	0.3985	0.3134	1.5708

With $\delta > 0.97$ we get $a > 1$, and the weakest parallel coincides with the equator, $\theta_0 = \pi/2 = 1.5708$. By using Eq. (3.102) we take $n = 0$ for the even mode, and $n = 1$ for the odd mode. The even and the odd modes are shown schematically in Fig. 3.10, and denoted by 1 and 2, respectively.

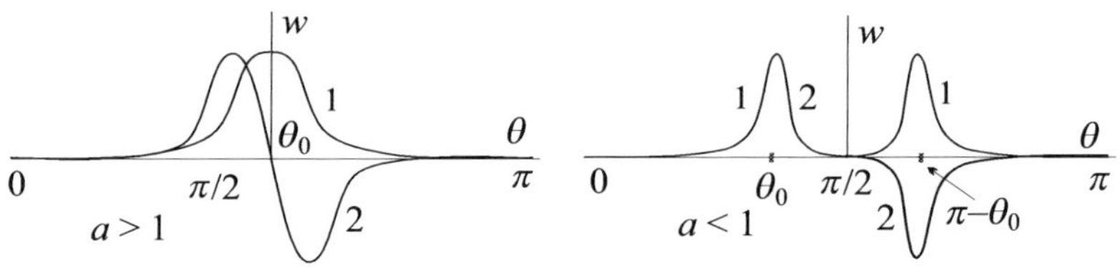

FIGURE 3.10: Even and odd modes of free vibrations (sketch).

With $\delta < 0.97$ we get $a < 1$, and in the shell surface there are two weakest parallels, θ_0 and $\pi - \theta_0$. If these parallels are sufficiently far from the equator, then for each of them Eq. (3.102) gives the identical values λ_a, as calculated with $n = 0$. A numerical solution shows that here there are also the even and the odd modes (see Fig. 4.9) with various values $\lambda^{(even)}$ and $\lambda^{(odd)}$. For larger $|t_0 - \pi/2|$ these values become closer to each other (see the columns 2 and 4 in Table 3.7). These eigenvalues are asymptotically double and satisfy the asymptotic estimate (3.98). From Table 3.7 it follows that with $\delta \leq 0.7$ the values $\lambda^{(even)}$ and $\lambda^{(odd)}$ differ in the 5th digit. In contrast, it should be noted that for $a > 1$

$$\Delta\lambda = \lambda^{(1)} - \lambda(0) = O(\mu). \tag{3.109}$$

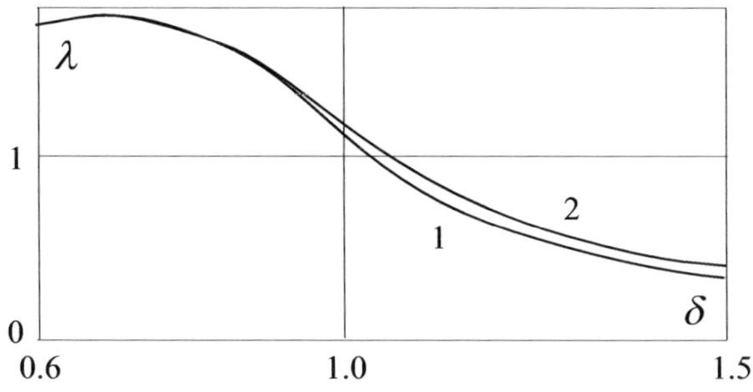

FIGURE 3.11: Functions 1 — $\lambda^{(even)}(\delta)$ and 2 — $\lambda^{(odd)}(\delta)$.

Figure 3.11 shows the numerically obtained graphs of the functions $\lambda^{(even)}(\delta)$ and $\lambda^{odd}(\delta)$. For $\delta < 1$ the curves 1 and 2 become closer. The difference $\Delta\lambda$ satisfies estimate (3.98); it changes its sign as δ changes (it is not seen in Fig. 3.11, because the difference δ is very small).

3.9 Vibration modes of a pre-stressed non-uniform cylindrical shell localized near a parallel

In Section 3.8, the reason of localization is a variable curvature of the generatrix. A circular cylindrical shell with a constant thickness does not have localized vibration modes. The localization is possible if the shell thickness is variable and/or (in the pre-stressed case) the initial stresses are non-uniform. We consider the case when the weakest parallel is far from the shell edges.

Consider a circular cylindrical shell of radius R and variable thickness $h(s)$ under the action of the momentless initial stress resultants T_1^0 and T_2^0. According to the thickness variation, Eqs. (3.48) requires a correction [222]

$$\mu^4\Delta_m(d\,\Delta_m w) - \lambda g w + 2\lambda_t\mu^2\left(t_1\frac{d^2w}{ds^2} - m^2 t_2 w\right) + \mu^2\frac{d^2\Phi}{ds^2} = 0,$$
$$\mu^4\Delta_m(g^{-1}\Delta_m\Phi) - \mu^2\frac{d^2w}{ds^2} = 0, \tag{3.110}$$

where

$$\Delta_m w = \frac{d^2w}{ds^2} - m^2 w, \qquad g(s) = \frac{h(s)}{h_0}, \quad d(s) = g^3(s), \quad \mu^4 = \frac{h_0^2}{12(1-\nu^2)R^2}. \tag{3.111}$$

Here, h_0 is the minimal value of thickness. We put

$$h(0) = h_0, \qquad h(s) = h_0 g(s), \quad g(s) = 1 + \eta s^2 + O(s^3), \quad \eta > 0, \tag{3.112}$$

and assume that the weakest parallel $s = 0$ lies far enough from the shell edges.

In this case the characteristic equation (3.99) reads as

$$\lambda = f(p,s) \equiv g^2(s)(p^2+\rho^2)^2 - 2\frac{\lambda_t}{g(s)}(t_1p^2+t_2\rho^2) + \frac{p^4}{(p^2+\rho^2)^2}. \quad (3.113)$$

Consider some particular cases in the dependence of parameters t_1, t_2, ρ in Eq. (3.113) (without loss of generality, we put $\lambda_t = 1$). We consider only the initial stresses at which the shell does not lose the stability. The shells of moderate length lose their stability under compression if $t_1 = 1+O(\mu)$ and at an external homogeneous pressure if $t_2 = O(\mu)$ [222]. That is why we assume that $t_1 < 1$, $t_2 \le 0$.

3.9.1 Axisymmetric vibrations

In this case $\rho = 0$ and Eq. (3.113) reads as

$$\lambda = f(p,s) \equiv g^2(s)p^4 - 2\frac{t_1}{g(s)}p^2 + 1. \quad (3.114)$$

The asymptotic formulas for natural frequencies and vibration modes are different in the cases $0 < t_1 < 1$ and $t_1 \le 0$.

At the axial compression ($0 < t_1 < 1$) the case (B) (see Eq.(3.94)) takes place. Formulas of Section 3.7 give

$$\begin{aligned} &\lambda^{(n,j)} = 1 - t_1^2 + \mu(n+1/2)\sqrt{64t_1^3\eta} + O(\mu^2), \qquad j = 1,2, \quad n = 0,1,2,\ldots, \\ &w^{(n,j)}(s) = \left(\mathcal{H}_n(z)\cos\Psi_j + O(\mu^{1/2})\right)\exp(-az^2/2), \\ &a = \sqrt{t_1\eta}, \quad z = \sqrt{a/\mu}\, s, \quad \Psi_j = \frac{p_0\, s}{\mu} + \Theta_j, \quad p_0 = \sqrt{t_1}. \end{aligned} \quad (3.115)$$

The eigenvalues $\lambda^{(n,j)}$ are asymptotically double (see Eq. (3.98)). The phases Θ_j of eigenfunctions $w^{(n,j)}(s)$ assume two definite values. If the shell is symmetric with respect the parallel $s = 0$, then one of the functions $w^{(n,j)}(s)$ is even, and the other is odd (or $\Theta_1 = 0$, $\Theta_2 = \pi/2$). The functions $w^{(0,1)}(s)$ and $w^{(0,2)}(s)$ are shown in Figs. 3.12(a) and 3.12(b).

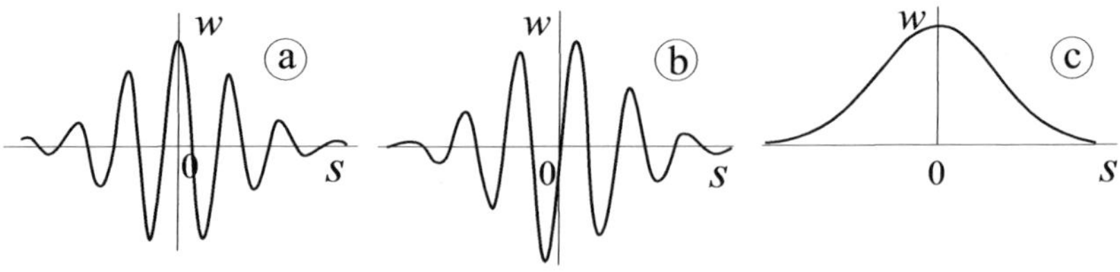

FIGURE 3.12: Vibration modes of a cylindrical shell with a variable thickness.

In the case $t_1 \leq 0$, it is impossible to use the results of Section 2.7, because the function (3.114) has a minimum at $p_0 = s = 0$, at this point $f^0_{ss} = 0$, and the condition (3.90) is not fulfilled. Therefore, addition analysis is required. Equations (3.110) are reduced to the equation

$$\mu^4 \frac{d^2}{ds^2}\left(g^3(s)\frac{d^2w}{ds^2}\right) + 2\mu^2 t_1 \frac{d^2w}{ds^2} - (\lambda - 1)gw = 0. \tag{3.116}$$

For $t_1 \leq 0$, the eigenvalues are simple, and they are close to $\lambda = 1$. The estimate $\lambda = 1+O(\mu^2 t_1)$ is valid. The eigenfunctions are not localized near the point $s = 0$. If for $0 < t_1 < 1$ the eigenvalues and eigenfunctions practically do not depend on the boundary conditions, then for $t_1 \leq 0$ this dependence exists.

Example 3.5. Consider a different-walled cylindrical shell with parameters

$$R/h_0 = 200, \quad g(s) = 1 + s^2, \quad -1.5 \leq s \leq 1.5, \quad \nu = 0.3. \tag{3.117}$$

We take the identical boundary condition at $s = \pm 1.5$. In this case the eigenfunctios are even or odd with respect the point $s = 0$.

TABLE 3.8: Frequency parameters of axisymmetric vibrations of a different-walled cylindrical shell.

t_1	$\lambda^{(0,1)}$	$\lambda^{(0,2)}$	$\lambda^{(a)}$
-1.0	1.0018	1.0059	
-0.5	1.0010	1.0034	
0.0	1.0003	1.0005	1.0000
0.2	0.9720	0.9720	0.9734
0.6	0.7098	0.7098	0.7123
1.0	0.1517	0.1517	0.1556
1.08	0.0041	0.0041	0.0082

In Table 3.8 for various values of an axial stress resultant t_1 we give the first two frequencies of bending vibrations of a shell with clamped edges, as found numerically. At $t_1 > 0$, the frequencies are asymptotically double, and for them the asymptotic values $\lambda^{(a)}$ are given. A shell of a constant thickness loses its stability with $t_1 = 1$, and the shell under consideration with variable thickness loses its stability with $t_1 = 1.082$. At $t_1 \leq 0.5$, the frequencies, as shown in Table 3.8, are not minimal, because the torsion frequency $\lambda = 0.56$, which does not depend on t_1, is smaller.

3.9.2 Non-axisymmetric vibrations

The asymptotic behavior depends substantially on the order of the number of waves in the circumferential direction, m.

The case of small m, in which the parameter $\rho = \mu m$ cannot be taken into account under using the above method, was described in Section 3.7. The function (3.113) has a minimum at $p_0 = 0$ or at small $p_0 = O(\rho)$; the value $f_{ss}^0 = O(\rho^2)$. We mention only that the localization does not take place, the frequency is essentially lower than at the axisymmetric vibrations, and the first vibration mode is of the form shown in Fig. 3.12(c).

Let the number of waves, m, be large, and $\rho = \mu m \sim 1$. Then for $t_1 < \rho^2$, $t_2 > -\rho^2$ the function (3.113) has a minimum at $p_0 = 0$, $s = 0$, and

$$\lambda_0 = \rho^4 - 2t_2\rho^2, \quad f_{pp}^0 = 4(\rho^2 - t_1), \quad f_{ss}^0 = 4\eta\rho^2(\rho^2 + t_2). \tag{3.118}$$

Equation (3.93) gives a set of simple eigenvalues

$$\lambda^{(n)} = \lambda_0 + \mu(n + 1/2)\sqrt{f_{pp}^0 f_{ss}^0} + O(\mu^2), \qquad n = 0, 1, 2, \ldots, \tag{3.119}$$

and the first vibration mode is shown in Fig. 3.12(c).

For $t_1 < \rho^2$, $t_2 < -\rho^2$, the function (3.113) describes two turning points $s = \pm s_0$ with $p = p_0 > 0$. This case is similar to vibrations of an oblate ellipsoid, which was studied in Section 3.8. Here we do not repeat this analysis.

For $\rho \sim 1$, $t_1 > \rho^2$, the function (3.113) has a minimum at $p = p_0 > 0$, where p_0 satisfies the equation

$$f_p(p, 0) = 4p\left(p^2 + \rho^2 - t_1 + \frac{p^2\rho^2}{(p^2 + \rho^2)^3}\right) = 0. \tag{3.120}$$

As before, the eigenvalues can be found from Eq. (3.119) with

$$\begin{aligned} \lambda_0 &= f(p_0, 0), \\ f_{pp}^0 &= 12p_0^2 + 4\rho^2 - 4t_1 - \frac{12p_0^2\rho^2(p_0^2 - \rho^2)}{(p_0^2 + \rho^2)^4}, \\ f_{ss}^0 &= 4\eta\left((p_0^2 + \rho^2)^4 + t_1 p_0^2 + t_2\rho^2\right). \end{aligned} \tag{3.121}$$

In contrast to the case $t_1 < \rho^2$, now the eigenvalues are asymptotically double, and the first vibration modes are of the form shown in Fig. 3.12(a) or 3.12(b).

Example 3.6. We consider a shell with parameters

$$g(s) = 1 + s^2, \quad -1.5 \le s \le 1.5, \quad \nu = 0.3. \tag{3.122}$$

TABLE 3.9: Frequency parameters of non-axisymmetric vibrations of a cylindrical shell with variable thickness.

N	m	R/h_0	t_1	t_2	$\lambda^{(0,1)}$	$\lambda^{(0,2)}$	$\lambda^{(a)}$	
1	2	200	−1.0	0.0	0.0452	0.2264		supported
2	2	200	−1.0	0.0	0.1056	0.2850		clamped
3	2	200	0.0	0.0	0.0413	0.2195		supported
4	2	200	0.0	0.0	0.1033	0.2772		clamped
5	2	200	1.0	1.0	0.0465	0.1604		supported
6	2	200	1.0	1.0	0.1089	0.1604		clamped
7	10	200	−1.0	0.0	0.0386	0.0706	0.0355	supported
8	10	200	0.0	0.0	0.0315	0.0529	0.0275	supported
9	25	200	0.0	0.0	0.9694	1.1310	0.9658	supported
10	10	200	1.0	0.0	0.0143	0.0178	0.0182	supported
11	25	500	1.0	0.0	0.0763	0.0763	0.0759	supported
12	25	500	1.0	-0.5	0.4494	0.4494	0.4503	supported

For a numerical solution we use Eqs. (3.5), which are more exact than Eqs. (3.110). In particular, Eqs. (3.5) contain the tangential inertia forces, which are omitted in Eqs. (3.110).

In Table 3.9, we use the same designations as in Table 3.6 (besides, "supported"corresponds to conditions $T_1 = v = w = M_1 = 0$ and "clamped"corresponds to conditions $u = v = w = \gamma_1 = 0$). The first two eigenvalues (the even $\lambda^{(0,1)}$ and odd $\lambda^{(0,2)}$ ones), as found from the numerical solution of Eqs. (3.5), are given. The identical clamped or the simply supported boundary conditions at the shell edges are studied.

For $m = 2$, the eigenfunctions are not localized that follows from the difference of eigenvalues for the clumped and the simply supported edges (see rows 1–6 of Table 3.9) (for the localized modes eigenvalues practically do not depend on the boundary conditions).

For $m = 10$ and for $m = 25$, the vibration modes are localized near $s = 0$. The difference between the values λ for the clumped and the simply supported edges is small and appears only in the 4th digit, and so in rows 7–12 of Table 3.9 the values λ are given only for the simply supported boundary conditions.

In rows 7 and 8, the eigenvalues are simple, and the even vibration mode is shown in Fig. 3.12(c). The asymptotic value $\lambda^{(a)}$ is given for the first eigenvalue $\lambda^{(0,1)}$.

In rows 9–12, the asymptotically double eigenvalues are given, and the corresponding vibration modes are shown in Figs. 3.12(a) and 3.12(b). The asymptotic value is common for both $\lambda^{(0,1)}$ and $\lambda^{(0,2)}$. For more thin shell (with larger R/h_0), the difference between the exact and the asymptotic values becomes smaller (see rows 11 and 12).

3.10 Localized modes of axisymmetric vibrations of an infinitely long non-uniform cylindrical shell resting on an elastic foundation

Here, we shall study localized axisymmetric vibrations of a long non-uniform cylindrical shell of the radius R in which the thickness $h(x)$, Young's modulus $E(x)$, Poisson's ratio $\nu(x)$ and the material density $\varrho(x)$ are functions of the longitudinal coordinate $x = Rs$. The shell rests on an inhomogeneous elastic foundation with the non-uniform coefficient of reaction $c_f(x)$. As above, it is assumed that the shell is pre-stressed by the membrane stress resultants T_1°, T_2°.

As the governing equations, we shall use the Flügge equations [57], which in the case of axisymmetric motion have the form:

$$
\begin{gathered}
\frac{\partial T_1}{\partial x} + T_1^\circ \frac{\partial^2 u_1}{\partial x^2} - \varrho h \frac{\partial^2 u_1}{\partial t^{*2}} = 0, \\
\frac{\partial^2 M_1}{\partial x^2} + \frac{1}{R} T_2 + T_1^\circ \frac{\partial^2 u_3}{\partial x^2} - \frac{1}{R^2} T_2^\circ u_3 + c_f u_3 + \varrho h \frac{\partial^2 u_3}{\partial t^{*2}} = 0.
\end{gathered}
\tag{3.123}
$$

Here, t^* is time, u_1, u_3 are the axial and normal (radial) displacements of the middle surface, and T_1, T_2, M_1 are additional membrane forces and moments induced by the shell vibration. For axisymmetric motion, these stress-resultants read

$$
\begin{gathered}
T_1 = \frac{Eh}{1-\nu^2}\left(\frac{\partial u_1}{\partial x} + \frac{\nu}{R} u_3\right), \quad T_2 = \frac{Eh}{1-\nu^2}\left(\nu \frac{\partial u_1}{\partial x} + \frac{1}{R} u_3\right), \\
M_1 = \frac{Eh}{12(1-\nu^2)} \frac{\partial^2 u_3}{\partial x^2}.
\end{gathered}
\tag{3.124}
$$

The value of spring constant $c_f(x)$ is influenced by elastic properties of a foundation (or filler) and the wave pattern on the shell surface (see Eq. (2.52)) as well. Here, we consider the Winkler model in accordance to which $c_f(x)$ depends only on the physical properties of an elastic foundation.

We introduce the dimensionless functions

$$
\begin{gathered}
d(s) = \frac{Eh^3}{E_0 h_0^3 (1-\nu^2)}, \quad g(s) = \frac{Eh}{E_0 h_0 (1-\nu^2)}, \quad \gamma(s) = \frac{\rho h}{\rho_0 h_0}, \\
c(s) = \frac{R^2 c_f}{E_0 h_0}, \quad \lambda = \frac{\rho_0 R^2}{E_0}\omega^2, \quad T_1^\circ = \mu_1^2 E_0 h_0 t_1, \quad T_2^\circ = E_0 h_0 t_2,
\end{gathered}
\tag{3.125}
$$

where h_0, E_0, ϱ_0 are the characteristic values of thickness, Young's modulus and material density, respectively, which will be chosen below, $\mu_1^4 = h_0^2/(12R^2)$ is a small parameter.

We assume that the applied loads do not result in buckling. A very long cylindrical shell of the length L buckles under the axial compression like a beam if $T_1^\circ = T_1^b \equiv \frac{1}{2}\pi^2 h_0 R_0^2/L^2$ [226]. Passing to dimensionless stresses, it is readily shown that if

$$t_1 < t_{cr} \equiv \frac{2}{\sqrt{1-\nu^2}} \qquad \frac{L^2}{R^2} < \frac{\pi^2\sqrt{3(1-\nu^2)}}{2}\frac{R}{h_0}, \tag{3.126}$$

then the stress-strain state of a shell will be subcritical.

From the known formula of Grashof-Bresse [38, 75], it is followed that a very long cylindrical shell buckles if $t_2 = \frac{h_0^2}{4R^2(1-\nu^2)} = O(\mu_1^4)$. In what follows, it is assumed that $t_1 < t_{cr}$, $t_2 \leq 0$.

The shell displacements are sought in the form:

$$u_1 = RU(s)\cos(\omega t^*), \qquad u_3 = RW(s)\cos(\omega t^*), \tag{3.127}$$

where ω is a required natural frequency of vibrations. The substitution of (3.127), (3.124) into Eqs. (3.123) leads to the coupled differential equations in the dimensionless form:

$$\begin{gathered}
\frac{d}{ds}\left[g(s)\left(\frac{dU}{ds}+\nu W\right)\right]+\mu_1^2 t_1\frac{d^2U}{ds^2}+\lambda\gamma(s)U=0,\\
\mu_1^4\frac{d^2}{ds^2}\left[d(s)\frac{d^2W}{ds^2}\right]+\nu g(s)\frac{dU}{ds}+\left[g(s)-t_2+c(s)\right]W\\
+\mu_1^2 t_1\frac{d^2W}{ds^2}-\lambda\gamma(s)W=0.
\end{gathered} \tag{3.128}$$

From all possible modes of axial symmetric vibrations, we shall consider bending vibrations which are accompanied by formation of a large number of short waves in the axial direction for which $\partial/\partial s \sim \mu_1^{-1}$ as $\mu_1 \to 0$. To analyze small bending vibrations, we assume that [154]

$$W = w \sim 1, \quad U = \mu u, \quad \text{where} \quad u \sim 1. \tag{3.129}$$

Let $s = s_0$ be the weakest parallel in the neighborhood of which it is observed localization of vibrations satisfying the conditions

$$w,\, u \to 0 \quad \text{as} \quad s \to \pm\infty. \tag{3.130}$$

Following the algorithm given in Section 3.7, we seek a solution of boundary-value problem (3.128), (3.130) in the form of asymptotic expansion

$$\mathbf{X}(s,\mu_1) = \sum_{k=0}^{\infty}\mu_1^{k/2}\mathbf{X}_k(\xi)\exp\left[\mathrm{i}\left(\mu_1^{-1/2}p\xi + 1/2b\xi^2\right)\right], \quad \Im b > 0, \tag{3.131}$$

where $\mathbf{X} = (u,\, w)$, $\mathbf{X}_k = (u_k,\, w_k)$ are two-dimensional vectors, $\xi = \mu_1^{-1/2}(s - s_0)$. The required eigenvalue λ is expanded into series (3.92).

The substitution of Eqs. (3.129), (3.131), (3.92) into (3.128) produces the sequence of algebraic equations

$$\sum_{j=0}^{k} \mathbf{L}_j \mathbf{X}_{k-j}^{\mathrm{T}} = \mathbf{0}, \quad k = 0, 1, 2, \ldots \tag{3.132}$$

where $\mathbf{L}_k$ are the 2×2-dimension matrices. The elements of $\mathbf{L}_0$ are as follows:

$$\begin{aligned} l_{11} &= -g(s_0)p^2, \quad l_{12} = \mathrm{i}\nu(s_0)g(s_0)p, \\ l_{21} = l_{12}, \quad l_{22} &= d(s_0)p^4 + g(s_0) - t_2 + c - t_1 p^2 - \lambda_0 \gamma(s_0). \end{aligned} \tag{3.133}$$

The matrices $\mathbf{L}_k$ with $k \geq 1$ are not given here.

Considering the homogeneous system of algebraic equations (3.132) with $k = 2$, we get the formulas:

$$u_0 = \mathrm{i}\frac{\nu(s_0)}{p} w_0(\xi), \tag{3.134}$$

$$\begin{aligned} \lambda_0 = f(p, s_0) = \gamma^{-1}(s_0)\{d(s_0)p^4 + [1 - \nu^2(s_0)]g(s_0) \\ +c(s_0) - t_2 - t_1 p^2\}. \end{aligned} \tag{3.135}$$

The resolvability condition of the inhomogeneous system of equations (3.132) for $k = 1$ is equivalent to the stationary condition of the function $f(p, s_0) : f_p = 0, f_s = 0$. Here and below, the indices p and s mean differentiation of the function $f(p, s_0)$ with respect to the corresponding variable. From these conditions one can find the wave number

$$p^\circ = \sqrt{\frac{t_1}{2d(s_0^\circ)}}, \tag{3.136}$$

and also get the equation for determining the weakest parallel $s = s_0^\circ$:

$$\gamma \left[\frac{d'}{4d^2} t_1^2 + c' - (\nu^2 g)' \right] - \gamma' \left[(1 - \nu^2)g + c - t_2 - \frac{t_1^2}{4d} \right] = 0. \tag{3.137}$$

The prime in (3.137) and below means the derivative of a function with respect to s.

Introduce the characteristic values of the thickness, Young's modulus and material density:

$$h_0 = h(s_0^\circ), \quad E_0 = E(s_0^\circ), \quad \rho_0 = \rho(s_0^\circ). \tag{3.138}$$

We note that apart from value (3.136), the equation $f_p = 0$ has also the trivial solution $p = 0$. However, taking into account (3.134), the root $p = 0$ should be omitted, and formula (3.136) is considered at $t_1 > 0$. Thus, the asymptotic solution in the form (3.131) is valid only in the case when a shell is subjected to compressive axial forces.

The substitution of (3.136) into (3.135) gives the zero-order approximation for a parameter λ in the asymptotic expansion (3.92):

$$\lambda_0^\circ = \frac{1}{\gamma(s_0^\circ)} \left\{ \left[1 - \nu^2(s_0^\circ)\right] g(s_0^\circ) - \frac{t_1^2}{4d(s_0^\circ)} + c(s_0^\circ) - t_2 \right\}. \tag{3.139}$$

The compatibility condition of the inhomogeneous system of algebraic equations (3.123) leads to the following equations:

$$\begin{gathered} f_{pp}^\circ b^2 + 2f_{ps}^\circ b + f_{ss}^\circ = 0, \\ \lambda_1^{(n)} = -\mathrm{i}(n+1/2)\left(f_{pp}^\circ b + f_{ps}^\circ\right) + \eta, \\ \eta = \frac{2d'(s_0^\circ)(p^\circ)^3}{\gamma(s_0^\circ)}, \qquad w_0 = \mathcal{H}_n(z), \\ z = \left[\frac{f_{ss}^\circ}{f_{pp}^\circ} - \left(\frac{f_{ps}^\circ}{f_{pp}^\circ}\right)\right]^{1/4} r\xi, \quad r = \left[\frac{f_{ss}^\circ}{f_{pp}^\circ} - \left(\frac{f_{ps}^\circ}{f_{pp}^\circ}\right)\right]^{1/4}, \end{gathered} \tag{3.140}$$

where $\mathcal{H}_n(z)$ is the nth degree Hermit polynomial.

As examples, consider some particular cases.

Example 3.7. Let parameters E, ρ, ν, c be constant, and the thickness $h(s) = h_0(1 + \kappa s^2/2)$ be variable. Here, the curve $s = s_0^\circ = 0$ is the weakest parallel, and the condition $\Im b > 0$ is equivalent to the inequality

$$\kappa\left[t_2 - c + (1-\nu^2)t_1^2\right] > 0. \tag{3.141}$$

We obtain the parameter

$$b = \frac{\mathrm{i}}{2}\sqrt{\frac{\kappa\left[t_2 - c + (1-\nu^2)t_1^2\right]}{t_1}}, \tag{3.142}$$

characterizing the degree of localization of eigenmodes and the required series $(n = 0, 1, \ldots)$ of natural frequencies

$$\begin{aligned} \omega = \sqrt{\frac{E}{R^2\rho}} &\left\{\left[1 - t_2 + c - 1/4(1-\nu^2)t_1^2\right]\right. \\ &\left. + 2\mu_1(n+1/2)\sqrt{\kappa\left[t_2 - c + (1-\nu^2)t_1^2\right] t_1} + O(\mu_1^2)\right\}^{1/2} \end{aligned} \tag{3.143}$$

as well.

It is seen from inequality (3.141) that the localization of eigenmodes is possible in a vicinity of the line $s = 0$ where the thickness is minimal and in the neighborhood of the parallel on which it is maximal. In the last case the stiffness of an elastic foundation or filler should be sufficiently large so that $c > t_2 + (1-\nu^2)t_1^2$.

Example 3.8. Consider the case when parameters h, ρ, ν, c are constant, and Young's modulus $E(s) = E_0(1 + e\, s^2/2)$ is variable. Here, the localization of free vibrations takes place in the vicinity of the parallel $s = s_0^\circ = 0$ where Young's modulus is minimal (for $e > 0$). One obtains

$$b = \frac{\mathrm{i}}{2}\sqrt{\frac{e\left[4 + (1-\nu^2)t_1^2\right]}{6t_1}}, \tag{3.144}$$

$$\begin{aligned} \omega = \sqrt{\frac{E}{R^2\rho}}\,\Big\{&\left[1 - t_2 + c - 1/4(1-\nu^2)t_1^2\right] \\ &+3\mu_1(n+1/2)\sqrt{1/6e\left[4+(1-\nu^2)t_1^2\right]t_1} + O(\mu_1^2)\Big\}^{1/2}. \end{aligned} \tag{3.145}$$

In this example, the parameter b as well as the correction to eigenfrequency accounting for inhomogeneity of Young's modulus do not depend on the stress-resultants t_2 and spring constant c.

Example 3.9. Let $c(s) = c_0 + c_1 s^2/2$, and remaining parameters be constants. In this case, one has a localization near the parallel $s = 0$ where the response of the Winkler foundation is minimal. We arrive the following relations:

$$b = \frac{\mathrm{i}}{2}\sqrt{\frac{c_1}{t_1}}, \tag{3.146}$$

$$\omega = \sqrt{\frac{E}{R^2\rho}}\left\{\left[1 - t_2 + c_0 - 1/4(1-\nu^2)t_1^2\right] + 2\mu_1(n+1/2)\sqrt{c_1 t_1} + O(\mu_1^2)\right\}^{1/2}. \tag{3.147}$$

From Eqs. (3.146), (3.147), it follows that increasing the spring constant (parameter c_0) leads to growth in the total stiffness of the mechanical system "shell-foundation"and, as a consequence, leads to an increase in the natural frequencies, and an increase in the inhomogeneity degree of a foundation results in more strong localization of vibration modes.

Completing this section, we point out to the papers [157, 159] dealing with a long pre-stressed nanotube embedded into an inhomogeneous elastic medium. Using the nonlocal version of the Flügge type theory for orthotropic cylindrical shells, the authors studied free vibrations localized near the weakest parallel on which the reaction of the surrounding medium is weakened by "soft inclusions."Moreover, in contrast to the above results related to the axially *compressed* cylindrical shell (for $t_1 > 0$), it was revealed the localized modes in the axially *stretched* cylindrical nanoscale shells (for $t_1 < 0$).

3.11 Transverse shear influence on the localized vibration modes of shells of revolution

Consider a thin elastic shell of revolution made of a transversely isotropic inhomogeneous in the thickness direction material. In particular, multi-layered shells and functionally graduated shells are considered. If the level of inhomogeneity is large, then the KL model is unacceptable. Instead, the TR model including a transverse shear should be used. In this section, we use the TR model, as described in Section 2.6. In the Section 2.7, it is explained how to replace approximately an inhomogeneous shell by an equivalent homogeneous one.

We write again Eqs. (2.64)–(2.66), (2.71) including a transverse shear, normal inertia forces, and initial stresses

$$\begin{aligned} &\Delta\Delta\Phi + K_0\Delta_R w = 0, \\ &D\Delta\Delta\Psi - \Delta_R\Phi - \lambda w - \Delta_T w = 0, \\ &w = \Psi - g\Delta\Psi, \\ &\Delta\Theta - g_1\Theta = 0. \end{aligned} \tag{3.148}$$

Then we separate the variables

$$\{w, \Phi, \Psi, \Theta\}(s, \varphi) = \{w, \Phi, \Psi, \Theta\}(s)\sin m\varphi, \tag{3.149}$$

and pass to the dimensionless variables in Eqs. (3.148) and in the boundary conditions (2.55). Then in the dimensionless form Eqs. (3.148) read as

$$\begin{aligned} &\Delta\Delta\Phi + \Delta_R w = 0, \qquad w = \Psi - g\Delta\Psi, \\ &\Delta\Delta\Psi - \Delta_R\Phi - \lambda w - \lambda_t\Delta_T w = 0, \\ &\Delta\Theta - g_1\Theta = 0, \end{aligned} \tag{3.150}$$

with the differential operators

$$\Delta = p^2 - q^2, \quad \Delta_R = k_2p^2 - k_1q^2, \quad \Delta_T = t_1p^2 - t_2q^2, \quad p = \frac{d}{d\hat{s}}, \tag{3.151}$$

and the notations

$$\begin{aligned} &\mu^4 = \frac{h^2}{12(1-\nu^2)R^2}, \quad s = R\hat{s}, \quad q = \frac{mR\mu}{B}, \quad k_1 = \frac{R}{R_1}, \quad k_2 = \frac{R}{R_2}, \\ &g = \frac{Eh\mu^2}{\Gamma}, \quad g_1 = \frac{2}{(1-\nu)g}, \quad \lambda = \frac{\rho R^2\omega^2}{E}, \quad \{T_1^0, T_2^0\} = -\lambda_t Eh\mu^2\{t_1, t_2\}. \end{aligned} \tag{3.152}$$

The dimensionless unknowns (with the hat sign) are introduced as

$$\{w, \Psi, \Theta\} = R\{\hat{w}, \hat{\Psi}, \hat{\Theta}\}, \quad \Phi = EhR^2\mu^2\hat{\Phi}, \quad \{u, v\} = \mu R\{\hat{u}, \hat{v}\},$$
$$\{T_1, T_2, S\} = Eh\{\hat{T}_1, \hat{T}_2, \hat{S}\}, \quad \{Q_1, Q_2\} = Eh\mu\{\hat{Q}_1, \hat{Q}_2\}, \tag{3.153}$$
$$\{\varphi_1, \varphi_2\} = \mu^{-1}\{\hat{\varphi}_1, \hat{\varphi}_2\}, \quad \{M_1, M_2, H\} = Eh\mu^2 R\{\hat{M}_1, \hat{M}_2, \hat{H}\}.$$

In what follows, the hat sign will be omitted.

Here, $\mu \ll 1$ is a small parameter, g is the transverse shear parameter, q is the wave parameter, λ is the frequency parameter. We shall study the localized non-uniform vibrations with the large variability of SSS. Then without lost of generality we may assume that the coefficients of the operators (3.151) are constant, and $R = B = R_2 = k_2 = 1$, $\lambda_t = 1$, and denote $q = \mu m$.

The general solution of Eqs. (3.150) reads as

$$Z(s) = \sum_{k=1}^{10} C_k Z_k e^{p_k s}, \tag{3.154}$$

where Z(s) denotes any unknown function, Z_k are the constant coefficients (see Eqs. (3.156) and (3.157) below), C_k are arbitrary coefficients, p_k are the roots of the characteristic 10th order equation

$$F_8(p)(\Delta - g_1) = 0, \quad F_8(p) = \Delta^4 + (1 - g\Delta)(\Delta_R^2 - \lambda\Delta^2 + \Delta^2\Delta_T), \tag{3.155}$$

where Δ, Δ_R, Δ_T are the same as in Eqs. (3.151). At $g = 0$ the equation $F_8(p) = 0$ coincides with Eq. (3.113), as obtained by using the KL model. Let the first group of roots p_k, $k = 1, 2, \ldots, 8$ be the roots of the equation $F_8(p) = 0$, and the roots of the second group be $p_{9,10} = \pm\sqrt{g_1 + q^2}$.

For the roots of the first group, the coefficients Z_k in Eq. (3.154) for the functions containing in the boundary conditions (2.55) are as follows:

$$\begin{aligned}
&w_k = 1, \quad \Psi_k = \frac{w_k}{1 - g\Delta_k}, \quad \Phi_k = -\frac{\Delta_{Rk} w_k}{\Delta_k^2}, \quad \Delta_k = p_k^2 - q^2,\\
&\Delta_{Rk} = k_2 p_k - k_1 q^2, \quad \varphi_{1k} = -p_k \Psi_k, \quad \varphi_{2k} = -q\Psi_k, \quad T_{1k} = -q^2\Phi_k,\\
&S_k = -p_k q \Phi_k, \quad M_{1k} = (p_k^2 - \nu q^2)\Psi_k, \quad H_k = (1 - \nu)p_k q \Psi_k,\\
&u_k = \frac{k_1 w_k - (q^2 + \nu p_k^2)\Phi_k}{p_k}, \quad v_k = \frac{k_2 w_k + (p_k^2 + \nu q^2)\Phi_k}{q},\\
&Q_{1k} = -p_k M_{1k} + qH_k - p_k t_1 w_k, \quad \Theta_k = 0, \quad k = 1, 2, \ldots, 8;
\end{aligned} \tag{3.156}$$

for the roots of the second group, they read as

$$\begin{aligned}
&\Theta_k = 1, \quad w_k = \Phi_k = \Psi_k = T_{1k} = S_k = u_k = v_k = 0,\\
&\varphi_{1k} = -q\Theta_k, \quad \varphi_{2k} = -p_k\Theta_k, \quad M_{1k} = p_k q(1 - \nu)\Theta_k,\\
&H_k = \frac{1 - \nu}{2}(p_k^2 + q^2)\Theta_k, \quad Q_{1k} = \frac{1 - \nu}{2} q(q^2 - p_k^2)\Theta_k, \quad k = 9, 10.
\end{aligned} \tag{3.157}$$

Let $0 \le s \le s_0$. We shall search for a localized near the edge $s = s_0$ the non-uniform vibration mode. Assume that the roots $\Re p_k > 0,\ k = 1, 2, 3, 4, 9$, and $\Re p_k < 0$ for the rest roots of Eq. (3.155). Then we seek the solution decreasing away from $s = s_0$ in the form

$$Z_j(s) = \sum_{k\in K} C_k Z_k^{(j)} e^{p_k(s-s_0)}, \quad K = \{1, 2, 3, 4, 9\}, \quad j = 1, \ldots, 5, \quad 0 \le s \le s_0, \tag{3.158}$$

where $Z_j(s)$ are the functions in the boundary conditions (2.55)

$$\begin{array}{llll} (1) & u = 0, & T_1 = 0, & (0), \\ (1) & v = 0, & S = 0, & (0), \\ (1) & w = 0, & Q_1 = 0, & (0), \\ (1) & \varphi_1 = 0, & M_1 = 0, & (0), \\ (1) & \varphi_2 = 0, & H = 0, & (0). \end{array} \tag{3.159}$$

Now the frequency equation reads as

$$\Delta_5(\lambda) = 0, \tag{3.160}$$

where $\Delta_5(\lambda)$ is the determinant of the 5th order that depends on the boundary conditions under consideration. For example, at a free edge, $T_1 = S = Q_1 = M_1 = H = 0$, we get

$$\Delta_5(\lambda, t_1, t_2, k_2, g, q) = \begin{vmatrix} T_{11} & T_{12} & T_{13} & T_{14} & 0 \\ S_{11} & S_{12} & S_{13} & S_{14} & 0 \\ Q_{11} & Q_{12} & Q_{13} & Q_{14} & Q_{19} \\ M_{11} & M_{12} & M_{13} & M_{14} & M_{19} \\ H_1 & H_2 & H_3 & H_4 & H_9 \end{vmatrix} = 0. \tag{3.161}$$

Non-uniform localized vibration modes exist not for all variants of boundary conditions (3.159) and not for all values of the parameters q, k_2, t_1, t_2, g.

We consider in more details a cylindrical shell under axial compression. We put $k_2 = t_2 = 0$; the parameters q, t_1, g are to be analyzed. At first, we find the critical value t_1^* of the compression parameter t_1 at which the shell buckling takes place. We seek vibration modes only for $t_1 < t_1^*$.

3.11.1 Effect of boundary conditions on stability of a circular cylindrical shell

Consider, at first, an internal stability of a shell under axial compression. A stability loss takes place if Eq. (3.155), $F_8(p) = 0$, has imaginary roots $p = i\hat{p}$ with $\lambda = 0$. We rewrite this equation as

$$t_1(\hat{p}, q) = \frac{1}{2}\left(\frac{(\hat{p}^2 + q^2)^2}{\hat{p}^2(1 + g(\hat{p}^2 + q^2))} + \frac{\hat{p}^2}{(\hat{p}^2 + q^2)^2}\right). \tag{3.162}$$

In contrast to the KL model with $g = 0$ where q remains indefinite [227, 251], in the framework of the TR model the function $t_1(\hat{p}, q)$ has a minimum at $q = 0$, namely, a buckling mode is axisymmetric. We get

$$\min_{\hat{p},q} t_1(\hat{p}, q) = \min_{\hat{p}} t_1(\hat{p}), \quad t_1(\hat{p}) = \frac{1}{2}\left(\frac{\hat{p}^2}{1 + g\hat{p}^2} + \frac{1}{\hat{p}^2}\right). \tag{3.163}$$

For $g < 1$, the minimum of the function $t_1(\hat{p})$ gives the critical value t_1^{cr} and the wave number $\hat{p}^{cr}$ of an axi-symmetric shell buckling under axial compression

$$t_1^{cr}(g) = 1 - \frac{g}{2}, \quad \hat{p}^{cr}(g) = \frac{1}{\sqrt{1-g}}. \tag{3.164}$$

As $g \to 1$, we get $\hat{p}_{cr} \to \infty$, therefore the wave length $\ell = 2\pi/p_{cr}$ goes to zero, the TR model (as any other 2D shell theories) becoming unacceptable. In [227, 251] it is shown that as $g \to 1$ the shell material loses its stability. Further we assume that $g < 1$.

For a homogeneous material, $g \ll 1$ and the KL model according to Eq. (3.164) gives $t_1^{cr} = 1$ (see [77, 222, 226]).

Near a weakly supported edge the non-uniform buckling modes are possible. In Section 3.5, for the KL model 8 variants of boundary conditions are indicated (see Fig. 3.7). For the TR model, a number of boundary conditions is equal to 5, and a number of studied variants of boundary conditions is 32.

The detailed analysis of these variants is performed in [251]. The variants are divided into four groups with a various behavior of the curves $t_1^*(q, g)$ for the fixed shear parameter g. Here, $t_1^*(q, g) < t_1^{cr}(g)$ is the root of Eq. (3.160) with $\lambda = 0$. The groups are as follows:

Group 1 contains 3 variants: (00000), (01001), (10000), for which the functions $t_1^*(q, g)$ have minimum at $q = q^* > 0$. For the free edge (00000), these functions are shown in Fig. 3.13.

Group 2 contains 10 variants: (00001), (00010), (00100), (00101), (1000), (10001), (10100), (10101), (11000), (11001), for that the minimum of functions $t_1^*(q, g)$ is reached at $q^* = 0$.

Group 3 contains 3 variants: (00011), (01010) and (11100). For these variants, some roots of the characteristic equation, $F_8(p) = 0$, have small real parts, and hence the localized buckling mode decreases very slowly.

Group 4 consists of the remaining 16 variants for which localized buckling modes are absent.

Here, a localization of buckling modes is considered near the edge $s = s_0$. It is necessary to bear in mind that the studied edge is to be supported more weakly than the opposite one ($s = 0$); namely, the value $t_1^*(q, g)$ for the edge $s = s_0$ is to be smaller than for the edge $s = 0$. In the opposite case, the roles of the edges change. The more complex case, when the values $t_1^*(q, g)$ for the both edges coincide with each other, was studied in [52].

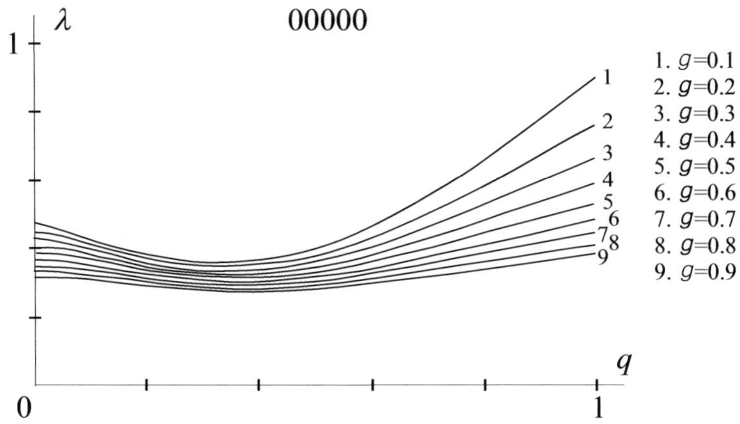

FIGURE 3.13: Functions $t_1^*(q,g)$ for the free edge (00000).

3.11.2 Localized non-uniform vibration modes

As it was mentioned earlier, a cylindrical shell may have localized non-uniform vibration modes only in the presence of a longitudinal compression. We assume that this compression, t_1, is smaller than the critical compression of a shell buckling with the same boundary conditions. In contrast to Section 3.5 (see Fig. 3.7), here we assume that for the chosen variant of boundary conditions and for the fixed g the initial compression satisfies the inequality

$$t_1 < t_1^{**}, \qquad t_1^{**} = \min_q t_1^*(g,q) = t_1^*(g,q^*). \tag{3.165}$$

To find the area of possible values of the frequency parameter, λ, we rewrite the characteristic equation (3.155) as follows:

$$\lambda = f(\hat{p}) = \frac{(\hat{p}^2+q^2)^2}{1+g(\hat{p}^2+q^2)} + \frac{\hat{p}^4}{(\hat{p}^2+q^2)^2} - t_1\hat{p}^2, \quad \hat{p} = ip, \quad \mathrm{i} = \sqrt{-1}. \tag{3.166}$$

Equation (3.155) has no imaginary roots if $\lambda < \min_{\hat{p}} f(\hat{p})$. The minimum is attained at $\hat{p} = 0$, therefore we will seek the roots of Eq. (3.155) and of Eqs. (3.160) for

$$\lambda < \lambda^* = \frac{q^4}{1+gq^2}. \tag{3.167}$$

For small q, the possible interval $0 < \lambda < \lambda^*$ is very narrow and independent of the boundary conditions variant and of the value t_1.

Now we shall search for non-uniform vibration frequencies as roots of Eq. (3.160). It seems that non-uniform vibration modes ought to correspond to each of 16 non-uniform buckling modes, but calculations do not support this assumption. There are only 6 variants of boundary conditions (00000, 00001,

00010, 01000, 01001, 10000) for which there exist localized non-uniform vibration modes for some values of parameters t_1, g, q.

We consider the localization mode near a free edge.

TABLE 3.10: Frequency parameters of non-uniform localized vibration modes.

	$g \to 0$				$g = 0.5$		
q	$t_1 = 0.38$	$t_1 = 0.3$	$t_1 = 0.15$	$t_1 = 0.1$	$t_1 = 0.31$	$t_1 = 0.15$	$t_1 = 0.1$
0.1	0.00004	0.0001	0.0001	0.00010	0.0001	0.0001	0.0001
0.2	0.00028	0.0010	0.0015	0.00155	0.0007	0.0015	0.0015
0.3	0.00002	0.0049	0.0077	0.00795	0.0017	0.0073	0.0076
0.4	0.0040	0.0176	0.0242	0.0263	0.0042	0.0227	0.0232
0.5	0.0326	0.0516	0.0620	− − −	0.0208	0.0538	− − −
0.6	0.1027	0.1197	0.1260	− − −	0.0671	0.1070	− − −
0.7	0.221	0.231	− − −	− − −	0.150	− − −	− − −
0.8	− − −	− − −	− − −	− − −	0.271	− − −	− − −
0.9	− − −	− − −	− − −	− − −	0.433	− − −	− − −

In Table 3.10, the values of the frequency parameter λ for some values of parameters t_1, g, q are presented. The columns with $g \to 0$ correspond to an isotropic shell with small g. Gaps denote the absence of localized modes. The parameter λ increases simultaneously with increasing the wave parameter q, and with decreasing the initial compression t_1. Localized modes are absent for large values of g and for small t_1. For small g, the real parts of some roots of Eq. (3.155) are very small, and localized modes decrease very slowly. The shell loses its stability with $t_1 = 0.38, q_* = 0.31$, this is why the value $\lambda = 0.00002$ in Table 3.10 at $q = 0.3$ is very small, because q is close to the critical value q_*.

Chapter 4

Localized vibration modes of cylindrical and conic shells

In this chapter, problems on vibrations of thin cylindrical and conical shells, and shells close to them as well are considered. In particular, shells with non-circular cross-section, shells with slant edge, and cylindrical panels are studied. In contrast to Chapter 3, here the problems do not admit an exact separation of variables. The asymptotic separation of variables, those with the asymptotic expansions lead to an approximate solution, is used. Natural frequencies and vibration modes localized near the weakest generatrix or near the rectilinear edge are found.

4.1 Equations of localized shells vibrations, and boundary conditions

In this chapter, low-frequency vibrations of shells with zero Gaussian curvature and of a close form are investigated using asymptotic expansions. A peculiarity of these problems consists in the presence of modes which correspond to the *semi-momentless* stress-strain state [173]. For these modes, the index of variation in the longitudinal direction is zero, and in the circumferential direction, 1/4. In the neighborhood of curvilinear shell edges the main semi-momentless solution is accompanied with the edge effect solutions, the indexes of variability of that in the longitudinal and in the circumferential directions are equal to 1/2 and 1/4, respectively.

The accuracy of Eqs. (2.41) of shallow shells is enough to construct the first two terms of asymptotic expansions. We rewrite this system in the dimensionless form after relating all linear values to the characteristic radius R of the neutral surface. On the cylindrical or conic shell surface, we introduce curvilinear orthogonal coordinates s, φ, where s is the length of generatrix, and φ is the coordinate in the circumferential direction with $0 \le \varphi < 2\pi$, $s_1(\varphi) \le s \le s_2(\varphi)$. The latter inequality allows us to study shells with slant edges. We choose the coordinates s, φ so that the first quadratic form of the surface reads as

$$I = R^2 ds^2 + B^2(s,\varphi)d\varphi^2, \qquad B(s,\varphi) = B(\varphi)(1-\eta s). \tag{4.1}$$

For a cylindrical shell $\eta = 0$, and for a conic shell, the distance between the point $s = 0$ and the vertex of the cone is $1/\eta$.

The curvatures of the surface are as follows:

$$\frac{1}{R_1} = 0, \qquad \frac{1}{R_2} = \frac{k(\varphi)}{B(s,\varphi)}. \tag{4.2}$$

In particular, for a circular cylindrical shell of radius R we get $k(\varphi) = 1$, and for a circular conic shell of radius R of the lower base and with angle α at the vertex we get $B(\varphi) = R$, $k(\varphi) = \cos\alpha$, $\eta = \sin\alpha$.

For a shell close to a cylindrical one, the curvilinear coordinates are introduced in Section 4.5.

In the dimensionless form Eqs. (2.41) read as

$$\mu^4 \Delta\Delta w - \lambda w + \Delta_R \Phi = 0, \qquad \Delta\Delta\Phi - \Delta_R w = 0, \tag{4.3}$$

where

$$\mu^4 = \frac{h^2}{12(1-\nu^2)R^2}, \qquad \lambda = \frac{\rho R^2 \omega^2}{E} \tag{4.4}$$

are the small thickness parameter, and the frequency parameter, respectively. The differential operators are as follows:

$$\Delta w = \frac{1}{\hat{B}}\frac{\partial}{\partial s}\left(\hat{B}\frac{\partial w}{\partial s}\right) + \frac{1}{\hat{B}}\frac{\partial}{\partial \varphi}\left(\frac{1}{\hat{B}}\frac{\partial w}{\partial \varphi}\right), \quad \Delta_R w = \frac{k(\varphi)}{\hat{B}(s)}\frac{\partial^2 w}{\partial s^2} \quad \hat{B} = \frac{B}{R}, \tag{4.5}$$

In what follows, the sign ^ is omitted.

The tangential stress resultants are expressed in terms of the stress function Φ as

$$T_1 = \frac{1}{B}\frac{\partial}{\partial \varphi}\left(\frac{1}{B}\frac{\partial \Phi}{\partial \varphi}\right) + \frac{1}{B}\frac{\partial B}{\partial s}\frac{\partial \Phi}{\partial s}, \quad T_2 = \frac{\partial^2 \Phi}{\partial s^2}, \quad S = -\frac{1}{B}\frac{\partial^2 \Phi}{\partial s \partial \varphi} + \frac{1}{B^2}\frac{\partial B}{\partial s}\frac{\partial \Phi}{\partial \varphi}. \tag{4.6}$$

We study thin shells of moderate length and investigate the low-frequency part of the spectrum for semi-momentless vibration modes [173]. For the semi-momentless SSS the unknown functions satisfy the strong inequality

$$\frac{\partial^2 F}{\partial s^2} \ll \frac{\partial^2 F}{\partial \varphi^2}. \tag{4.7}$$

For these vibrations the following simplification of Eqs. (4.3) is possible. The first summand in the Laplace operator (4.5) has no effect on the construction of the first and the second terms of asymptotic expansions, and so Eqs. (4.3) assume the form

$$\frac{\mu^4}{B^4(s)}\frac{\partial^4 w}{\partial \varphi^4} - \lambda w + \frac{k(\varphi)}{B(s)}\frac{\partial^2 \Phi}{\partial s^2} = 0, \qquad \frac{1}{B^4(s)}\frac{\partial^4 \Phi}{\partial \varphi^4} - \frac{k(\varphi)}{B(s)}\frac{\partial^2 w}{\partial s^2} = 0. \tag{4.8}$$

As a result, the system (4.8) has the 4th order in s, and it is necessary to find two main boundary conditions which the semi-momentless solution satisfies at the edges $s = s_1(\varphi)$ and $s = s_2(\varphi)$.

The problem of finding the main boundary conditions can be solved by excluding the edge effect solutions. For the shell buckling with semi-momentless modes, this problem was solved in [222], [226]. It was shown that for both buckling and vibrations, the main boundary conditions coincide, and here only the final results are presented.

We consider 16 variants of boundary conditions, and use the notation introduced in Eqs. (2.23) and (2.27). The result is different for the straight and for the slant edge $s = s_0(\varphi)$.

For the straight edge with $ds_0/d\varphi = 0$, the 16 considered variants of the boundary conditions can be split into four groups: the clamped support group, the simple support group, the weak support group, and the free edge group. Table 4.1 shows the main boundary conditions and the corresponding variants of full boundary conditions, containing in the group.

TABLE 4.1: Main boundary conditions in the case of a straight edge for different groups of boundary conditions: C(clamped), S(simply supported), W(weakly supported), F(free edge).

Group	Main boundary conditions	Full boundary conditions
C	$w = \dfrac{\partial w}{\partial s} = 0$	1111, 1110, 1101, 1100, 1011, 1010
S	$w = \Phi = 0$	0111, 0110, 0101, 0100, 0011, 0010
W	$\dfrac{\partial}{\partial s}\left(\dfrac{w}{B}\right) = \dfrac{\partial}{\partial s}\left(\dfrac{\Phi}{B}\right) = 0$	1001, 1000
F	$\Phi = \dfrac{\partial \Phi}{\partial s} = 0$	0001, 0000

The clamped support group (C) contains 6 variants of boundary conditions. Among conditions corresponding to the clamped support group, there is the condition $u = 0$ and at least one of conditions $v = 0$ or $w = 0$. The group of simple support differs from the clamped support group by the presence of the condition $T_1 = 0$ instead of the condition $u = 0$. The group of weak support contains 2 variants. In this group, there is the constraint $u = 0$ and no constraints in the circumferential and transversal directions, $S = Q_1^* = 0$. In the group of free edge, the constraint $u = 0$ is replaced by the condition $T_1 = 0$. The constraint for an angle of rotation, $\gamma_1 = 0$, or its absence has no effect on the belonging to a group.

The slant edge with $ds_0/d\varphi \neq 0$. In this case, to construct the edge effect integrals, we use the orthogonal coordinates s_m, φ_t (see [45]), and instead of the tangential deflections u, v we introduce the projections u_m, v_t of deflection onto the normal and tangent to the edge. This circumstance influences the result of excluding the edge effect integrals from the boundary conditions. As a result, 16 variants of boundary conditions (see Eqs. (2.27) are divided into the following four groups (see [222, 226]): the clamped support group, the special group, the group of simple support, and the group of free edge. The distribution of the boundary conditions in these groups is shown in Table 4.2.

TABLE 4.2: Groups of boundary conditions in the case of a slant edge.

Group	Full boundary conditions
Clamped	1111, 1110, 1101, 1100
Special	1011, 1010
Simply supported	0111, 0110, 0101, 0100, 0011, 0010, 1001, 1000
Free edge	0001, 0000

The main boundary conditions for three groups (clamped, simple support, and free edge) are the same as in Table 4.1 for the straight edge. For the special group, the main boundary conditions are as follows

$$w = 0, \quad \frac{\partial w}{\partial s} - a_i \Phi = 0, \tag{4.9}$$

and the constants a_i are different for the variants 1011 and 1010 [222, 226].

4.2 Low-frequency vibrations of circular cylindrical shells

We consider low-frequency vibrations of a thin circular cylindrical shell of radius R and length L for various boundary conditions at the edges $s = 0$ and $s = l = L/R$. The corresponding vibration modes

$$w(s, \varphi) = w(s) \cos m\varphi \tag{4.10}$$

are not localized, and they occupy the entire shell surface. The results, which are given below, will be used in the following analysis. They are also of an independent interest.

In the case $B(s) \equiv 1$, and $\kappa(\varphi) \equiv 1$, Eqs. (4.8) after the separation of variables (4.10) are recast into the beam equation

$$\frac{d^4w}{ds^4} - \alpha^4 w = 0, \qquad \lambda = \mu^4 m^4 + \frac{\alpha^4}{m^4}. \tag{4.11}$$

The minimum value of the frequency parameter, $\lambda \simeq 2\mu^2\alpha^2$, is attained at $m \simeq \sqrt{\alpha/\mu}$.

Values of a parameter α are calculated by satisfying the boundary conditions at $s = 0$ and for $s = l$ (see Table 4.3, where the derivative is denoted by the stroke).

TABLE 4.3: Main boundary conditions for a circular cylindrical shell.

Group	Main boundary conditions
Clamped	$w = w' = 0$
Simply supported	$w = w'' = 0$
Weakly supported	$w' = w''' = 0$
Free edge	$w'' = w''' = 0$

For various combinations of the main boundary conditions at the edges $s = 0$ and $s = l$, Table 4.4 shows the first two values of parameter $\beta_n = \alpha_n l$ and their asymptotic expressions as $n \to \infty$.

TABLE 4.4: Eigenvalues of the parameter $\beta_n = a_n l$.

N	$s = 0$	$s = l$	β_1	β_2	β_n
1	Clamped	Clamped	4.730	7.853	$\pi(n + 1/2)$
2	Clamped	Simply supported	3.927	7.069	$\pi(n + 1/4)$
3	Simply supported	Simply supported	π	2π	πn
4	Clamped	Weakly supported	2.365	5.498	$\pi(n - 1/4)$
5	Clamped	Free edge	1.876	4.694	$\pi(n - 1/2)$
6	Simply supported	Weakly supported	1.571	4.712	$\pi(n - 1/2)$
7	Simply supported	Free edge	0.000	3.926	$\pi(n - 3/4)$
8	Weakly supported	Weakly supported	0.000	3.142	$\pi(n - 1)$
9	Weakly supported	Free edge	0.000	2.365	$\pi(n - 5/4)$
10	Free edge	Free edge	0.000	0.000	$\pi(n - 3/2)$

Next, we consider the effect of additional boundary conditions on the natural frequency. This influence is of the relative order μ or $h_*^{1/2}$. The accuracy of Eqs. (4.8) and (4.11) is not enough to examine this effect, because in these equations the terms of the order of μ were omitted. That is why for the numerical solution we use the more accurate equations (3.5).

Example 4.1. We consider a circular cylindrical shell with parameters $R/h = 200$, $l = L/R = 3$, $\nu = 0.3$. We shall calculate the first natural frequencies for all 16 variants of boundary conditions in the assumption that at the both shell edges $s = 0$ and $s = l$ the boundary conditions are identical and the number of waves m in the circumferential direction is 6. The results are presented in Table 4.5.

In columns 1, 3, 5, the boundary conditions are given, and in columns 2, 4, 6, the corresponding values of λ are presented. The columns 1 and 2 correspond to the clamped support group with $\beta = 4.73$, the columns 3, 4 correspond to the simple support group with $\beta = \pi$, and the columns 5, 6 correspond to the weakly supported group and to the free edge group with $\beta = 0$. In the second row, the values of λ, as found from the approximate equation (4.11), are given for the corresponding values of parameter β. The accuracy of Eq. (4.11) is not high (especially for the boundary conditions of the clamped support group, but the results of Table 4.5 demonstrate that the values of λ for the boundary conditions of each group are close to each other.

TABLE 4.5: Dependence of the frequency parameter, λ, on the boundary conditions.

1	2	3	4	5	6
$\beta = 4,73$	0.00774	$\beta = \pi$	0.00389	$\beta = 0$	0.00297
1111	0.00634	0111	0.00386	1001	0.00273
1110	0.00629	0110	0.00375	1000	0.00272
1101	0.00628	0101	0.00380	0001	0.00273
1100	0.00627	0100	0.00375	0000	0.00272
1011	0.00625	0011	0.00386		
1010	0.00597	0010	0.00372		

From Table 4.5, it follows the well-known result that larger values of λ correspond to more strong constraints. More hard constraints can be obtained if we change 0 by 1 in the symbolic four digital designation of a variant.

4.3 Localized vibrations modes of cylindrical shells

Here, localized low-frequency vibrations modes of thin (in the general case) non-circular cylindrical shells with slant edges are studied. We shall use Eqs. (4.3) which can be rewritten as

$$\varepsilon^4 \Delta\Delta w - \hat{\lambda} w + \Delta_R \Phi = 0, \quad \varepsilon^4 \Delta\Delta \Phi - \Delta_R w = 0 \tag{4.12}$$

with

$$\varepsilon = \sqrt{\mu} \sim h_*^{1/4}, \qquad \hat{\lambda} = \varepsilon^{-4}\lambda, \qquad \hat{\Phi} = \varepsilon^{-4}\Phi, \tag{4.13}$$

where ε, λ, and Φ are the new thickness small parameter, the frequency parameter, and the stress function, respectively. The operators Δw and $\Delta_R w$ for a cylindrical shell (with $B(s) \equiv 1$) read as

$$\Delta w = \frac{\partial^2 w}{\partial s^2} + \frac{\partial^2 w}{\partial \varphi^2}, \qquad \Delta_R w = k(\varphi)\frac{\partial^2 w}{\partial s^2}, \tag{4.14}$$

where $k(\varphi) > 0$ is the curvature of the shell middle surface.

The boundary conditions (2.27) are set at the edges $s = s_1(\varphi)$ and $s = s_2(\varphi)$.

We seek semi-momentless vibrations modes for which the function $w(s, \varphi)$ rapidly varies in the circumferential direction (with the index of variation equal to $\varsigma = 1/4$). For the conditions that will be indicated below and which depend on the functions $k(\varphi)$, $s_1(\varphi)$, $s_2(\varphi)$, the deflection $w(s, \varphi)$ is localized near the weakest generatrix $\varphi = \varphi_0$. To find it, we assume temporary that the functions $k(\varphi)$, $s_1(\varphi)$, $s_2(\varphi)$ are the constants equal to values of these functions at $\varphi = \varphi_0$. Then we pass to the problem on vibrations (studied in Section 4.2) of a circular cylindrical shell with radius $R = k^{-1}(\varphi_0)$ and constant length $l(\varphi_0) = s_2(\varphi_0) - s_1(\varphi_0)$. As in Section 4.2, we search for a periodic in φ solution of Eqs. (4.12) in the form

$$w(s, \varphi) = w(s)\sin(p\varphi/\varepsilon), \quad \hat{\Phi}(s, \varphi) = \hat{\Phi}(s)\sin(p\varphi/\varepsilon). \tag{4.15}$$

Substituting Eqs. (4.15) into Eqs. (4.12) and omitting the summands of order of ε^2, we obtain

$$\begin{aligned} &\frac{k^2(\varphi_0)}{p^4}\frac{d^4 w}{ds^4} + (p^4 - \hat{\lambda})w = 0, \\ &\hat{\lambda} = f(p, \varphi_0) \equiv p^4 + \frac{\beta^4 k^2(\varphi_0)}{p^4 l^4(\varphi_0)}, \end{aligned} \tag{4.16}$$

where the parameter β is the same as in Section 4.2 (it depends on the boundary conditions at $s = s_1$ and $s = s_2$). The following analysis shows that the vibration mode (4.15) and the frequency parameter (4.16) give the approximations for the exact solution only in the case when the values $p = p_0$ and $\varphi_0 = \varphi_{00}$ give the minimum of the function $f(p, \varphi_0)$, defined by Eq. (4.16). Besides,

$$\begin{aligned} &\hat{\lambda} = \lambda_0 = \min_{p,\varphi_0} f = \frac{2\beta^2 k(\varphi_{00})}{l^2(\varphi_{00})}, \\ &p_0^4 = \frac{\lambda_0}{2}, \qquad \min_{\varphi_0} \frac{k(\varphi_0)}{l^2(\varphi_0)} = \frac{k(\varphi_{00})}{l^2(\varphi_{00})}. \end{aligned} \tag{4.17}$$

Equations (4.17) give the approximate value of the frequency parameter $\hat{\lambda} = \lambda_0$ and the weakest generatrix φ_{00}. The asymptotic solution of Eqs. (4.12), as given below, is based on Maslov's algorithm [131].

The formal asymptotic solution of Eqs. (4.12) will be sought in the form (see also [233], [232], [222], [226]) which is an extension of Eq. (3.91) to the two dimensional case

$$
\begin{aligned}
& w(s,\varphi,\varepsilon) = w_* \exp\{\mathrm{i}\varepsilon^{-1/2}p\xi - (1/2)b\xi^2\}, \qquad \Re b > 0, \\
& w_* = \sum_{n=0}^{\infty} \varepsilon^{n/2} w_n(s,\xi), \qquad \xi = \varepsilon^{-1/2}(\varphi - \varphi_0), \\
& \hat{\lambda} = \lambda_0 + \varepsilon\lambda_1 + \varepsilon^2\lambda_2 + \dots .
\end{aligned}
\tag{4.18}
$$

The function Φ will be sought in the same form. Here, $w_n(s,\xi)$ are the polynomials in ξ, the wave number $p > 0$, and the condition $\Re b > 0$ is responsible for the decrease of the vibrations amplitudes away from the weakest generatrix $\varphi = \varphi_0$.

To find the unknown functions w_n and Φ_n, and the numbers p, b, φ_0, λ_n, we substitute the function (4.18) into Eqs. (4.12) and equalize the coefficients by the same powers of of $\varepsilon^{1/2}$. We expand the coefficients of Eqs. (4.12) depending on φ in power series of $\varphi - \varphi_0 = \varepsilon^{1/2}\xi$.

First, by using the second equation (4.12), we express the function Φ_* in terms of the function w_*. As a result, we get

$$
\begin{aligned}
\Phi_* = & -\frac{\partial^2}{\partial s^2}\left\{\frac{w_*}{p^4} - \frac{4\varepsilon^{1/2}}{p^5}\left(b\xi w_* - \mathrm{i}\frac{\partial w_*}{\partial \xi}\right) + \frac{10\varepsilon}{p^6}\left[(b^2\xi^2 - \mathrm{i}b)w_* \right.\right. \\
& \left.\left. -2\mathrm{i}b\xi\frac{\partial w_*}{\partial \xi} - \frac{\partial^2 w_*}{\partial \xi^2}\right]\right\} + O(\varepsilon^{3/2}).
\end{aligned}
\tag{4.19}
$$

Now the first equation (4.12) gives a sequence of equations for determining functions w_n

$$
\mathbf{L}_0 w_0 = 0, \quad \mathbf{L}_0 w_1 + \mathbf{L}_1 w_0 = 0, \quad \mathbf{L}_0 w_2 + \mathbf{L}_1 w_1 + \mathbf{L}_2 w_0 = 0, \dots \tag{4.20}
$$

with the operators

$$
\begin{aligned}
\mathbf{L}_0 z = & \frac{k^2(\varphi_0)}{p^4}\frac{\partial^4 z}{\partial s^4} + (p^4 - \lambda_0)z, \\
\mathbf{L}_1 z = & \left(b\frac{\partial \mathbf{L}_0}{\partial p} + \frac{\partial \mathbf{L}_0}{\partial \varphi_0}\right)\xi z - \mathrm{i}\frac{\partial \mathbf{L}_0}{\partial p}\frac{\partial z}{\partial \xi}, \\
\mathbf{L}_2 z = & \frac{1}{2}\left(b^2\frac{\partial^2 \mathbf{L}_0}{\partial p^2} + 2b\frac{\partial^2 \mathbf{L}_0}{\partial p \partial \varphi_0} + \frac{\partial^2 \mathbf{L}_0}{\partial \varphi_0^2}\right)\xi^2 z \\
& -\mathrm{i}\left(b\frac{\partial^2 \mathbf{L}_0}{\partial p^2} + \frac{\partial^2 \mathbf{L}_0}{\partial p \partial \varphi_0}\right)\xi\frac{\partial z}{\partial \xi} \\
& -\frac{1}{2}\frac{\partial^2 \mathbf{L}_0}{\partial p^2}\left(\mathrm{i}z + \frac{\partial^2 z}{\partial \xi^2}\right) - \frac{1}{2}\frac{\partial^2 \mathbf{L}_0}{\partial p \partial \varphi_0}z + \mathbf{N}z, \\
\mathbf{N} = & -\lambda_1.
\end{aligned}
\tag{4.21}
$$

The solution of Eq. (4.20) in the leading approximation

$$\mathbf{L}_0 w_0 = 0 \tag{4.22}$$

with the corresponding main boundary conditions read as

$$w_0 = P(\xi) w^0(s, p, \varphi_0), \qquad \lambda_0 = f(p, \varphi_0), \tag{4.23}$$

where $w^0(s, p, \varphi_0)$ is the solution of Eq. (4.16), the function $f(p, \varphi_0)$ is defined by Eq. (4.16), and the polynomial $P(\xi)$ is to be found in the next approximations. Here and below, the main boundary conditions are not fixed. It is only important that the obtained boundary value problems are self-adjoin.

In the first-order approximation, Eq. (4.20) reads as

$$\mathbf{L}_0 w_1 + G_1 = 0, \quad G_1 = (b\xi P_0(\xi) - iP_0'(\xi)) \frac{\partial \mathbf{L}_0}{\partial p} w^0 + \xi P_0(\xi) \frac{\partial \mathbf{L}_0}{\partial \varphi_0} w^0. \tag{4.24}$$

The compatibility condition of the corresponding boundary value problems is as follows:

$$\int_{s_1}^{s_2} w^0 G_1 \, ds = 0. \tag{4.25}$$

Differentiating Eq. (4.21) with respect to the parameters q and φ_0, we obtain the boundary value problems

$$\begin{aligned} \mathbf{L}_0 w_p + \frac{\partial \mathbf{L}_0}{\partial p} w^0 - \frac{\partial \lambda_0}{\partial p} w^0 = 0, \\ \mathbf{L}_0 w_\varphi + \frac{\partial \mathbf{L}_0}{\partial \varphi_0} w^0 - \frac{\partial \lambda_0}{\partial \varphi_0} w^0 = 0. \end{aligned} \tag{4.26}$$

The compatibility conditions of these problems are as follows:

$$\begin{aligned} \int_{s_1}^{s_2} w^0 \mathbf{L}_0 w_p \, ds = \int_{s_1}^{s_2} w_p \mathbf{L}_0 w^0 \, ds = 0, \\ \int_{s_1}^{s_2} w^0 \mathbf{L}_0 w_\varphi \, ds = \int_{s_1}^{s_2} w_\varphi \mathbf{L}_0 w^0 \, ds = 0. \end{aligned} \tag{4.27}$$

Now, due to Eqs. (4.24), (4.26) and (4.27), we can rewrite Eq. (4.25) as follows:

$$\left[(b\xi P_0(\xi) - \mathrm{i}P_0'(\xi)) \frac{\partial \lambda_0}{\partial p} + \xi P_0(\xi) \frac{\partial \lambda_0}{\partial \varphi_0} \right] \int_{s_1}^{s_2} \left(w^0\right)^2 ds = 0, \tag{4.28}$$

From Eq. (4.28), one can conclude that

$$\frac{\partial \lambda_0}{\partial p} = 0, \qquad \frac{\partial \lambda_0}{\partial \varphi_0} = 0. \tag{4.29}$$

If equalities (4.29) hold, then the solution of Eq. (4.13) reads as

$$w_1 = P_1(\xi) w^0 + \xi P_0(\xi)(b w_p + w_\varphi) - \mathrm{i}P_0'(\xi) w_p, \tag{4.30}$$

where w_p and w_φ are the solutions of the boundary value problems (4.26) and the function $P_1(\xi)$ is not defined now (it is to be defined from the following approximations).

The compatibility condition of the boundary value problem (4.30) in the second-order approximation leads to the second order differential equation with respect to $P_0(\xi)$,

$$-\frac{1}{2}f_{pp}P_0'' - \mathrm{i}(bf_{pp} + f_{p\varphi})\left(\xi P_0' + \frac{1}{2}P_0\right) - \lambda_1 P_0 + c\xi^2 P_0 = 0 \tag{4.31}$$

with

$$2c = b^2 f_{pp} + 2bf_{p\varphi} + f\varphi\varphi, \qquad f_{pp} = \frac{\partial^2 f}{\partial p^2}, \quad f_{p\varphi} = \frac{\partial^2 f}{\partial p \partial \varphi_0}, \quad f_{\varphi\varphi} = \frac{\partial^2 f}{\partial \varphi_0^2}, \tag{4.32}$$

and the function $f(p, \varphi_0)$ defined by Eq. (4.26).

The condition $c = 0$ is necessary for the existence of a polynomial solution of Eq. (4.31). From the square equation $c = 0$, we get

$$b = \frac{-f_{p\varphi} + \mathrm{i}r}{f_{pp}}, \qquad r = \sqrt{d}, \qquad d = f_{pp}f_{\varphi\varphi} - (f_{p\varphi})^2. \tag{4.33}$$

From Eq. (4.18) it follows that for the decreasing of the vibrations amplitudes away from the weakest generatrix $\varphi = \varphi_{00}$ it is enough to carry out the conditions

$$f_{pp} > 0, \qquad d > 0. \tag{4.34}$$

With the fulfillment of these conditions the function $f(p, \varphi_0)$ has at the point $p = p_0, \varphi_0 = \varphi_{00}$ a strong minimum, the quadratic form $f_{pp}(p-p_0)^2 + 2f_{p\varphi}(p-p_0)(\varphi_0 - \varphi_{00}) + f_{\varphi\varphi}(\varphi_0 - \varphi_{00})^2$ is positively definite.

For

$$\lambda_1 = \lambda_1^{(n)} = r\left(n + \frac{1}{2}\right), \qquad n = 0, 1, 2, \ldots \tag{4.35}$$

Equation (4.31) has the solution

$$P(\xi) = \mathcal{H}_n(\theta), \qquad \theta = \sqrt{\frac{r}{f_{pp}}}\,\xi, \tag{4.36}$$

where $\mathcal{H}_n(\theta)$ is the n-th degree Hermite polynomial.

Therefore, we find the set of eigenvalues

$$\hat{\lambda}^{(n)} = \lambda_0 + \varepsilon r\left(n + \frac{1}{2}\right) + O(\varepsilon^2), \qquad n = 0, 1, 2, \ldots \tag{4.37}$$

As in Section 3.7, these eigenvalues are asymptotically double. For the fixed n, there are two very close to each other exact eigenvalues $\hat{\lambda}^{(n,1)}$ and $\hat{\lambda}^{(n,2)}$, and

$$|\hat{\lambda}^{(n,1)} - \hat{\lambda}^{(n,2)}| = O(\varepsilon^N) \tag{4.38}$$

for any N. The corresponding exact eigenfunctions are equal to linear combinations of the real and imaginary parts of expression (4.18).

The construction of the following approximations of the series (4.18) is discussed in [222], [226], and here it is not presented by the following reasons. At first, the leading approximation for a vibration mode, and the zero- and first-order approximations for the frequency parameter give the correct qualitative picture. Secondly, the accuracy of Eqs. (4.12) is not enough to construct the following approximations. And finally, the following approximations depend not only on the main boundary conditions but also on the additional ones. For the buckling problems the influence of the additional boundary conditions was discussed in [222], [226].

The asymptotically double eigenvalues and eigenfunctions are met in a lot of problems on free vibrations of a rigid body. We mention the problem on the high-frequency free vibrations of an elliptic membrane that is described by the Helmholtz equation [17]. In this problem, there is a set of eigenfunctions localized near a short diameter of an ellipse. Similar forms are inherent in eigenfunctions corresponding to high-frequency thickness vibrations of isotropic [235] or anisotropic [234] plates of variable thickness. The last problem has application in designing quartz resonators [26]. In these problems, the eigenfunction may be localized near points at which the plate has maximum or minimum thickness. The same picture occurs for the high-frequency thickness vibrations of a rod of variable thickness having the form of a body of revolution [99].

We consider two examples.

Example 4.2. Consider free low-frequency vibrations of a thin circular cylindrical shell of moderate length with slant edge. Let the shell edges

$$s_1(\varphi) = 0, \qquad s_2(\varphi) = L + R \sin\chi(\cos\varphi - 1) \tag{4.39}$$

be simply supported. The angle χ is shown in Fig. 4.1.

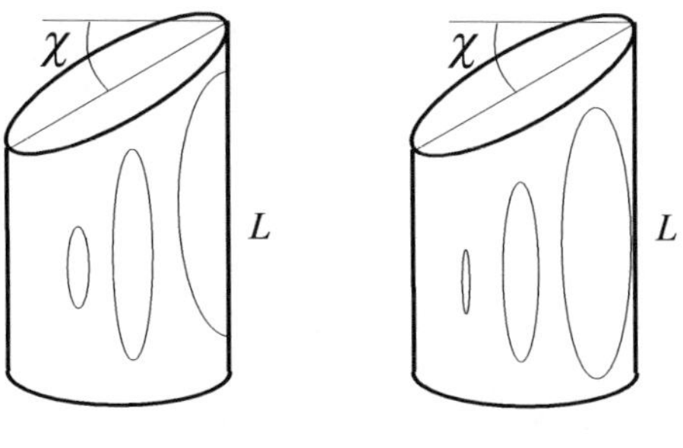

FIGURE 4.1: Cylindrical shell with the slant edge.

Here, the longest generatrix with $\varphi = \varphi_{00} = 0$ is the weakest. Calculations by using Eqs. (4.16), (4.32), (4.37) with $\beta = \pi$ give

$$\lambda_0 = \frac{2\pi^2}{l^2}, \quad p_0^4 = \frac{\pi^2}{l^2}, \quad f_{pp} = \frac{32\pi}{l}, \quad f_{\varphi\varphi} = \frac{4\pi^2 \sin\chi}{l^3}, \quad l = \frac{L}{R}. \tag{4.40}$$

The approximate values of the dimensionless frequency parameters $\hat{\lambda}^{(n)}$ and the natural frequencies $\omega^{(n)}$ in the initial designations are as follows:

$$\hat{\lambda}^{(n)} = \frac{2\pi^2}{l^2}\left(1 + 4\varepsilon\sqrt{\frac{2\sin\chi}{\pi}} + O(\varepsilon^2)\right), \qquad n = 0, 1, 2, \ldots,$$

$$\left(\omega^{(n)}\right)^2 = \frac{\pi^2 Eh}{\rho R L^2\sqrt{3(1-\nu^2)}}\left(1+4\varepsilon\sqrt{\frac{2\sin\chi}{\pi}}+O(\varepsilon^2)\right), \quad \varepsilon^8 = \frac{h^2}{12(1-\nu^2)R^2}. \tag{4.41}$$

Frequencies (4.41) are asymptotically double. The corresponding even and odd with respect to line $\varphi = 0$ vibration modes are schematically shown in Fig. 4.1.

The detailed analysis of low-frequency vibrations of cylindrical shells with a slant edge is contained in the book by Filippov [55]. In this monograph as well as in papers [56, 93, 94], the higher-order asymptotic approximations are built, and the vibrations of two cylindrical shells attached to each other by an angle are investigated.

Example 4.3. Consider free low-frequency vibrations of a thin cylindrical shell of constant length L with elliptical cross-section with semi-axes a_0 and b_0 ($a_0 < b_0$) (see Fig. 4.2). The shell edges $s = 0$ and $s = L$ are simply supported. This example is interesting because here the separation of variables is possible, and we may compare the asymptotic solution with the numerical solution. Numerical computations can be performed with high level of accuracy (say, with an accuracy up to 10^{-12}) that is impossible for the finite element methods.

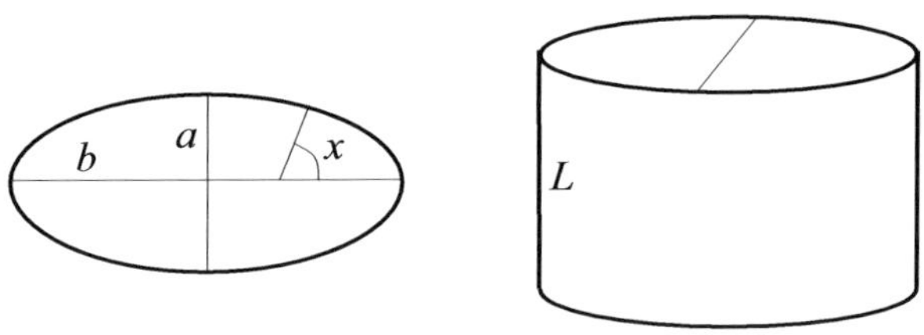

FIGURE 4.2: Elliptical cylindrical shell.

We perform the separation of variables $w(s, \varphi) = w(\varphi)\sin \pi s/L$, and use instead of φ the new independent variable x (the angle between the shell normal and the diameter, Fig. 4.2). Then omitting small summands in the operator Δ we obtain the boundary value problem

$$\varepsilon_0^4 k^4(x)\frac{d^4 w}{dx^4} - \lambda w + k(x)\Phi = 0, \qquad \varepsilon_0^4 k^4(x)\frac{d^4 \Phi}{dx^4} - k(x) w = 0 \tag{4.42}$$

with

$$\varepsilon_0^8 = \frac{h^2 L^4}{12(1-\nu^2)a^6}, \qquad \lambda = \frac{\rho\omega^2 L^4}{\pi^2 \varepsilon_0^4 E a^2},$$
$$k(x) = \frac{dx}{d\varphi} = e^{-2}\left(\sin^2 x + e^2\cos^2 x\right)^{3/2}, \qquad e = \frac{b}{a}. \tag{4.43}$$

Here, ε_0 is the new small parameter, λ is the frequency parameter, $k(x)$ is the curvature of directrix (ellipse). We seek a periodic solution of Eqs. (4.42) at the segment $0 \le x \le 2\pi$.

In this problem, there are two weakest generatrices, $x = \pi/2$ and $x = 3\pi/2$, each of them generates two close to each other natural frequencies, as a result, the natural frequencies are *asymptotically fourfold*. For them, the common asymptotic formula (4.41) is valid:

$$\lambda^{(n)} = \frac{2}{e^2}\left(1 + \varepsilon_0\sqrt{\frac{48(e^2-1)}{e^3}}\left(n + \frac{1}{2}\right) + O(\varepsilon_0^2)\right), \qquad n = 0, 1, 2, \ldots \tag{4.44}$$

To distinguish numerically the close frequencies, we mention that the eigenfunctions are even or odd with respect to two ellipse diameters. These functions are schematically shown in Fig. 4.3.

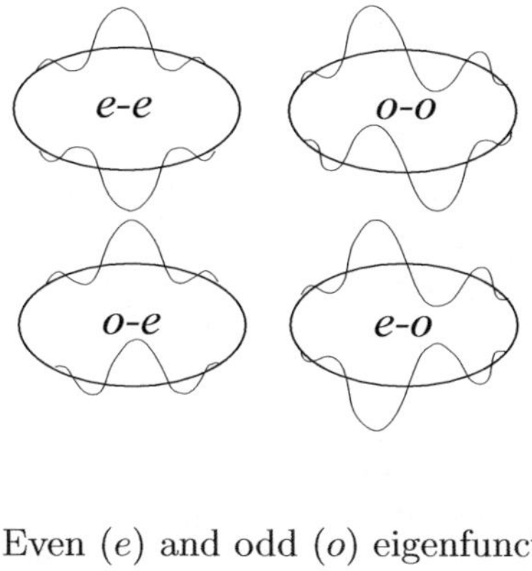

FIGURE 4.3: Even (e) and odd (o) eigenfunctions (sketch).

To calculate the corresponding eigenvalues of the parameter λ, we perform numerical integration in the quarter of the ellipse (in the interval $0 \le x \le \pi/2$), and satisfy the boundary conditions:

$$\begin{aligned} &w' = w''' = \Phi' = \Phi''' = 0 \qquad \text{for even end,} \\ &w = w'' = \Phi = \Phi'' = 0 \qquad \text{for odd end.} \end{aligned} \tag{4.45}$$

For the semi-axes ratio $e = b/a = 1.4$, we give in Table 4.6 some numerical results for asymptotically fourfold eigenvalues depending on the small parameter ε_0. Only the case $n = 0$ is considered.

TABLE 4.6: Asymptotically fourfold eigenvalues.

ε_0	λ^a	λ^{ee}	λ^{oo}	λ^{eo}	λ^{oe}
0.01	1.0413	1.04145	=	=	=
0.02	1.0622	1.06255	=	=	=
0.05	1.125	1.1267209	=	1.1267217	=
0.1	1.229	1.235766405641	5654	1.233745567910	7918
0.2	1.439	1.413183	1.413186	1.472816	1.472835

In Table 4.6, the values λ^a are found from Eq. (4.44) for $n = 0$, and the asymptotically fourfold eigenvalues λ^{ee}, λ^{oo}, λ^{eo} and λ^{oe} are found numerically with an accuracy up to 10^{-12}. For $\varepsilon_0 = 0.01$ and $\varepsilon_0 = 0.02$, the equalities $\lambda^{ee} = \lambda^{oo} = \lambda^{eo} = \lambda^{oe}$ hold with an accuracy up to 10^{-12}. For larger values of ε_0, the eigenvalues are split into two groups $\lambda^{ee}, \lambda^{oo}$ and $\lambda^{eo}, \lambda^{oe}$, and the distances between eigenvalues within the group is much smaller than other distances. For $\varepsilon_0 = 0.05$, the equalities $\lambda^{ee} = \lambda^{oo}$ and $\lambda^{eo} = \lambda^{oe}$ hold with an accuracy up to 10^{-12}. For $\varepsilon_0 = 0.1$, it may be seen the small difference of the order 10^{-10} between eigenvalues within groups (for λ^{oo} and λ^{oe} the last four digits are given), and with increasing ε_0 the difference between eigenvalues becomes larger.

4.4 Localized vibration modes of a cylindrical panel near a free or a weakly supported rectilinear edge.

Let us examine small free low-frequency vibrations of a thin elastic cylindrical panel of radius R and length L. Assume that one of the rectilinear edges, for instance $y = 0$, is free or weakly supported and there exist non-uniform low-frequency vibration modes localized near this edge and exponentially decreasing away from this edge. We will not take into account the boundary conditions at the opposite edge $y = y_0$ assuming that it is more rigid than at the edge $y = 0$, and the edges are far enough from each other so that it is possible to neglect their mutual influence.

We consider arbitrary boundary conditions at the curvilinear edges $x = 0$ and $x = -l = -L/R$ with the following restriction. In Section 4.1, we have considered semi-momentless vibration modes with the separation of boundary conditions into four groups shown in Table 4.1. Here, we take only the combinations of boundary conditions for which the parameter β_1 is non-zero (see Table 4.4). These combinations are contained in the first 6 rows of Table 4.4.

This problem is of interest, because it is simpler than the problem on vibrations of a rectangular plates. Indeed, for the problem under consideration, the following asymptotic formula is obtained [54], [52]:

$$\hat{\lambda} = \lambda_0 + \varepsilon^2 \lambda_2 + O(\varepsilon^3). \tag{4.46}$$

On the other hand, for low-frequency vibrations of a rectangular plate (see Fig. 4.4) with arbitrary boundary conditions, the analytic expressions for natural frequencies are absent. For high-frequency vibrations of a rectangular plate the asymptotic expressions for frequencies and modes may be found in [35].

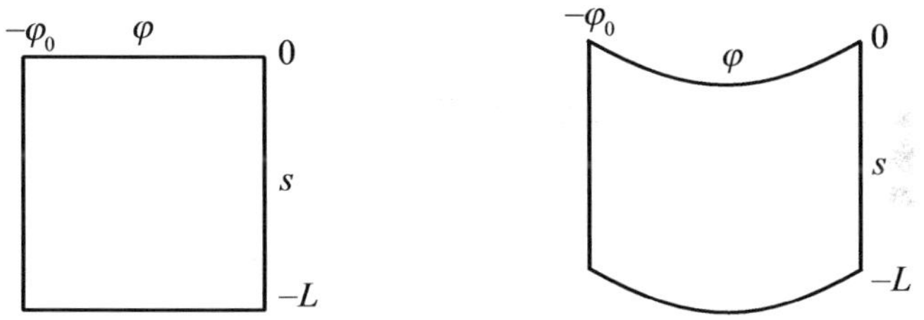

FIGURE 4.4: Rectangular plate (left) and cylindrical panel (right).

From the asymptotic point of view, at $\varepsilon \to 0$ the panel deflection can be represented in the form of the sum of three components:

$$w(x, y, \varepsilon) = w^0(x, \tau, \varepsilon) + w^e(\xi, \tau, \varepsilon) + w^c(\xi, \zeta, \varepsilon), \tag{4.47}$$

where $x = s/R$ is the dimensionless length of generatrix, y is the angle in circumferential direction, $\tau = y/\varepsilon$, $\xi = x/\varepsilon^2$, $\zeta = \varphi/\varepsilon^2$ are the auxiliary variables obtained by scaling. The functions w^0, w^e and w^c regularly depend on ε. The first summand $w^0 \sim 1$ describing the semi-momentless state is the principal one. It satisfies all the boundary conditions at the rectilinear edge $y = 0$ and the main boundary conditions at the curvilinear edges $x = 0$ and $x = -l$. The second term $w^e = O(\varepsilon^2)$ quickly decreasing together with x is the edge effect integral near the edge $x = 0$ and in the sum with w^0 satisfies the additional boundary condition at $x = 0$. Finally, the third summand $w^c = O(\varepsilon^2)$ quickly attenuates in both the x- and y- directions away from the angle point $x = y = 0$; in the sum with w^e, it satisfies the boundary conditions at $y = 0$ near the point $x = 0$. The similar summand appears near the angle $x = -l$, $y = 0$.

The first- and second-order approximations for the function $w^0 = w_0^0 + \varepsilon^2 w^2$ as well as the formulas for λ_0 and λ_2 in Eq. (4.46) were found in [52]. There, an estimate of $O(\varepsilon^3)$ was also obtained, which owes its appearance to the influence of the term w^c. Here, we limit ourself with the construction of function w_0^0 and find the main term λ_0 in (4.46), which depends on the main boundary conditions at the curvilinear edges and on all boundary conditions at the rectilinear edge $y = 0$. The next term $\varepsilon^2\lambda_2$, as found in [52], depends on all boundary conditions at the curvilinear edges, and the accuracy of Eqs. (4.12) is not enough to find it.

We put $k(\varphi) = 1$ in Eqs. (4.12) and exclude the function Φ. As a result, we arrive at the equation

$$\varepsilon^8\Delta\Delta\Delta\Delta w - \varepsilon^4\hat{\lambda}\Delta\Delta w + \frac{\partial^4 w}{\partial x^4} = 0, \qquad \Delta w = \frac{\partial^2 w}{\partial x^2} + \frac{\partial^2 w}{\partial y^2}. \tag{4.48}$$

In view of the accepted accuracy, we put approximately

$$\Delta w = \frac{\partial^2 w}{\partial y^2}. \tag{4.49}$$

Now the variables are separated,

$$w_0^0(x, y) = X(x)Y(y), \tag{4.50}$$

where

$$\frac{d^4X}{dx^4} - \alpha^4 X = 0, \qquad \varepsilon^8\frac{d^8Y}{dy^8} - \varepsilon^4\lambda_0\frac{d^4Y}{dy^4} + \alpha^4 Y = 0. \tag{4.51}$$

We choose the parameter α in such a way that the main boundary conditions at the curvilinear edges $x = 0$ and $x = -l$ be satisfied. Then, we obtain

$$\alpha = \frac{\beta}{l}, \tag{4.52}$$

where the values of β for various boundary conditions are given in Table 4.4 (we consider only the cases with $\beta > 0$).

The second equation in (4.51) after scaling reads as

$$\frac{d^8Y}{d\eta^8} - 2\lambda'\frac{d^4Y}{d\eta^4} + Y = 0, \qquad \eta = \frac{y}{\varepsilon}\sqrt{\frac{\beta}{l}}, \quad \lambda_0 = 2\lambda'\frac{\beta^2}{l^2}. \tag{4.53}$$

We seek the solution of this equation satisfying the given boundary conditions at $\eta = 0$ and the attenuation condition $Y \to 0$ at $\eta \to -\infty$ in the form

$$Y(\eta) = \sum_{k=0}^{4} C_k e^{q_k\eta}, \quad \mathrm{Re}(q_k) > 0, \tag{4.54}$$

where q_k are the roots of equation $q^8 - 2\lambda' q^4 + 1 = 0$.

Below, the boundary conditions at $y = 0$ or $\eta = 0$ in the leading approximation are expressed through the function Y [52]:

$$\begin{aligned}
&v = 0 \quad \Longrightarrow \quad Y^{(-1)} = 0; \qquad && T_2 = 0 \quad \Longrightarrow \quad Y^{(-4)} = 0;\\
&u = 0 \quad \Longrightarrow \quad Y^{(-2)} = 0; && S = 0 \quad \Longrightarrow \quad Y^{(-3)} = 0;\\
&w = 0 \quad \Longrightarrow \quad Y^{(0)} = 0; && Q_2^* = 0 \quad \Longrightarrow \quad Y^{(3)} = 0;\\
&\gamma_2 = 0 \quad \Longrightarrow \quad Y^{(1)} = 0; && M_2 = 0 \quad \Longrightarrow \quad Y^{(2)} = 0,
\end{aligned} \tag{4.55}$$

where $Y^{(k)}(\eta)$ are the successive derivatives and integrals of the function $Y(\eta)$:

$$Y^{(0)}(\eta) = Y(\eta), \quad Y^{(k+1)}(\eta) = \frac{dY^{(k)}}{d\eta}, \quad Y^{(k-1)}(\eta) = \int_{-\infty}^{\eta} Y^{(k)}\, d\eta. \quad (4.56)$$

At a rectilinear edge, there are 6 (of 16) variants of weak constraints for which there exist localized vibration modes. These variants are as follows: free edge (0000), 4 variants with one constraint (1000, 0100, 0010, 0001) and one variant with two constraints (1001), $u = \gamma_2 = 0$:

$$\begin{array}{lll} T_2 = S = Q_{2*} = M_2 = 0, & (0000), & \lambda'_0 = 0,113, \quad \lambda'_0 = 0,973; \\ T_2 = S = Q_{2*} = \gamma_2 = 0, & (0001), & \lambda'_0 = 0.223; \\ T_2 = u = Q_{2*} = M_2 = 0, & (0100), & \lambda'_0 = 0.223; \\ T_2 = u = Q_{2*} = \gamma_2 = 0, & (0101), & \lambda'_0 = 0.419; \\ T_2 = S = w = M_2 = 0, & (0010), & \lambda'_0 = 0.809; \\ v = S = Q_{2*} = M_2 = 0, & (1000), & \lambda'_0 = 0.809. \end{array} \quad (4.57)$$

The corresponding vibration modes, as normalized by the condition

$$\int_{-\infty}^{0} Y(\eta)\, d\eta = 1, \quad (4.58)$$

are shown in Fig. 4.5.

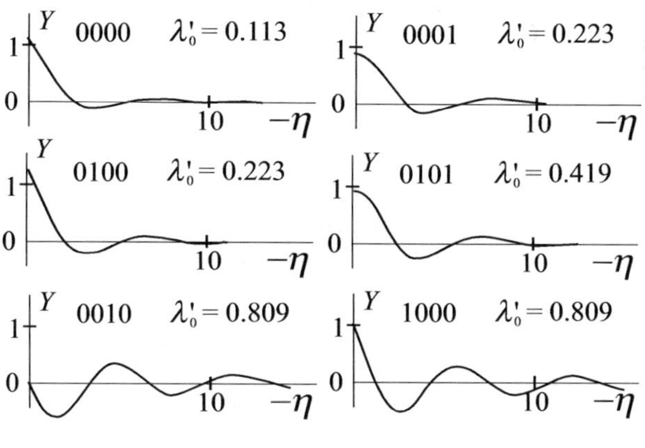

FIGURE 4.5: Vibration modes localized near a rectilinear edge.

The value $\lambda' = 1$ corresponds to a cylindrical shell, closed in the circumferential direction, and also to a cylindrical panel with the more hard constraint of rectilinear edges than in Eqs. (4.57).

The effect of localization is accompanied by lowering the frequency parameter λ'. The largest decreasing (ninefold) is observed for the free edge. Note that in this case there is the second frequency of localized vibrations $\lambda_0' = 0.973$ for which the eigenfunction $Y_2(\eta)$ decreases much slower than the first one. This function is shown in Fig. 4.6 together with the eigenfunction $Y_1(\eta)$ for $\lambda_0' = 0.113$.

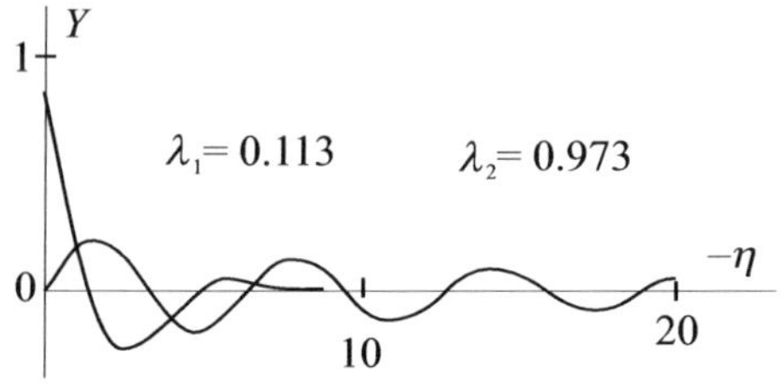

FIGURE 4.6: Vibration modes localized near a free edge.

In the leading approximation, the frequency ω of localized vibrations of cylindrical panel in the initial designations reads as

$$\omega^2 = \frac{Eh\beta^2\lambda_0'}{\rho RL^2\sqrt{3(1-\nu^2)}}\left(1 + O(\varepsilon^2)\right), \tag{4.59}$$

where the parameter β depends on the main boundary conditions at the curvilinear edges (see Table 4.4), and the parameter λ_0' depends on the boundary conditions at the rectilinear edges. For a weak constraint, the parameter λ_0' is given by Eqs. (4.57), and for a hard constraint, one has $\lambda_0' = 1$.

Example 4.4. Consider a cylindrical panel $-\pi/2 \le y \le 0$ with simply supported curvilinear edges with parameters $l = 1$, $\nu = 0.3$. We take three values of the relative thickness $h_* = h/R = 0.01,\ 0.002,\ 0.001$. Calculations were performed by Z. G. Ershova, and the results are presented below. The natural frequency is written in the form

$$\omega = \sqrt{\frac{E}{\rho R^2}}\,\Omega. \tag{4.60}$$

To discuss the accuracy of the asymptotic approach used, the dimensionless frequency Ω is presented in four various ways (see Table 4.7):

- Ω_0 is found from the approximate asymptotic Eq. (4.59),
- $\Omega^{(a)}$ is found from the more accurate asymptotic Eq. (4.59), including the summand $O(\varepsilon^2)$,
- $\Omega^{(e)}$ is the exact value obtained as a result of numerical solution of Eq. (4.48) after separation of variables,
- $\Omega^{(f)}$ is found using the finite elements method.

TABLE 4.7: Frequency parameters of vibration of a cylindrical panel.

boundary conditions	0000	0001	0100	0101	0010	1000
$h_* = 0.01$						
Ω_0	0.0820	0.1153	0.1153	0.1582	0.2198	0.2198
$\Omega^{(a)}$	0.1021	0.1262	0.1314	0.1620	0.2191	0.2238
$\Omega^{(e)}$	0.1004	0.1244	0.1315	0.1613	0.2202	0.2261
$\Omega^{(f)}$	0.1014	0.1266	0.1276	0.1605	0.2245	0.2222
$h_* = 0.002$						
Ω_0	0.0367	0.0516	0.0516	0.0707	0.0983	0.0983
$\Omega^{(a)}$	0.0407	0.0536	0.0549	0.0714	0.0983	0.0994
$\Omega^{(e)}$	0.0405	0.0534	0.0549	0.0714	0.0984	0.0996
$\Omega^{(f)}$	0.0400	0.0541	0.0510	0.0699	0.0975	0.0990
$h_* = 0.001$						
Ω_0	0.0259	0.0365	0.0365	0.0500	0.0695	0.0695
$\Omega^{(a)}$	0.0279	0.0374	0.0382	0.0503	0.0696	0.0702
$\Omega^{(e)}$	0.0280	0.0375	0.0381	0.0503	0.0695	0.0701
$\Omega^{(f)}$	0.0272	0.0380	0.0342	0.0488	0.0682	0.0702

The opposite rectilinear panel edge is clamped. The accuracy of the asymptotic values Ω_0 and $\Omega^{(a)}$ is seen to increase as the thickness parameter h_* decreases for all variants of boundary conditions.

4.5 Localized vibration modes of shells close to cylindrical ones

To construct localized vibration modes of shells close to cylindrical ones, we use the asymptotic method described in Section 3.3. Low-frequency localized vibrations of shells close in shape to shells with zero Gaussian curvature were investigated in [164], [165]. Similar results for the buckling of initial momentless state of such shells can be found in [222], [226].

Now, we shall describe the geometry of the neutral surface of a shell close to the reference cylindrical surface which may be generally non-circular. Let a surface be specified by the vector equation

$$\mathbf{r}(s, \varphi) = \mathbf{r}_0(s, \varphi) + \delta\, g(s, \varphi)\mathbf{n}, \tag{4.61}$$

where the vector $\mathbf{r}_0(s, \varphi)$ describes the reference cylindrical shell, $\mathbf{n}$ is the external unit normal to it, δ is the small parameter, and the smooth slowly varying function $g(s, \varphi) \sim 1$ describes the deviation form from the reference surface.

Let the curvilinear coordinates s, φ be chosen so that the first quadratic form of the reference surface is $ds^2 + d\varphi^2$. Then coefficients of the first quadratic form of surface (4.61) and its radii of curvature read as follows:

$$A_{11} = 1 + \delta^2 (g'_s)^2, \quad A_{12} = \delta^2 g'_s g'_\varphi, \quad A_{22} = 1 + 2\delta\, k\, g + \delta^2 \left((g'_\varphi)^2 + k^2 g^2 \right),$$
$$\frac{1}{R_1} = -\delta\, g''_{ss} + O(\delta^2), \quad \frac{1}{R_{12}} = \delta\, g''_{s\varphi} + O(\delta^2), \quad \frac{1}{R_2} = k - \delta \left(g''_{\varphi\varphi} k^2 g \right) + O(\delta^2), \tag{4.62}$$

where $k = k(\varphi)$ is the curvature of the reference cylindrical shell.

We seek a vibration mode localized near the weakest generatrix in the same form (4.18). We take the order of deviation δ such that the deviation is included in the leading approximation, and put

$$\delta = \varepsilon^2, \qquad \varepsilon^8 = \frac{h^2}{12(1-\nu^2)R^2}. \tag{4.63}$$

Then the deviation from the reference surface is of the order $\sqrt{Rh}$.

The curvilinear coordinates s, φ on surface (4.61) (in general case) are not orthogonal. But the non-orthogonality is small and it is possible to use Eqs. (4.12) to construct the leading approximation of the vibration mode and two first approximations for the eigen-frequency. For a convenience, we rewrite Eq. (4.12) as follows:

$$\varepsilon^4 \Delta\Delta w - \hat{\lambda} w + \Delta_R \hat{\Phi} = 0, \qquad \varepsilon^4 \Delta\Delta\hat{\Phi} - \Delta_R w = 0. \tag{4.64}$$

Here contrast to Eq. (4.14) we take

$$\Delta_R w = \frac{\partial}{\partial\varphi}\left(\frac{1}{R_1}\frac{\partial w}{\partial\varphi} \right) + k(\varphi)\frac{\partial^2 w}{\partial s^2}, \qquad \frac{1}{R_1} = -\varepsilon^2 \frac{\partial^2 g}{\partial s^2}. \tag{4.65}$$

In the general case, $R_1 = R_1(s, \varphi)$. For simplicity we study the case when each line $\varphi =$ const on the shell surface is an arc of a circle of large radius $R_1 \sim \varepsilon^{-2}$. Hence

$$\Delta_R w = \varepsilon^2 \frac{\partial}{\partial\varphi}\left(k_1(\varphi)\frac{\partial w}{\partial\varphi} \right) + k(\varphi)\frac{\partial^2 w}{\partial s^2}, \qquad k_1(\varphi, s) = -\frac{\partial^2 F}{\partial s^2}. \tag{4.66}$$

Note that for a shell of moderate length $l = L/R \sim 1$ the radius of curvature of the parabola $g = ax(l-x)$ with the same maximum deviation as a circle is equal to the radius of a circle (with an accuracy up to the values of order ε). Therefore, Eq. (4.66) can be also applied to deviations of paraboloidal form.

After excluding the function Φ in Eqs. (4.64), we arrive at the equation

$$\varepsilon^8 \frac{\partial^8 w}{\partial\varphi^8} - \varepsilon^4 \hat{\lambda} \frac{\partial^4 w}{\partial\varphi^4} + \left(\varepsilon^2 \frac{\partial}{\partial\varphi}\left(k_1(\varphi)\frac{\partial}{\partial\varphi} \right) + k(\varphi)\frac{\partial^2}{\partial s^2} \right)^2 w = 0. \tag{4.67}$$

We seek an asymptotic solution of this equation in the same form: (4.18)

$$w(s,\varphi,\varepsilon) = w_* \exp\{\mathrm{i}\varepsilon^{-1/2}p\xi - (1/2)b\xi^2\}, \qquad w_* = \sum_{n=0}^{\infty} \varepsilon^{n/2} w_n(s,\xi),$$

$$\hat{\lambda} = \lambda_0 + \varepsilon\lambda_1 + \varepsilon^2\lambda_2 + \ldots, \qquad \xi = \varepsilon^{-1/2}(\varphi - \varphi_0), \qquad \Re b > 0. \tag{4.68}$$

Arguing as in Section 4.3, we find the eigenfunctions localized near the weakest generatrix $\varphi = \varphi_{00}$ as well as the asymptotically double eigenvalues

$$\hat{\lambda}^{(n)} = \lambda_0 + \varepsilon\, r\left(n + \frac{1}{2}\right) + O(\varepsilon^2), \quad \lambda_0 = \min_{p,\varphi_0} f(p,\varphi_0), \quad f(p,\varphi_0) = p^4 + \left(\frac{k\beta^2}{p^2 l^2} + k_1\right)^2 \tag{4.69}$$

with $r = \sqrt{d}$, $d = f_{pp}f_{\varphi\varphi} - (f_{p\varphi})^2$. Here, we assume that conditions (4.34) guaranteeing decreasing the wave amplitudes far away from the generatrix $\varphi = \varphi_{00}$ are fulfilled.

The function w_0 in expansion (4.68) can be found from the boundary value problem consisting of the equation

$$\frac{k^2}{p^4}\frac{d^4 w_0}{ds^4} - 2\frac{k\,k_1}{p^2}\frac{d^2 w_0}{ds^2} + (p^4 + k_1^2 - \lambda_0)w_0 = 0, \quad 0 \le s \le l, \tag{4.70}$$

and of the main boundary conditions similar to those given in Table 3.1. The values $k_1 = k_1(\varphi_0)$, $k = k(\varphi_0)$, and $l = l(\varphi_0)$ depend on the parameter φ_0.

After changing of variables

$$\lambda_0 = \frac{k}{l^2}\lambda_*, \quad \chi = \frac{k_1 l}{\sqrt{k}}, \quad p^4 = \frac{k_1 p_*^4}{l^2}, \quad s = l\, s_*, \tag{4.71}$$

one obtains

$$\lambda_0 = f(p,\varphi_0) = \frac{k(\varphi_0)}{l^2(\varphi_0)} f_*\left(p_*, \chi(\varphi_0)\right), \quad f_* = \lambda_*, \tag{4.72}$$

where the frequency parameter λ_* can be found from the equation

$$\frac{1}{p_*^4}\frac{d^4 w_0}{ds_*^4} - 2\frac{\chi}{p_*^2}\frac{d^2 w_0}{ds_*^2} + (p_*^4 + \chi^2 - \lambda_*)w_0 = 0, \quad 0 \le s_* \le 1, \tag{4.73}$$

that contains the parameter $\chi(\varphi_0)$ depending on the curvature k_1 of the generatrix.

First, we consider a shell of revolution with curvature k_1 of the generatrix and consider three variants of the main boundary conditions: simply supported–simply supported (SS), clamped–clamped (CC), and clamped–free edge (CF). The main boundary conditions for the simply supported and the clamped groups are the same as for a cylindrical shell (see Table 4.1). For a group of free edge the main boundary conditions are $T_1 = S = 0$ or

$$w'' - p^2\chi w = 0, \qquad w''' - p^2\chi w' = 0. \tag{4.74}$$

For these variants of boundary conditions and for the five values of the curvature parameter $\chi = 2, 1, 0, -1, -2$, Figure 4.7 shows the plots of the functions $\lambda_*(p)$. Here, $p = \varepsilon m$ is the wave parameter, where m is the number of waves in the circumferential direction. For the first two variants of boundary conditions (SS and CC) the frequency decreases with the parameter χ, and in the last variant, CF, the frequency demonstrates the opposite behavior.

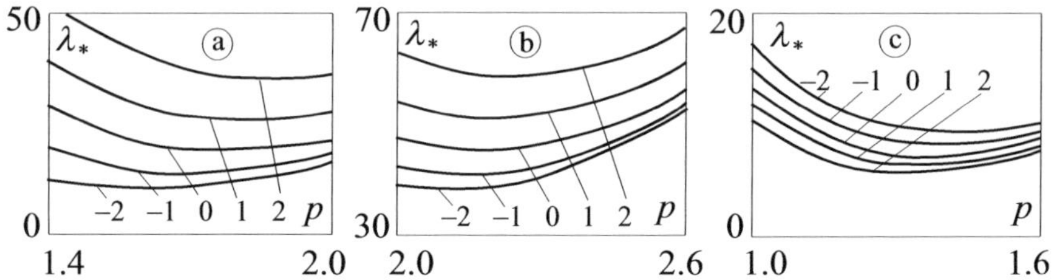

FIGURE 4.7: Plots of functions $\lambda_*(p)$ at $\chi = 2, 1, 0, -1, -2$ ((a) SS, (b) CC, (c) CF).

In Table 4.8, for the same boundary conditions and the same values of χ, the minimum values of the functions $f_*(p, \chi)$ and the corresponding values $p = p_0$ are given.

TABLE 4.8: Minimum frequency parameters for a near-cylindrical shell.

	SS		CC		CF	
χ	p_0	λ_*	p_0	λ_*	p_0	λ_*
−2	1.61	9.98	2.11	37.89	1.41	9.36
−1	1.70	14.18	2.14	40.44	1.39	8.06
0	1.77	19.74	2.17	44.75	1.37	7.03
1	1.84	26.79	2.21	50.85	1.35	6.21
2	1.90	35.44	2.24	58.76	1.32	5.54

For a shell of revolution no localized modes exist. If the parameters k and/or l and/or χ depend on the angle φ_0, then for the values $p = p_0$, $\varphi_0 = \varphi_{00}$ giving the minimum of function $f(p, \varphi_0)$, there appear the localized modes near the generatrix φ_{00}. The value $p = p_0$ depends only on the main boundary conditions and on the parameter χ (see Table 4.8). The value $\varphi_0 = \varphi_{00}$ depends also on the shell form, and in the general case it cannot be found. Consider some examples.

Example 4.5. Consider vibrations of a cylindrical shell with slightly bent axis (see Fig. 4.8). Let be $k = 1$, $l = \text{const}$, and let the deviation from the cylinder be $g(s,\varphi) = a\,s(l - s)\cos\varphi$. Then the functions $k_1(\varphi)$ and $\chi(\varphi)$ in Eq. (4.72) are as follows:

$$k_1(\varphi) = 2a\cos\varphi, \qquad \chi(\varphi) = 2al\cos\varphi. \tag{4.75}$$

The maximum deviation of the shell surface from the cylindrical reference surface is $\varepsilon^2 Ral^2/4$.

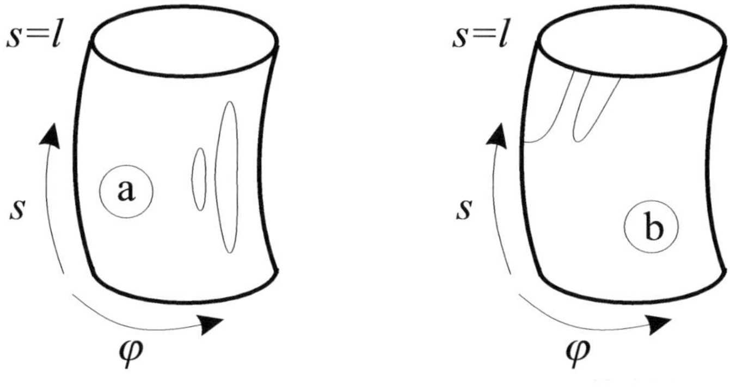

FIGURE 4.8: Vibration modes of a cylindrical shell with a slightly bent axis (a — SS or CC, b — CF).

Consider the same 3 variants of boundary conditions (SS, CC, CF). The vibration mode is localized near the line φ_{00} on which $f(p,\varphi)$ has a minimum. From Table 4.8 it follows that for boundary conditions SS and CC the vibration mode is localized near $\varphi_{00} = \pi$, where the parameter χ is minimal, namely in the region with negative Gaussian curvature (see Fig. 4.8a). In the case of CF, the line $\varphi_{00} = 0$ is the weakest one (see Fig. 4.8b).

In all cases the eigenvalues are asymptotically double. To calculate them, we use Eq. (4.69) with the notation (4.71) and (4.72):

$$\hat{\lambda}^{(n)} = \frac{\lambda_*}{l^2} + \varepsilon\left(n + \frac{1}{2}\right)\sqrt{f_{pp}f_{\varphi\varphi}} + O(\varepsilon^2), \tag{4.76}$$

with

$$f_{p\varphi} = 0, \quad f_{pp} = \frac{1}{l\sqrt{2a}}\frac{\partial^2 f_*}{\partial p_*^2}, \quad f_{\varphi\varphi} = \frac{1}{l}\frac{\partial f_*}{\partial \chi}\frac{d^2 k_1}{d\varphi^2}. \tag{4.77}$$

Here, all the functions are calculated at the point giving the minimum for the function f_*.

Consider a numerical data. Let be $l = 2$, $a = 0.5$. Integrating Eq. (4.72), we find:

(case SS)

$$\varphi_{00} = \pi, \quad \chi = -2, \quad \lambda_* = 9.98, \quad \frac{\partial^2 f_*}{\partial p_*^2} = 1.07, \quad \frac{\partial f_*}{\partial \chi} = 3.62, \tag{4.78}$$

(case CC)

$$\varphi_{00} = \pi, \quad \chi = -2, \quad \lambda_* = 37.89, \quad \frac{\partial^2 f_*}{\partial p_*^2} = 1.52, \quad \frac{\partial f_*}{\partial \chi} = 1.67, \tag{4.79}$$

(case FC)

$$\varphi_{00} = 0, \quad \chi = 2, \quad \lambda_* = 9.36, \quad \frac{\partial^2 f_*}{\partial p_*^2} = 0.61, \quad \frac{\partial f_*}{\partial \chi} = -6.07. \quad (4.80)$$

Example 4.6. Consider a barrel-like shell with a slant edge as shown in Fig. 4.9. We assume that $k = 1$, $F(s, \varphi) = k_1\, s(l-s)/2$, $l = l_0-(1-\cos\varphi)\sin\alpha$, where k_1 is the dimensionless curvature of the generatrix, α is the angle of inclination of the upper edge. The edges are simply supported.

Then the solution of Eq. (4.70) is of the form $w(s) = w_0 \sin(\pi s/l)$ and

$$f(p, \varphi_0) = p^4 + \left(\frac{\pi^2}{p^2 l^2(\varphi_0)} + k_1\right)^2. \quad (4.81)$$

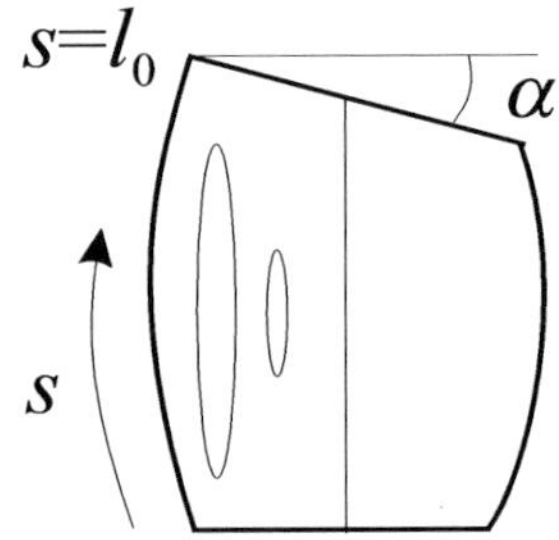

FIGURE 4.9: A barrel-like shell with a slant edge.

For $k_1 > 0$, the function (4.81) has a minimum when $\varphi_0 = 0$ and $p = p_0$ that satisfies the equation

$$\frac{c}{p^2} + k_1 = \frac{p^6}{c}, \qquad c = \frac{\pi^2}{l_0^2}. \quad (4.82)$$

The derivatives in Eqs. (4.69) read as

$$f_{p\varphi} = 0, \quad f_{pp} = \frac{8c^2}{p_0^6}, \quad f_{\varphi\varphi} = \frac{4p_0^4 \sin\alpha}{l_0}. \quad (4.83)$$

The vibration mode is localized near the longest generatrix.

4.6 Localized vibration modes of conic shells

For a conic shells, the asymptotic relations describing localized modes and frequencies are the same as given in Section 4.3 for a cylindrical shell. The single difference consists of the expression of a differential operator $\mathbf{L}_0 w_0$ in Eq. (4.21). For a conic shell the equation of zero approximation has variable coefficients and reads as

$$\frac{k^2(\varphi)}{p^4 B(s,\varphi)} \frac{d^2}{ds^2}\left(B^3(s,\varphi)\frac{d^2 w_0}{ds^2}\right)+\left(\frac{p^4}{B^4(s,\varphi)}-\lambda_0\right) w_0 = 0, \quad s_1(\varphi) \le s \le s_2(\varphi), \tag{4.84}$$

with

$$\lambda = \varepsilon^4(\lambda_0 + \varepsilon\lambda_1 + \dots), \qquad B(s,\varphi) = B_0(\varphi)(1-\eta s), \quad 0 \le \eta \le 1. \tag{4.85}$$

The remaining notation is the same as in Section 4.3.

Shell of revolution. Consider a conic shell of revolution with straight edges. In this case $B_0(\varphi) = 1$, $p = m\varepsilon$, and the values $k = \cos\alpha$, $\eta = \sin\alpha$ are constant (α is the angle at the cone vertex, m is the number of waves in the circumferential direction). The variable s in Eq. (4.84) changes in the range $0 \le s \le s_2 = L/R \le 1/\eta$. After scaling $s = s_2 x$ we get $0 \le x \le 1$, and now Eq. (4.84) reads as

$$\frac{a}{p^4 B_1} \frac{d^2}{dx^2}\left(B_1^3 \frac{d^2 w_0}{dx^2}\right) + \left(\frac{p^4}{B_1^4} - \lambda_0\right) w_0 = 0, \tag{4.86}$$

with

$$a = \frac{\cos^2\alpha}{s_2^4}, \qquad B_1 = 1 - \zeta x, \quad \zeta = s_2 \sin\alpha \le 1. \tag{4.87}$$

After substitutions $p^4 = \sqrt{a}\, p_1^4$, $\Lambda = \lambda_0/\sqrt{a}$, the parameter a is excluded,

$$\frac{1}{p_1^4 B_1} \frac{d^2}{dx^2}\left(B_1^3 \frac{d^2 w_0}{dx^2}\right) + \left(\frac{p_1^4}{B_1^4} - \Lambda\right) w_0 = 0, \quad 0 \le x \le 1, \tag{4.88}$$

and Eq. (4.88) contains a single geometric parameter $\zeta \le 1$, which depends on the ratio of the radii of edge circles.

Figure 4.10 shows the values of the frequency parameter Λ (found by numerical solution) as a function of the geometric parameter ζ and the wave parameter p_1. The value $\zeta = 0$ corresponds to a cylindrical shell, and the value $\zeta = 1$, to a full cone. The simply supported and clamped edges are studied. In the initial designations the natural frequency ω and the number of waves m in the circumferential direction read as

$$\omega^2 = \frac{Eh\cos\alpha}{\rho R L^2\sqrt{12(1-\nu^2)}}\Lambda, \qquad m = \frac{p}{\varepsilon} = \left(\frac{12(1-\nu^2)R^6}{L^4 h^2}\right)^{1/8} p_1. \tag{4.89}$$

Consider an example with a localization of vibration modes near the weakest generatrix.

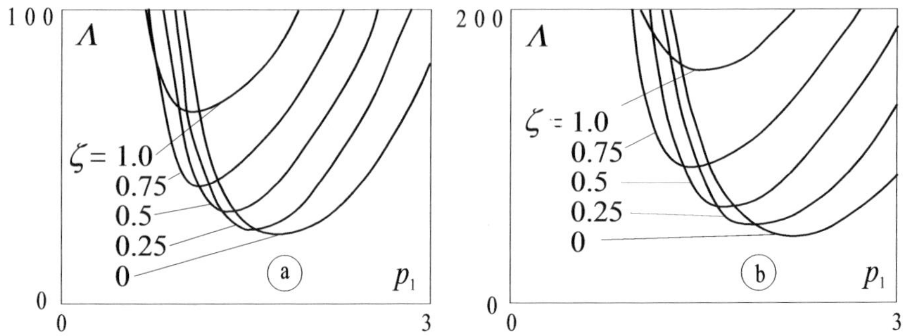

FIGURE 4.10: Frequency parameter for conic shell of revolution ((a) — simple supported edges, (b) — clamped edges).

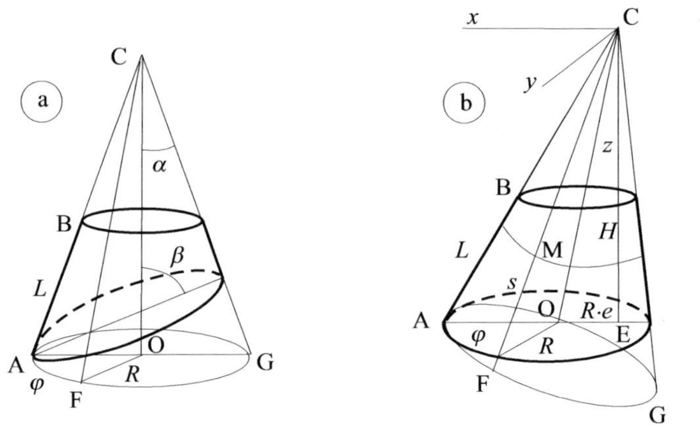

FIGURE 4.11: Cut-off conic shell with a slant edge (a), non-straight cone (b).

Example 4.7. We find the natural frequency of a truncated conic shell of revolution with a slant edge. The lower edge is inclined to the axis of cone by an angle β $(\alpha < \beta < \pi/2)$ (see Fig. 4.11(a)).

As before, we take $B_0(\varphi) = 1$, $k = \cos\alpha$, $\eta = \sin\alpha$ in Eq. (3.81), but now the interval of integration is as follows:

$$s_1(\varphi) \leq s \leq s_2, \qquad s_1(\varphi) = \frac{\xi(1-\cos\varphi)}{\sin\alpha(1-\xi\cos\varphi)}, \quad \xi = \frac{\operatorname{tg}\alpha}{\operatorname{tg}\beta}, \qquad s_2 = \frac{L}{R}, \tag{4.90}$$

where the co-ordinate in the generatrix s is counted away from the circle. The asymptotically double vibration modes are localized near the longest generatrix AB with $\varphi = \varphi_{00} = 0$ and the summand λ_0 in (4.85) is the same as for a conic shell with straight edges. Note that for some values of the problem parameters and for the angle $\varphi > 0$, the inequality $s_1(\varphi) < s_2$ can be violated (in this case the lower foundation crosses the upper edge), but due to a localization near the longest generatrix $\varphi = 0$ the following calculations remain correct.

After changing of variables

$$s = (s_2 - s_1)x + s_1, \quad p_1 = a_p(\varphi)p, \quad \lambda = a_\lambda(\varphi)\Lambda(p_1, \zeta), \quad \zeta(\varphi) = \frac{(s_2 - s_1)\sin\alpha}{1 - s_1 \sin\alpha}, \tag{4.91}$$

with

$$a_\lambda(\varphi) = \frac{\cos\alpha(s_2 - s_1)^2}{1 - s_1 \sin\alpha}, \qquad a_p(\varphi) = \left(\frac{(s_2 - s_1)^2}{\cos\alpha(1 - s_1 \sin\alpha)^3}\right)^{1/4}, \tag{4.92}$$

Eq. (4.84) is transformed to the form (4.87). Now using Eqs. (4.33) and (4.35) with $n = 0$ we obtain

$$\lambda_1 = \frac{1}{2}\sqrt{f_{pp} f_{\varphi\varphi}}, \qquad f(p, \varphi) = a_\lambda(\varphi)\Lambda\left(a_p(\varphi)p, \zeta(\varphi)\right), \tag{4.93}$$

with

$$f_{pp} = a_\lambda(0)a_p^2(0)\frac{\partial^2 \Lambda}{\partial p_1^2}, \qquad f_{\varphi\varphi} = \frac{d^2 a_\lambda}{d\varphi^2}\Lambda + a_\lambda(0)\frac{d^2\zeta}{d\varphi^2}\frac{\partial\Lambda}{\partial\zeta}, \tag{4.94}$$

where all the functions are calculated with $\varphi = 0$, $p_1 = p_{10}$ for which the function $f(p, \varphi)$ has a minimum.

In Fig. 4.12(b), the non-straight conic shell is shown. The projection E of a vertex C onto the foundation does not coincide with the center O of foundation. Again, the weakest generatrix AB coincides with the longest one. In more details, localized vibrations are discussed in [141].

4.7 Low-frequency vibrations of a Timoshenko–Reissner circular cylindrical shell

In Sections 4.1–4.2, some localization problems based on the KL model for cylindrical and conic shells were studied. In the next three sections, some of these problems will be considered in the framework of the TR model. Assume that the shell material is transversely isotropic with small transversal shear parameter g. We begin with the low-frequency vibrations problem of a circular cylindrical shell. Here, the vibration modes localization do not have a place, but the obtained results are used to solve the localization problems.

As in Sections 2.7 and 3.10, we shall consider a circular cylindrical shell of dimensionless length $l \sim 1$ bounded by the parallels $s = 0$ and $s = l$. The low-frequency vibrations are described approximately by Eqs. (2.65) or (3.148), and in the dimensionless form these equations read as

$$\begin{aligned} &\Delta\Delta\Phi + \Delta_R w = 0, \\ &D\Delta\Delta\Psi - \Delta_R\Phi - \lambda w = 0, \\ &w = \Psi - g\Delta\Psi, \\ &\Delta\Theta - g_1\Theta = 0 \end{aligned} \tag{4.95}$$

with

$$\begin{aligned} &\Delta = \frac{\partial^2}{\partial s^2} + \frac{\partial^2}{\partial \varphi^2}, \quad \Delta_R = \frac{\partial^2}{\partial s^2}, \quad D = \frac{h_*^2}{12(1-\nu^2)} = \mu^4, \quad h_* = \frac{h}{R}, \\ &\lambda = \frac{\rho R^2\omega^2}{E}, \quad g = \frac{DK_0}{\Gamma} = O(h_*^{2-\varsigma_3}), \quad \varsigma_3 \geq 0, \quad g_1 = \frac{2}{(1-\nu)g}. \end{aligned} \tag{4.96}$$

We seek the solution of Eqs. (4.95) satisfying at the each edge $s = 0$ and $s = l$ the following five ($j = 1, \ldots, 5$) boundary conditions (2.55) (here, for convenience the boundary conditions are numbered):

$$\begin{array}{llllll} 1 & u_1 = 0 & (1) & \text{or} & T_1 = 0 & (0), \\ 2 & u_2 = 0 & (1) & \text{or} & S_1 = 0 & (0), \\ 3 & w = 0 & (1) & \text{or} & Q_1 = 0 & (0), \\ 4 & \varphi_1 = 0 & (1) & \text{or} & M_1 = 0 & (0), \\ 5 & \varphi_2 = 0 & (1) & \text{or} & H = 0 & (0). \end{array} \tag{4.97}$$

For a fixed number m of waves in the circumferential direction, we seek the solution of Eqs. (4.95) in the form

$$Z(s, \varphi) = Z\, e^{ps+im\varphi}, \tag{4.98}$$

where Z is any unknown function in the boundary conditions (3.94).

The general solution of Eqs. (4.95) reads as

$$Z(s, \varphi) = e^{im\varphi} \sum_{n=1}^{10} C_n Z_n e^{p_n s}, \tag{4.99}$$

where C_n are arbitrary constants, and p_n are the roots of characteristic equation

$$F_8(p)(p^2 - m^2 - g_1) = 0, \;\; F_8(p) = \mu^4(p^2 - m^2)^4 + (1 + g(m^2 - p^2))(p^4 - \lambda(p^2 - m^2)^2). \tag{4.100}$$

Here, we study the part of the spectrum disposed near the minimum natural frequency. That is why we take

$$m = O(h_*^{-\varsigma_2}), \quad 0 < \varsigma_2 < 1/2. \tag{4.101}$$

For simplicity we consider the case $\varsigma_2 = 1/4$.

The equation $F_8(p) = 0$ has four roots $p = p_s = O(1)$ for which $|p_s|^2 \ll m^2$. These roots correspond to a semi-momentless solution and approximately satisfy the equation

$$p_s^4 = \lambda m^4 - \frac{\mu^4 m^8}{1 + g m^2}. \tag{4.102}$$

Also, the equation $F_8(p) = 0$ has four roots $p = p_e = O(h_*^{-\varsigma_e})$ for which $|p_e|^2 \gg m^2$. These roots correspond to an edge effect integral and approximately satisfy the equation

$$\mu^4 p_e^4 - g p_e^2 + 1 = 0. \tag{4.103}$$

If the shear parameter $g \ll \mu^2$, then Eq. (4.103) can be simplified, $\mu^4 p_e^4 + 1 = 0$. If $g \sim \mu^2$, then all summands in Eq. (4.103) have the same order. In the case of a very small transverse shear $g \gg \mu^2$, Eq. (4.103) has real roots of different orders (approximately):

$$\begin{aligned} p_{e1}^2 &= g\mu^{-4}, \qquad & p_{e1} &= O(h_*^{-\varsigma_3/2}), \\ g p_{e2}^2 &= 1, \qquad & p_{e2} &= O(h_*^{\varsigma_3/2-1}). \end{aligned} \tag{4.104}$$

The above condition $|p_e|^2 \gg m^2$ is fulfilled if $\varsigma_3 < 3/2$. We restrict ourselves to the case $\varsigma_3 < 3/2$ or $g \ll \mu$, $\Gamma \gg K_0 h_*^{3/2}$, because only in this case the above semi-momentless solution exists.

Finally, Eq. (4.100) has two large real roots

$$p_b^2 = m^2 + g_1, \quad p_b = O(h_*^{-1+\varsigma_3/2}), \tag{4.105}$$

corresponding to a boundary layer solution.

The orders ς_e and ς_b of the roots p_e and p_b versus the order ς_3 of the shear parameter η are shown in Fig. 4.12.

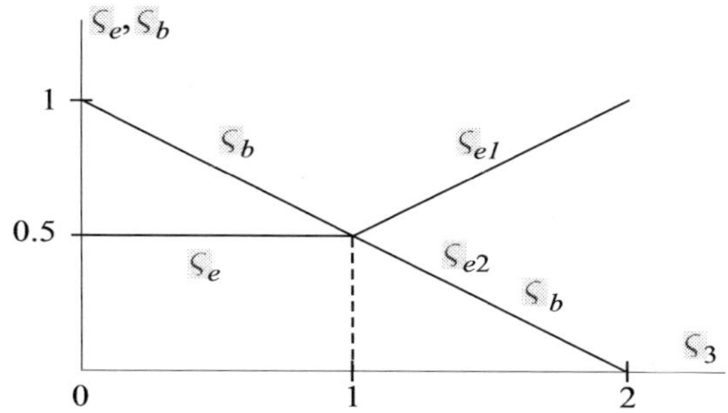

FIGURE 4.12: Functions $\varsigma_e(\varsigma_3)$ and $\varsigma_b(\varsigma_3)$.

If the value p_s^4 is given, then the frequency parameter λ can be found from Eq. (4.102):

$$\lambda = \frac{p_s^4}{m^4} + \frac{\mu^4 m^4}{1 + gm^2}. \tag{4.106}$$

For $gm^2 \ll 1$, Eq. (4.106) coincides with Eq. (4.11) obtained by using the KL model. The case $gm^2 \geq 1$ is not considered here because the inequality $g \ll \mu$ does not hold. In this case it is necessary to solve the full boundary value problem, which leads to a determinant of 10th order.

Now we seek the value p_s^4. As in Section 4.2 for the KL model, the semi-momentless solution is reduced to the fourth-order equation

$$\frac{\partial^4 w_s}{ds^4} - p_s^4 w_s = 0. \tag{4.107}$$

To find two main boundary conditions at each edge ($s = 0$ and $s = l$) from five conditions (4.97), we exclude the edge effect solutions and the boundary layer solution that exponentially decrease away from the edges. It occurs that there are the same four groups of main boundary conditions as in the KL model (see Table 3.3). To exclude the decreasing solution, we present all ten unknown functions involved in conditions (4.97) (see Table 4.9, based on Eqs. (3.156), (3.157)).

TABLE 4.9: Functions involved in the boundary conditions.

1	2	3	4	5
Z		Z_s	$Z_e(p_e)$	Z_b
u	$\dfrac{p(m^2+\nu p^2)}{(p^2-m^2)^2}$	$\dfrac{w'}{m^2}+\dfrac{(2+\nu)w'''}{m^4}$	$\dfrac{\nu}{p_e}$	0
v	$\dfrac{m(m^2-(2+\nu)p^2}{(p^2-m^2)^2}$	$\dfrac{w}{m}-\dfrac{\nu w''}{m^3}$	$-\dfrac{2+\nu}{p_e^2}$	0
w	1	1	1	0
γ_1	$-p$	$-w'$	$-p_e$	m
γ_2	$-m$	$-m\,w$	$-m$	p_b
T_1	$\dfrac{m^2p^2}{(p^2-m^2)^2}$	$\dfrac{w''}{m^2}+\dfrac{2p_s^4 w}{m^4}$	$\dfrac{m^2}{p_e^2}$	0
S	$\dfrac{mp^3}{(p^2-m^2)^2}$	$\dfrac{w'''}{m^3}+\dfrac{2p_s^4 w'}{m^5}$	$\dfrac{m}{p^e}$	0
Q_1	$\dfrac{p(m^2-p^2)}{1+g(m^2-p^2)}$	$m^2w'-w'''$	$-p_e^3$	$\dfrac{1-\nu}{2}m(m^2-p_b^2)$
M_1	$\dfrac{p^2-\nu m^2}{1+g(m^2-p^2)}$	$-\nu m^2 w+w''$	p_e^2	$(1-\nu)mp_b$
H	$\dfrac{(1-\nu)pm}{1+g(m^2-p^2)}$	$(1-\nu)w''$	$(1-\nu)m\,p_e$	$\dfrac{1-\nu}{2}(m^2-p_b^2)$

Column 2 in Table 4.9 contains solutions of the three first Eqs. (4.95). Column 3 contains semi-momentless solutions in the zero and first approximations, as obtained from column 2 under the assumption $p \ll m$ on replacing p^k by $w^{(k)}$ (the prime denotes the derivative). Also, due to the assumption $g \ll \mu$, the denominator $1 + g(m^2 - p^2)$ for the semi-momentless solution is replaced approximately by 1. Column 4 contains the coefficients of the edge effect solutions in the leading approximations, as obtained from column 2 under the assumption $p \gg m$. Column 5 contains coefficients of the boundary layer solutions.

For a fixed m, we write a solution near the edge $s = 0$ as

$$Z(s) = Z_s(s) + Z_e(s) + Z_b(s), \tag{4.108}$$

with

$$Z_e(s) = C_1 Z_e(p_{e1}) e^{p_{e1} s} + C_2 Z_e(p_{e2}) e^{p_{e2} s}, \quad Z_b(s) = \Theta Z_b e^{p_b s}. \tag{4.109}$$

Here, Z is the any function in column 1, Z_s is the semi-momentless solution in column 3, Z_e and Z_b are the edge effect and the boundary layer solutions decreasing away from the edge $s = 0$, real parts of roots p_{e1}, p_{e2}, p_b being negative.

Let us chose some variant of boundary conditions (of 32 possible variants)

$$Z^{(1)}(0) = \cdots = Z^{(5)}(0) = 0. \tag{4.110}$$

These equations can be written as

$$\begin{array}{l} \alpha_0^{(j)} w_s(0) + \alpha_1^{(j)} w_s'(0) + \alpha_2^{(j)} w_s''(0) + \alpha_3^{(j)} w_s'''(0) \\ +C_1 Z_e^{(j)}(p_{e1}) + C_2 Z_e^{(j)}(p_{e2}) + \Theta Z_b^{(j)} = 0 \end{array} \tag{4.111}$$

with $j = 1, \ldots, 5$. Equations (4.111) form a linear system with respect to the seven unknowns w_s, w_s', w_s'', w_e''', C_1, C_2, Θ. Excluding the values C_1, C_2, Θ, we get two boundary conditions for the function $w_s(s)$, which can be used to solve Eq. (4.107).

As for the KL model, there are four groups of main boundary conditions. The result of the edge effect and boundary layer exclusion depends on the group (see Table 4.10).

TABLE 4.10: Results of exclusion of edge effect for groups (Gr.) of boundary conditions (C — clamped, S — simply supported, W — weakly supported, F — free edge).

Gr.	Main bound.con.	Results	of exclusion
C	$w = w' = 0$	$w + a_2^c w'' + a_3^c w''' = 0,$	$w' + b_2^c w'' + b_3^c w''' = 0$
S	$w = w'' = 0$	$w + a_1^s w' + a_3^s w''' = 0,$	$w'' + b_1^s w' + b_3^s w''' = 0$
W	$w' = w''' = 0$	$w' + a_0^w w + a_2^w w'' = 0,$	$w''' + b_0^w w + b_2^w w'' = 0$
F	$w'' = w''' = 0$	$w'' + a_0^f w + a_1^f w' = 0,$	$w''' + b_0^f w + b_1^f w' = 0$

For the given variant (4.110) of boundary conditions, the choice of main boundary conditions is correct if the corresponding coefficients a_i^k, b_i^k are small compared with 1. The maximum value of these coefficients estimates the error of using of the semi-momentless approach.

Solving Eqs. (4.111) we get the following results.

The clamped support group ($w_s = w_s' = 0$) includes 14 variants of boundary conditions:

1111(0,1), 1110(0,1), 1101(0,1), 1100(0,1), 1011(0,1), 1010(0,1), 10011, 10001.

The simple support group ($w_s = w_s'' = 0$) also includes 14 variants of boundary conditions:

0111(0,1), 0110(0,1), 0101(0,1), 0100(0,1), 0011(0,1), 0010(0,1), 00011, 00001.

The weak support group ($w_s' = w_s''' = 0$) includes 2 variants of boundary conditions:

10010, 10000.

Finally, the free edge group ($w_s'' = w_s''' = 0$) also includes 2 variants of boundary conditions:

00010, 00000.

Here, for brevity we write 1111(0,1) in place of 11111, 11110, and so on, namely, the both variants with the boundary conditions $\gamma_2 = 0$ and $H = 0$ are included in the same group.

The simple support (or free edge) group may be obtained from the clamped support (or weak support) group by replacing the condition $u = 0$ by $T_1 = 0$.

In the cases with (0,1), the groups for the TR model are the same as for the KL model. The difference between the TR and the KL models appears in the cases 10011, 10001 and 00011, 00001 with the boundary condition $\gamma_2 = 0$ (which is absent in the KL model). For the TR model, these variants belong to the clamped and simple support groups, respectively, and the corresponding variants 1001, 1000 and 0001, 0000 for the KL model are included in the weak support group and in the free edge group, respectively.

To discuss the error

$$\varepsilon = \max_i \{|a_i^k|,\ |b_i^k|\} \tag{4.112}$$

of this semi-momentless approach, we assume that

$$m = O(h_*^{-1/4}), \quad \{|p_{e1}|,\ |p_{e2}|\} = O(h_*^{-1/2}), \quad g = O(h_*). \tag{4.113}$$

Solutions of Eqs. (4.111) show that for the most of the 32 variants the error is $\varepsilon = O(h_*^{1/2})$. For the 11 variants (1010(0,1), 1011(0,1), 00101, 0011(0,1), 10011, 10010, 00011, 00001) the error is larger, $\varepsilon = O(h_*^{1/4})$.

The variant 01101 ($T_1 = v = w = M_1 = \gamma_2 = 0$) is exactly satisfied by the conditions $w_s = w_s' = 0$ and the edge effect and the boundary layer solutions do not appear.

Further calculations are exactly the same as for the KL model (see Section 3.2). To find p_s in Eq. (4.107), it is necessary to consider boundary conditions at two edges $s = 0$ and $s = l$ simultaneously. Then the value p_s is equal to $p_s = \beta/l$, where β depends on the groups of the main conditions at the edges $s = 0$ and $s = l$ (see Table 4.4).

Consider the numerical data:

$$h_* = 0.005, \quad \nu = .3, \quad g = 0.002, \quad m = 7. \tag{4.114}$$

Some solutions of Eqs. (4.111) are given below:

$$\begin{aligned}
&11111 \quad w_s = 0.0062w_s'' + 0.00001w_s''', \\
&\qquad\quad\ w_s' = 0.0147w_s'' - 0.0456w_s''', \\
&10100 \quad w_s = -0.0003w_s'' + 0.085w_s''' + 0.003w_s''''', \\
&\qquad\quad\ w_s' = 0.096w_s''' + 0.0058w_s''''', \\
&00100 \quad w_s = 0.0006w_s' + 0.085w_s''' + 0.0034w_s''''', \\
&\qquad\quad\ w_s'' = -0.001w_s' + 0.197w_s''' - 0.04w_s'''' + 0.008w_s''''', \\
&00000 \quad w_s'' = 0.0043w_s - 0.025w_s' - 0.001w_s'''', \\
&\qquad\quad\ w_s''' = 0.264w_s - 0.068w_s' + 0.0007w_s'''' - 0.041w_s'''''.
\end{aligned} \tag{4.115}$$

Here, by using Eq. (4.107), the derivatives w_s'''' and w_s''''' can be expressed as $w'''' = p_s^4 w$ and $w''''' = p_s^4 w'$, respectively. But the value p depends on the main boundary conditions at $s = l$. For example, in the case 00000 for the minimum root of Eq. (4.107), we get $p_s = 1.876/l$ for the free edge–clamped (FC) combination of groups, and $p_s = 0$ for the rest three combinations (FS, FW, FF) of the main boundary conditions (see Table 4.4). The remaining last cases are not considered here.

For the KL model a more detailed exclusion of edge effect conditions can be found in [53].

In the next two sections we generalize some results of Sections 4.3 and 4.4 to the TR shell model with transversal shear.

4.8 Localized vibration modes of a Timoshenko–Reissner non-circular cylindrical shell

Consider a low-frequency vibrations of a non-circular cylindrical shell of constant length L. As curvilinear coordinates s, y on a cylindrical surface we shall take the length s on a generatrix $(0 \le s \le L)$, and the arc length y on the directrix $(0 \le y \le y_0)$. All linear values are related to a scale factor R,

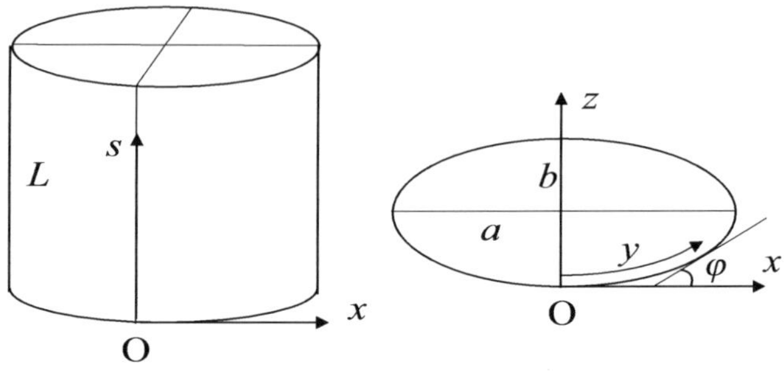

FIGURE 4.13: Cylindrical shell.

which will be chosen later. The Cartesian coordinates x, z of a point on the directrix satisfy the relations

$$\frac{dx}{dy} = \cos\varphi, \quad \frac{dz}{dy} = \sin\varphi, \quad k(y) = \frac{d\varphi}{dy}, \tag{4.116}$$

where φ is the angle between the tangent to the directrix and the x-axis (see Fig. 4.13). The directrix is assumed to be a sufficiently smooth closed convex ($k(s) > 0$) curve. We take as the origin the point O on the directrix at which the curvature is minimal, and take $R^{-1} = k(0) = \min_y k(y)$. We assume that $k_0'' = d^2k/dy^2|_{y=0} > 0$.

For an elliptic cylindrical shell with ellipse semi-axes a and b ($a > b$) (Fig. 4.13), we get

$$\begin{aligned} k(y) &= \frac{d\varphi}{dy} = \frac{1}{R}\left(e^2 + (1-e^2)\cos^2\varphi\right)^{3/2}, \\ e &= \frac{a}{b} > 1, \quad R = \frac{a^2}{b}, \quad k_0'' = 3R^{-3}(e^2-1). \end{aligned} \tag{4.117}$$

The vibration modes are localized near the generatrix $y = 0$, and in the case of an elliptical shell they are localized near the two generatrices with $\varphi = 0$ and $\varphi = \pi$.

In the previous section, it was established that for the TR model a semi-momentless approach is acceptable for any boundary conditions if

$$g \ll \mu. \tag{4.118}$$

We recall that for isotropic material $g = O(\mu^4)$; therefore, the restriction (4.118) is not very limiting. We assume that the restriction (4.118) is fulfilled with the exception of simple support boundary conditions at the both edges, because in this case the edge effect and the boundary layer are absent (the solution $w(s, y) = w(y)\sin\pi x/l$) satisfies all equations and boundary conditions).

The system (4.95)

$$\begin{aligned} &\Delta\Delta\Phi - \Delta_R w = 0, \\ &w = \Psi - g\Delta\Psi, \\ &\mu^4\Delta\Delta\Psi + \Delta_R\Phi - \lambda w = 0, \\ &\Delta = \frac{\partial^2}{\partial s^2} + \frac{\partial^2}{\partial y^2}, \quad \Delta_R = k(y)\frac{\partial^2}{\partial s^2} \end{aligned} \tag{4.119}$$

under the assumption $\partial^2/ds^2 \ll \partial^2/dy^2$ gives the system,

$$\begin{aligned} &\varepsilon^4\frac{\partial^4\hat{\phi}}{\partial y^4} - k(y)\frac{\partial^2 w}{\partial s^2} = 0, \\ &w = \psi - g\frac{\partial^2\Phi}{\partial y^2}, \\ &\varepsilon^4\frac{\partial^4\Phi}{\partial y^4} + \frac{\partial^2\hat{\Phi}}{\partial s^2} - \hat{\lambda}w = 0 \end{aligned} \tag{4.120}$$

with

$$\varepsilon = \sqrt{\mu} \sim h_*^{1/4}, \qquad \hat{\lambda} = \varepsilon^{-4}\lambda, \qquad \hat{\Phi} = \varepsilon^{-4}\Phi, \tag{4.121}$$

which is similar to Eqs. (4.12). Here, ε and $\hat{\lambda}$ are the new small parameter and the frequency parameter.

As in Section 4.3, we seek a formal asymptotic solution of Eqs. (4.120) in the form (4.18):

$$\begin{aligned} &w(s, y, \varepsilon) = w_* \exp\{\mathrm{i}\varepsilon^{-1/2}q\xi - (1/2)b\xi^2\}, \qquad \Re b > 0, \\ &w_* = \sum_{n=0}^{\infty} \varepsilon^{n/2} w_n(s, \xi), \qquad \xi = \varepsilon^{-1/2}(y - y_0), \\ &\hat{\lambda} = \lambda_0 + \varepsilon\lambda_1 + \varepsilon^2\lambda_2 + \dots. \end{aligned} \tag{4.122}$$

The functions Ψ and $\hat{\Phi}$ will be sought in the same form. Here, $w_n(s, \xi)$ are the polynomials in ξ, the wave number $q > 0$, and the condition $\Re b > 0$ is responsible for the decrease of the vibration mode away from the weakest generatrix $y = y_0$.

Substituting (4.122) into Eqs. (4.120), we arrive in the leading approximation (as $\varepsilon \to 0$) at the following ODE:

$$\frac{k^2(y_0)}{q^4}\frac{d^4 w_0}{ds^4} + \frac{q^4 w_0}{1 + \tilde{g}q^2} - \lambda_0 w_0 = 0, \quad g = \varepsilon^2\tilde{g}. \tag{4.123}$$

As in Section 4.3, we take

$$\frac{d^4 w_0}{ds^4} = p_s^4 w_0, \quad p_s = \frac{\beta}{l}, \tag{4.124}$$

where the parameter β depends on the main boundary conditions at the shell edges and is given in Table 4.4. The function $f(q, y)$

$$\hat{\lambda} = f(q, y) = \frac{q^4}{1 + \tilde{g}q^2} + \frac{p_s^4 k^2(y)}{q^4} \tag{4.125}$$

plays an important role in the following calculations. The set of eigenvalues in the zero and the first approximations, corresponding to localized vibration modes, depends only on f (see Section 4.3):

$$\hat{\lambda}^{(n)} = \lambda_0 + \varepsilon r \left(n + \frac{1}{2}\right) + O(\varepsilon^2), \quad n = 0, 1, \ldots, \tag{4.126}$$

where

$$\begin{aligned} &\lambda_0 = \min_{q,y}\{f(q, y)\} = f(q_0, y_0), \quad r = \sqrt{f_{yy} f_{qq}}, \\ &f_{qq} = \frac{\partial^2 f}{\partial q^2}, \quad f_{yy} = \frac{\partial^2 f}{\partial y^2}, \quad t_{qy} = 0. \end{aligned} \tag{4.127}$$

Here, the derivatives f_{yy} and f_{qq} are calculated with $q = q_0$, $y = y_0$. The eigenvalues $\hat{\lambda}^{(n)}$ are asymptotically double.

Taking into account that the weakest generatrix $y_0 = 0$, and $k(0) = 1$, we get

$$\lambda_0 = \min_q \left\{\frac{q^4}{1 + \tilde{g}q^2} + \frac{p_s^4}{q^4}\right\}. \tag{4.128}$$

Calculating the minimum, we obtain

$$\lambda_0 = p_s^2 f_1(g_*), \qquad q_0 = \sqrt{p_s} f_2(g_*), \qquad g_* = p_s \tilde{g} = \mu g p_s. \tag{4.129}$$

Here, for the KL model ($g = 0$) $f_1(0) = 2$, $f_2(0) = 1$. For the TR model with $g > 0$, the correcting functions $f_1(g_*)$ and $f_2(g_*)$ are shown in Fig. 4.14. For a small g_* we get

$$f_1(g_*) = 2 - g_* + O(g_*^2), \quad f_2(g_*) = 1 - \frac{3}{16} g_* + O(g_*^2). \tag{4.130}$$

According to restriction (4.118) for the arbitrary boundary conditions at the shell edges, the constructed semi-momentless solution is acceptable only for $\tilde{g} \ll 1$, and for a shell with the simply supported edges this restriction is absent.

We calculate the value $\lambda_1^{(n)} = r((1/2) + n))$ in Eq. (4.122) for an elliptical shell. We get

$$r = \left(\frac{12(e^2 - 1)[10p_s^4(1 + \tilde{g}q_0^2)^3 + 8(6 + 3\tilde{g}q_0^2 + \tilde{g}^2 q_0^4)]}{q_0^{10}(1 + \tilde{g}q_0^2)^3}\right)^{1/2} \tag{4.131}$$

with $q_0 = \sqrt{p_s} f_2(g_*)$, $e = a/b$.

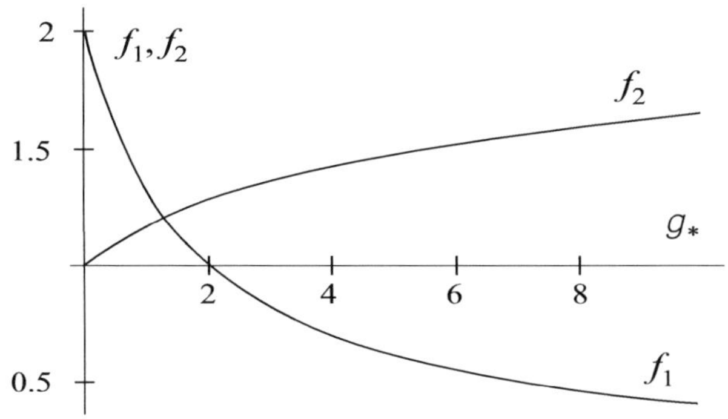

FIGURE 4.14: The correcting functions $f_1(g_*)$ and $f_2(g_*)$.

The localized natural frequencies $\omega^{(n)}$, $n = 0, 1, \ldots$, for an elliptical cylindrical shell in the initial designations read as

$$(\omega^{(n)})^2 = \frac{E\mu^2}{\rho R^2}\left(\lambda_0 + \varepsilon\lambda_1^{(n)} + O(\varepsilon^2)\right) \tag{4.132}$$

with

$$\begin{aligned} &R = \frac{a^2}{b}, \quad \mu^4 = \frac{h^2}{12(1-\nu^2)R^2}, \quad \varepsilon = \sqrt{\mu}, \\ &p_s = \frac{R\beta}{L}, \quad \lambda_0 = p_s^2 f_1(g_*), \quad q_0 = \sqrt{p_s} f_2(g_*), \quad g_* = \mu g p_s. \end{aligned} \tag{4.133}$$

As for the KL model, all the frequencies are asymptotically fourfold. The localized vibration modes for $n = 0$ are shown in Fig. 4.3.

4.9 Localized vibration modes of Timoshenko–Reissner cylindrical panel with a free or a weakly supported edge

We consider the same cylindrical panel as in Section 4.4 (with $0 \le s \le l$, $0 \le \varphi \le \varphi_0$), and suppose that the edge $\varphi = 0$ is free or weakly supported. We shall seek vibration modes which are localized near this edge and decrease exponentially with distance from it. In Section 4.4, this problem was solved in the framework of the KL model, and also 6 variants of boundary conditions for which the sought-for modes exist were found (see Fig. 4.6). Here, we use the TR model.

We begin with Eqs. (4.95) accounting for the transverse shear

$$\begin{aligned} &\Delta\Delta\Phi + \Delta_R w = 0, \\ &\mu^4\Delta\Delta\Psi - \Delta_R\Phi - \lambda w = 0, \\ &w = \Psi - g\Delta\Psi, \\ &\Delta\Theta - g_1\Theta = 0, \qquad g_1 = \frac{2}{(1-\nu)g}, \end{aligned} \tag{4.134}$$

and consider the following the boundary conditions at the edge $\varphi = 0$:

$$\begin{array}{llllll} 1 & v = 0 & (1) & \text{or} & T_2 = 0 & (0), \\ 2 & u = 0 & (1) & \text{or} & S = 0 & (0), \\ 3 & w = 0 & (1) & \text{or} & Q_2 = 0 & (0), \\ 4 & \varphi_2 = 0 & (1) & \text{or} & M_2 = 0 & (0), \\ 5 & \varphi_1 = 0 & (1) & \text{or} & H = 0 & (0). \end{array} \tag{4.135}$$

The boundary conditions on the remaining edges will be discussed later.

The complete solution $Z = Z_s + Z_b$ near the edge $\varphi = 0$ consists of the semi-momentless solution Z_s and of the boundary layer solution Z_b, where Z stays instead any unknown function in Eqs. (4.135). We restrict ourselves to the leading approximation and will not take into account the influence of the solution near the angles $s = 0, l,\ \varphi = 0$ of the panel.

To construct the semi-momentless solution, we put

$$y = \frac{\hat{y}}{\varepsilon}, \quad \varepsilon = \sqrt{\mu}, \quad \lambda = \mu^2\hat{\lambda}, \quad \Phi = \mu^2\hat{\Phi}. \tag{4.136}$$

Then in the leading approximation Eqs. (4.134) yield

$$\begin{aligned} &\frac{\partial^4\hat{\Phi}}{\partial\hat{y}^4} + \frac{\partial^2 w}{\partial s^2} = 0, \\ &\frac{\partial^4\Psi}{\partial\hat{y}^4} - \frac{\partial^2\hat{\Phi}}{\partial s^2} - \hat{\lambda}w = 0, \\ &w = \Psi - \tilde{g}\frac{\partial^2\Psi}{\partial\hat{y}^2}, \qquad \tilde{g} = \frac{g}{\mu}. \end{aligned} \tag{4.137}$$

Excluding the functions w and Φ, one arrives at the following equation for the function Ψ:

$$\frac{\partial^8\Psi}{\partial\hat{y}^8} + \hat{\lambda}\tilde{g}\frac{\partial^6\Psi}{\partial\hat{y}^6} - \hat{\lambda}\frac{\partial^4\Psi}{\partial\hat{y}^4} - \tilde{g}\frac{\partial^6\Psi}{\partial\hat{y}^2\partial s^4} + \frac{\partial^4\Psi}{\partial s^4} = 0. \tag{4.138}$$

If the both edges, $s = 0$ and $s = l$, are simply supported, then the exact separation of variables, $\Psi(s, \hat{y}) = Y(\hat{y})\sin(\pi s/l)$ is possible, and Eq. (4.138) is acceptable for any $\tilde{g}$. For the arbitrary boundary conditions this equation is acceptable only for $\tilde{g} \ll 1$ (or $g \ll \mu$), and we assume that this restriction is fulfilled. We put $\tilde{g} = 0$ and $\Psi = w$ in Eq. (4.138). Then it coincides with the equation

$$\frac{\partial^8 w}{\partial \hat{y}^8} - \hat{\lambda}\frac{\partial^4 w}{\partial \hat{y}^4} + \frac{\partial^4 w}{\partial s^4} = 0. \tag{4.139}$$

obtained in the framework of the KL model, and for the semi-momentless solution we may use the analysis of Section 4.4.

At first, we perform the separation of variables

$$w(s, \hat{y}) = Y(\hat{y})X(s), \tag{4.140}$$

with

$$\frac{d^8 Y}{d\hat{y}^8} - \hat{\lambda}\frac{d^4 Y}{d\hat{y}^4} + p_s^4 Y + 0, \qquad \frac{d^4 X}{ds^4} - p_s^4 X = 0, \quad p_s = \frac{\beta}{l}. \tag{4.141}$$

Here, β depends on the groups of the main boundary conditions at the $s = 0$ and $s = l$ (see Table 4.4). After scaling, Eq. (4.141) can be reduced to the following equation:

$$\frac{d^8 Y}{d\eta^8} - 2\lambda'\frac{d^4 Y}{d\eta}^4 + Y = 0, \qquad 2\lambda' = \frac{\hat{\lambda}}{\sqrt{p_s}}, \quad \eta = \frac{\hat{y}}{\sqrt{p_s}}. \tag{4.142}$$

The general solution of Eq. (4.142), which decreases away from the edge $\eta = 0$, reads as

$$Y(\eta) = \sum_{k=1}^{4} C_k Y_k e^{q_k \eta} + C_5 Y_5 e^{q_5 \eta}, \qquad \mathrm{Re}(q_k) < 0, \tag{4.143}$$

where Y stays instead of the unknown functions involved in the boundary conditions (4.135), C_k are arbitrary constants, q_k are the roots of equation $q^8 - 2\lambda' q^4 + 1 = 0$, and $q_5 = -\varepsilon\sqrt{g_1/p_s}$. The condition $\Re q_k < 0$ for $k = 1, 2, 3, 4$, holds only for $\lambda' < 1$. We normalize the functions Y_k so that $Y_k(0) = O(1)$. From column 2 in Table 4.9 we find the coefficients Y_k in the zero approximation (see Table 4.11, columns 2 and 5). In columns 3 and 6 of Table 4.11, the coefficients for the boundary layer solutions are given.

TABLE 4.11: Coefficients Y_k.

1	2	3	4	5	6
v	q_k^{-1}	0	T_2	q_k^{-4}	0
u	q_k^{-2}	0	S	q_k^{-3}	0
w	1	0	Q_2	q_k^3	$-(1-\nu)(\sqrt{p_s})^{-1}q_5^2/2$
φ_2	q_k	$-\sqrt{p_s}$	M_2	q_k^2	$-(1-\nu)q_5$
φ_1	1	$-q_5 p_s^{-1}$	H	q_k	$-(1-\nu)(\sqrt{p_s})^{-3}q_5^2/2$

For the chosen 5 boundary conditions $Y^{(1)} = \cdots = Y^{(5)} = 0$ we get the frequency equation in the form of a determinant of 5th order:

$$\Delta_5(\lambda') = \begin{vmatrix} Y_1^{(1)} & Y_2^{(1)} & Y_3^{(1)} & Y_4^{(1)} & Y_5^{(1)} \\ Y_1^{(2)} & Y_2^{(2)} & Y_3^{(2)} & Y_4^{(2)} & Y_5^{(2)} \\ Y_1^{(3)} & Y_2^{(3)} & Y_3^{(3)} & Y_4^{(3)} & Y_5^{(3)} \\ Y_1^{(4)} & Y_2^{(4)} & Y_3^{(4)} & Y_4^{(4)} & Y_5^{(4)} \\ Y_1^{(5)} & Y_2^{(5} & Y_3^{(5)} & Y_4^{(5)} & Y_5^{(5)} \end{vmatrix} = 0. \tag{4.144}$$

In this determinant, the 5th column and the 5th row appear due to the TR model. For the KL model, the 5th column and the 5th row are absent and Q_2 is changed by $Q_{2*} = Q_2 - p_s H$.

We compare the values λ' obtained by the TR and the KL models. To this aim we expand the determinant $\Delta_5(\lambda')$ in the elements of the 5th column. We note that $|q_5| \gg 1$, while the remaining values in the determinant are of order 1.

In the TR model, the boundary conditions $\varphi_1 = 0$ or $H = 0$ are additional ones, compared with the KL model.

If the condition $H = 0$ is imposed, then the TR model gives the same frequencies as the KL model (with the relative error of order p_5^{-1}). Indeed, if the condition $Q_2 = 0$ is not imposed, then the element $Y_5^{(5)} = H$ is the largest one, and the result is evident. If the both conditions $H = 0$ and $Q_2 = 0$ are imposed, then two elements $Y_5^{(3)} = Q_2$ and $Y_5^{(5)} = H$ are of the maximum orders of q_5^2, but using the linear combination $Q_2 - p_s H$ for which $Y_5^{(3)} - p_s Y_5^{(5)} = O(1)$, we again return to the KL model.

Let now at the edge $y = 0$ the restriction $\varphi_1 = 0$ is imposed. This condition may be fulfilled only in the framework of the TR model. According to the KL hypotheses $\varphi_1 = -\partial w/\partial y$, and in the TR model the angle φ_1 is an independent variable. We consider 6 variants of boundary conditions, for which in the KL model the localized near the edge $y = 0$ vibration modes exist. We expand the determinant (4.144) in the elements of the 5th column. For the determinants of 4th order we use the designation

$$\Delta(\alpha_1, \alpha_2, \alpha_3, \alpha_4) = \begin{vmatrix} q_1^{\alpha_1} & q_2^{\alpha_1} & q_3^{\alpha_1} & q_4^{\alpha_1} \\ q_1^{\alpha_2} & q_2^{\alpha_2} & q_3^{\alpha_2} & q_4^{\alpha_2} \\ q_1^{\alpha_3} & q_2^{\alpha_3} & q_3^{\alpha_3} & q_4^{\alpha_3} \\ q_1^{\alpha_4} & q_2^{\alpha_4} & q_3^{\alpha_4} & q_4^{\alpha_4} \end{vmatrix}. \tag{4.145}$$

Firstly, we recall the 6 variants of boundary conditions (see Eqs. (4.57)), which admit the localization modes in the framework of the KL model, and give the corresponding approximate frequency equations (see Table 4.12).

TABLE 4.12: The values λ' for the KL molel and the approximate equations.

N	Boundary conditions		λ_0	Equations
1	$T_2 = S = Q_2 = M_2 = 0$	(0000)	0.113	$\Delta(-4,-3,3,2)=0$
2	$T_2 = u = Q_2 = M_2 = 0$	(0100)	0.223	$\Delta(-4,-2,3,1)=0$
3	$T_2 = S = Q_2 = \varphi_2 = 0$	(0001)	0.223	$\Delta(-4,-3,3,1)=0$
4	$T_2 = u = Q_2 = \varphi_2 = 0$	(0101)	0.413	$\Delta(-4,-2,3,1)=0$
5	$v = S = Q_2 = M_2 = 0$	(1000)	0.809	$\Delta(-1,-3,3,2)=0$
6	$T_2 = S = w = M_2 = 0$	(0010	0.809	$\Delta(-4,-3,0,2)=0$

For the TR model we add the condition $\varphi_1 = 0$ to 4 conditions in Table 4.12, and consider the determinant (4.144) of 5th order. It occurs that only two variants of boundary conditions

$$\begin{array}{lll} T_2 = S = Q_2 = M_2 = \varphi_1 = 0 & (00001) & \lambda' = 0.809 \\ T_2 = S = w = M_2 = \varphi_1 = 0 & (00101) & \lambda' = 0.809 \end{array} \tag{4.146}$$

lead to a decrease of the frequency parameter λ' compared with $\lambda' = 1$ and to an appearance of localized vibration modes. We notice that the constrain $\varphi_1 = 0$ is not typical for 2D shell theories.

As for the KL model, according to Eqs. (4.143), we construct solutions satisfying the condition

$$Y(\eta) \to 0 \quad \text{at} \quad \eta \to \infty. \tag{4.147}$$

The effect of the opposite edge $y = y_0$ is not investigated.

Concluding remarks. The asymptotic analysis is based on the assumption that the shell is of the moderate length, and the values L/R and $p_s = \beta R/L$ are of the order of 1.

The transverse shear G_{13}/E varies in the wide range from $1/[2(1+\nu)]$ for an isotropic shell to very small values for a transversely isotropic shell. In Sections 4.8 and 4.9, the semi-momentless approximation is used. This approximation is based on the assumptions which lead to the restriction on the transverse shear

$$g \ll \mu \quad \text{or} \quad \frac{G_{13}}{E} \gg h_*^{3/2}. \tag{4.148}$$

If this restriction does not hold, then the semi-momentless approximation may be used only if the edges $s = 0$ and $s = l$ are simply supported (in this case the exact separation of variables $w(s, \varphi) = w(\varphi)\sin(n\pi s/L)$ is valid. In the remaining cases, the semi-momentless simplification is unacceptable, and the full boundary value problem (2.14), (2.54), (2.55) is to be solved.

Chapter 5

Localized Parametric Vibrations of Thin Shells

In this chapter, localized parametric vibrations of thin shells are considered. Thin non-circular conical, cylindrical and nearly cylindrical shells with oblique edges under different variants of load periodically varying in time are studied. In particular, cylindrical shells subjected to the axial periodic forces as well as conical and cylindrical shells experiencing pulsing radial pressure are considered. The effect of transverse shear on localized forms of parametric vibrations of laminated shells is analyzed in detail. Using the asymptotic approach developed in Chapter 4 in combination with the multiple scale method with respect to time, solutions of the governing equations are constructed in the form of functions which are exponentially decay far from some generatrix or meridian and growing with time in the case of parametric resonance. The system of two differential equations with periodic in time coefficients is derived to determine the amplitude of parametric vibrations. The main region of parametric instability invariant with respect to the shell shape and the type of loading is found.

5.1 Localized parametric vibrations of cylindrical shells under periodic axial forces

Consider a non-circular cylindrical shell of thickness h and length $L = Rl$, where R is the characteristic dimension of the shell (to be introduced below). The curvilinear coordinate system s, φ is the same as in the previous chapter.

Let the shell be under the combined non-uniform axial load

$$T_1^\circ = \mu^2 EhT_1, \quad T_1(\varphi, t) = f_0(\varphi) + \mu f_1(\varphi)\cos(\Omega t), \tag{5.1}$$

where $\hat{t} = t_c t$ and $\hat{\Omega} = \Omega/t_c$ are time and the circular frequency of excitation, t and Ω are their dimensionless counterparts, respectively, $t_c = R\sqrt{\rho/E}$ is the characteristic time, E, ρ are Young's modulus and the material density, respectively, and μ is the small parameter introduced by Eq. (2.29). In addition, it is assumed that

$$f_j, \frac{\partial f_j}{\partial \varphi}, \Omega \sim 1 \quad \text{at} \quad \mu \to 0. \tag{5.2}$$

In this case, the axial force (5.1) results in the membrane time-dependent stress state. Now Eqs. (2.48) may be used as the governing ones. We rewrite them in the dimensionless form:

$$\begin{aligned} &\mu^4\Delta^2 w^* - \mu^2 k(\varphi)\frac{\partial^2 \Phi^*}{\partial s^2} + \mu^2 T_1(\varphi, t)\,\frac{\partial^2 w^*}{\partial s^2} + \frac{\partial^2 w^*}{\partial t^2} = 0, \\ &\mu^2\Delta^2\Phi^* - k(\varphi)\frac{\partial^2 w^*}{\partial s^2} = 0, \end{aligned} \tag{5.3}$$

where $w^* = w/R$, $\Phi^* = \Phi/(\mu^2 EhR^2)$ are the dimensionless counterparts of the normal displacement w and stress function Φ, respectively, and $k(\varphi) = R/R_2(\varphi)$ is the dimensionless curvature of the shell. In what follows, the asterisk * corresponding to dimensionless magnitudes is omitted.

We consider the simplest variant of boundary conditions, the simple support conditions called Navier's conditions:

$$w = \frac{\partial^2 w}{\partial s^2} = \Phi = \frac{\partial^2 \Phi}{\partial s^2} = 0. \tag{5.4}$$

In the general case, when $T_1(t)$ is a function of time, Eqs. (5.3) describe the shell motion in the vicinity of momentless dynamic stress state. These equations have been repeatedly used by many authors (see, among many others, [34,238,246]) to study parametric vibrations of circular cylindrical shells. Note that their use for solving such problems is justified (from the point of view of the correctness of a required asymptotic solution), if the index of variation of the axial force $T_1(\varphi, t)$ in time (the index of dynamism) does not exceed the index of dynamism of the required solution. In our case, if the last condition in (5.2) is satisfied, these indexes coincide. However, due to low intensity of the periodic component of the stress resultant $T_1(\varphi, t)$, the effect of the latter on the asymptotic solution to be constructed below will appear only in the second approximation.

When assuming $f_1 = 0$, then under a variable curvature and/or not uniform distribution of the stationary component $f_0(\varphi)$ of force (5.1), it is possible the localization of eigenmodes near some generatrix $\varphi = \varphi_0$. We show here that for $f_1 \neq 0$, Eqs. (5.3) admit nonstationary solutions describing the forms of parametric vibrations localized in the neighborhood of this generatrix [115,151].

5.1.1 Asymptotic approach

To satisfy the boundary conditions (5.4), we assume

$$w = w_n(\varphi, t)\sin(\mu^{-1}q_n s), \quad \Phi = \Phi_n(\varphi, t)\sin(\mu^{-1}q_n s),$$
$$q_n = \mu n \pi l^{-1}, \quad n = 1, 2, 3, ... \tag{5.5}$$

Substituting (5.5) into Eqs. (5.3), one obtains the coupled differential equations

$$\mu^4 \frac{\partial^4 w_n}{\partial \varphi^4} - 2\mu^2 q_n^2 \frac{\partial^2 w_n}{\partial \varphi^2} + q_n^2 k(\varphi)\Phi_n + [q_n^4 - q_n^2 T_1(\varphi, t)]w_n + \frac{\partial w_n}{\partial t^2} = 0,$$
$$\mu^4 \frac{\partial^4 \Phi_n}{\partial \varphi^4} - 2\mu^2 q_n^2 \frac{\partial^2 \Phi_n}{\partial \varphi^2} + q_n^4 \Phi_n - q_n^2 k(\varphi) w_n = 0 \tag{5.6}$$

with respect to w_n, Φ_n. Further, in order to avoid cumbersome entries, the index n is omitted.

To construct a solution rapidly decreasing far away from the weakest generatrix $\varphi = \varphi_0$ and increasing during time in the case of parametric instability, we use the asymptotic method outlined in Chapter 4 in combination with the multiple scale method by time. The required solution of Eqs. (5.6) is assumed to be of the form:

$$w(\varphi, t, \mu) = \sum_{j=0}^{\infty} \mu^{j/2} w_j(\xi, t_0, t_1, t_2, ...) \exp\{\mathrm{i}(\mu^{-1/2} p\xi + 1/2 b\xi^2)\},$$
$$\mathrm{i} = \sqrt{-1}, \quad \xi = \mu^{-1/2}(\varphi - \varphi_0), \quad t_m = \mu^m t, \quad m = 0, 1, 2, ... \tag{5.7}$$

where $\Im b > 0$, w_j are polynomials in ξ with coefficients depending on time t_0 and the *slow time* t_m for $m > 0$. The function Φ is sought in the same form.

It is assumed that on some limited time interval there are the estimates:

$$p, \ |b| \sim 1, \quad |y|, \ \left|\frac{\partial y}{dx}\right| = O(1) \quad \text{at} \quad \mu \to 0, \tag{5.8}$$

where y is any of the functions w_j, Φ_j, and x is any of arguments.

Ansatz (5.7) differs from similar decomposition (4.18) in the assumed amplitude dependence on the *slow time.*

The functions $k(\varphi)$, $f_0(\varphi)$, $f_1(\varphi)$ are expanded into series in the powers of $\mu^{1/2}\xi$ in the neighborhood of the weakest generatrix $\varphi = \varphi_0$.

Substituting (5.7) into equations (5.6) and excluding the function Φ, one arrives at the sequence of differential equations

$$\sum_{m=0}^{j} \mathbf{L}_m w_{j-m} = 0, \ j = 0, 1, 2, ... \tag{5.9}$$

Here

$$\mathbf{L}_0 z = \frac{\partial^2 z}{\partial t_0^2} + H^2(p, q, \varphi_0), \tag{5.10}$$

$$H(p, q, \varphi_0) = \sqrt{(p^2 + q^2)^2 + \frac{k^2(\varphi_0) q^4}{(p^2 + q^2)^2} - f_0(\varphi_0) q^2}, \tag{5.11}$$

and the operators $\mathbf{L}_j$ for $j \geq 1$ are introduced in (4.21), where

$$\mathbf{N} z = 2\frac{\partial^2}{\partial t_0 \partial t_1} z - q^2 f_1(\varphi_0) \cos(\Omega t_0) z. \tag{5.12}$$

Consider Eqs. (5.9) step by step.

In the leading approximation, one has homogeneous differential equation

$$\mathbf{L}_0 w_0 = \frac{\partial^2 w_0}{\partial t_0^2} + H^2(p, q, \varphi_0) w_0 = 0. \tag{5.13}$$

Its solution may be written as follows

$$w_0 = P_{0,c}(\xi, t_1, t_2, ...) \cos \omega_0 t_0 + P_{0,s}(\xi, t_1, t_2, ...) \sin \omega_0 t_0, \tag{5.14}$$

where

$$\omega_0 = H(p, q, \varphi_0), \tag{5.15}$$

and $P_{0,c}, P_{0,s}$ are unknown polynomials in ξ with coefficients depending on $t_1, t_2, \ldots$

In the first-order approximation (for $j = 1$), Eq. (5.9) is non-homogeneous. Under condition (5.15), the terms on the right of this equation gives rise to secular solutions with respect to t_0. The conditions for the absence of these terms turn out to be the same as those for the localization of natural modes in the neighborhood of the weakest generatrix $\varphi = \varphi_0^\circ$ but different depending on q. These conditions read as

$$2k'(\varphi_0) - f_0'(\varphi_0) = 0, \quad p^2 = k^{1/2}(\varphi_0)q - q^2 \quad \text{for} \quad q < 1, \tag{5.16}$$

$$2k(\varphi_0)k'(\varphi_0) - q^2 f_0'(\varphi_0) = 0, \quad p = 0 \quad \text{for} \quad q > 1. \tag{5.17}$$

The appropriate dimensionless frequencies are

$$\omega_0 = q\sqrt{2k(\varphi_0) - f_0(\varphi_0)} \tag{5.18}$$

and

$$\omega_0 = \sqrt{q^4 - f_0(\varphi_0)q^2 + k^2(\varphi_0)} \tag{5.19}$$

for the cases $q < 1$ and $q > 1$, respectively.

Equations (5.16) - (5.19) serve to define φ_0°, p°, ω_0°. Now the value $R = R_2(\varphi_0^\circ)$ may be accepted by the characteristic dimension of the shell.

We note that $\mathbf{L}_1 w_0 \equiv 0$ at $\varphi = \varphi_0 = \varphi_0^\circ$, $p = p^\circ$, $\omega = \omega_0^\circ$. Then Eq. (5.9) for $j = 1$ becomes homogeneous and admits the solution

$$w_1 = P_{1,c}(\xi, t_1, t_2, ...) \cos \omega_0 t_0 + P_{1,s}(\xi, t_1, t_2, ...) \sin \omega_0 t_0, \tag{5.20}$$

where $P_{1,c}, P_{1,s}$ are unknown polynomials in ξ.

It is seen from formulas (5.18), (5.19) that $f_0 < f_b$, where $f_b = 2k(\varphi_0^\circ)$ for case (5.16) and $f_b = q^2 + k^2(\varphi_0^\circ)q^{-2}$ in case (5.17). Here f_b is the leading approximation of the critical axial force resulting in buckling of the shell in the vicinity of the weakest line $\varphi = \varphi_0^\circ$ [226]. If the shell curvature and axial force are not dependent of the circumferential coordinate φ, then f_b corresponds to the classical axial buckling force calculated by the well-known formula of Lorentz [123,124].

In what follows, the superscript $^\circ$ is omitted so that $p^\circ, \varphi_0^\circ, \omega_0^\circ$ are replaced by p, φ_0, ω_0, respectively.

Theoretically, parametric instability is observed in the case when the ratio of the loading frequency Ω to the natural frequency ω is equal or close to one of the following values [247]

$$\frac{\Omega}{\omega} = \frac{2}{1}, \frac{2}{2}, \frac{2}{3}, \frac{2}{4}, \ldots \tag{5.21}$$

However, in practice there are usually observable only cases when the ratio is 2/1 or 2/2, less often 2/3, the condition $\Omega/\omega = 2/2$ corresponding to the usual resonance.

Consider here the case of the most interest when $\Omega \approx 2\omega_0$. Let be

$$\Omega = 2\omega_0 + \mu\sigma, \quad \sigma \sim 1 \quad \text{at} \quad \mu \to 0, \tag{5.22}$$

where σ is the frequency detuning parameter.

The second-order approximation. When taking into account (5.14)-(5.20) and (5.22), then Eq. (5.9) for $j = 2$ takes the form

$$\begin{gathered} \mathbf{L}_0 w_2 = -(N_c \cos \omega_0 t_0 + N_s \sin \omega_0 t_0) \\ +r[P_{0,c} \cos(3\omega_0 t_0 + \sigma t_1) + P_{0,s} \sin(3\omega_0 t_0 + \sigma t_1], \end{gathered} \tag{5.23}$$

where

$$N_c = -\frac{1}{2}\frac{\partial^2 H}{\partial p^2}\frac{\partial^2 P_{0,c}}{\partial \xi^2} + a\xi \frac{\partial P_{0,c}}{\partial \xi} + \left(c\xi^2 + \frac{1}{2}a\right) P_{0,c} + 2\omega_0 \frac{\partial P_{0,s}}{\partial t_1}$$

$$+r(P_{0,s} \sin \sigma t_1 - P_{0,c} \cos \sigma t_1),$$

$$N_s = -\frac{1}{2}\frac{\partial^2 H}{\partial p^2}\frac{\partial^2 P_{0,s}}{\partial \xi^2} + a\xi \frac{\partial P_{0,s}}{\partial \xi} + \left(c\xi^2 + \frac{1}{2}a\right) P_{0,s} + 2\omega_0 \frac{\partial P_{0,c}}{\partial t_1}$$

$$+r(P_{0,s} \cos \sigma t_1 - P_{0,c} \sin \sigma t_1), \quad a = -\mathrm{i}\left(b\frac{\partial^2 H}{\partial p^2} + \frac{\partial^2 H}{\partial \varphi_0 \partial p}\right),$$

$$c = \frac{1}{2}\left(b^2 \frac{\partial^2 H}{\partial p^2} + 2b\frac{\partial^2 H}{\partial \varphi_0 \partial p} + \frac{\partial^2 H}{\partial \varphi_0^2}\right), \quad r = \frac{1}{2}q^2 f_1(\varphi_0).$$

A particular solution of the inhomogeneous equation (5.23) contains secular terms generated by the first two summands on the right hand side of the equation. The condition for their absence is the equalities

$$N_c = 0 \quad N_s = 0, \tag{5.24}$$

leading to the differential equation

$$\begin{gathered} -\frac{1}{2}\frac{\partial^2 H}{\partial p^2}\frac{\partial^2 \mathbf{X}}{\partial \xi^2} + a\xi \frac{\partial \mathbf{X}}{\partial \xi} + \left(c\xi^2 + \frac{1}{2}a\right)\mathbf{X} \\ +2\omega_0 \left(\begin{smallmatrix} 0 & -1 \\ 1 & 0 \end{smallmatrix}\right) \frac{\partial \mathbf{X}}{\partial t_1} + r(\varphi_0)\begin{pmatrix} \cos \sigma t_1 & \sin \sigma t_1 \\ \sin \sigma t_1 & -\cos \sigma t_1 \end{pmatrix} \mathbf{X} = 0 \end{gathered} \tag{5.25}$$

with respect to the vector $\mathbf{X} = (P_{0,s}, P_{0,c})^{\mathrm{T}}$, where the superscript ^{T} means the transposition.

For the vector equation (5.25) to have a solution in the form of a polynomial in ξ, one needs to put $c = 0$. From here, one obtains

$$b = \frac{1}{8}q(q - q^2)^{-1/2}\{f_0'(\varphi_0) + \mathrm{i}\left[4k''(\varphi_)) - 2f_0''(\varphi_0)\right]^{1/2}\} \tag{5.26}$$

and

$$b = \frac{1}{2}\mathrm{i}q(q^4 - 1)^{-1/2}\left[2k''(\varphi_0) + 2k'^2(\varphi_0) - q^2 f_0''(\varphi_0)\right]^{1/2} \tag{5.27}$$

for cases (5.16) and (5.17), respectively.

It is seen that $\Im b > 0$ if

$$q < 1, \quad 2k''(\varphi_0) > f_0''(\varphi_0) \tag{5.28}$$

or

$$q > 1, \quad 2k''(\varphi_0) + 2k'^2(\varphi_0) - q^2 f_0''(\varphi_0) > 0. \tag{5.29}$$

Hence, in the case $q < 1$ the generatrix $\varphi = \varphi_0^\circ$ is found from conditions (5.16), (5.28), and for $q > 1$ it is determined from (5.17), (5.29). It can be seen that when $q \approx 1$, then the requirement $|b| = O(1)$ does not hold. This case requires the reconstruction of asymptotic expansions and will be considered below as the special case.

When taking into account (5.26), (5.27), the vector equation (5.25) admits the solution in the form

$$\mathbf{X} = \mathcal{H}_m(\theta\xi)\mathbf{Y}_m, \quad \theta^2 = a\left(\frac{\partial^2 H}{\partial p^2}\right)^{-1}, \tag{5.30}$$

where $\mathbf{Y}_m = (S_m(t_1)), C_m(t_1))^{\mathrm{T}}$, and $\mathcal{H}_m(x)$ is the Hermite polynomial of degree m. Substituting (5.30) into Eq. (5.25), one arrives at the homogeneous system of differential equations

$$\dot{\mathbf{Y}}_m(t_1) = \mathbf{A}_m(t_1)\,\mathbf{Y}_m(t_1) \tag{5.31}$$

with the periodic matrix

$$\mathbf{A}_m(t_1) = \begin{pmatrix} -a_0 \sin \sigma t_1 & -a_{2,m} + a_0 \cos \sigma t_1 \\ a_{2,m} + a_0 \cos \sigma t_1 & a_0 \sin \sigma t_1 \end{pmatrix} \tag{5.32}$$

with

$$a_0 = \frac{q^2 f_1(\varphi_0)}{4\omega_0}, \quad a_{2,m} = a\frac{2m+1}{4\omega_0}. \tag{5.33}$$

Finally, for the normal deflection we get the formula

$$w = \sin\left(\frac{\pi n s}{l}\right) \exp\left\{\mathrm{i}\mu^{-1}\left[p(\varphi - \varphi_0) + \frac{1}{2}b(\varphi - \varphi_0)^2\right]\right\}$$
$$\times \left\{\mathcal{H}_m\left[\mu^{-1/2}(\varphi - \varphi_0)\theta\right]\left[S_m(\mu t)\sin(\omega_0 t) + C_m(\mu t)\cos(\omega_0 t)\right] + O\left(\mu^{1/2}\right)\right\}. \tag{5.34}$$

Note that the forms of parametric oscillations (5.34) are fundamentally different in cases (5.16) and (5.17): for

$$q_n = \frac{\pi n h_*^{1/2}}{l[12(1-\nu^2)]^{1/4}} < 1, \tag{5.35}$$

one has $p \neq 0$, and for $q_n > 1$, one obtains $p = 0$. Properties of the solutions of system (5.31) will be discussed below.

Remark 5.1. The form of the localized parametric oscillations (5.34) is found with an accuracy of a certain factor, functionally dependent on the *slow time* $t_k = \mu^k t$ for $k \geq 2$. Indeed, if $\mathbf{Y}_m(t_1)$ is a solution of equations (5.31), then $F(t_k)\mathbf{Y}_m(t_1)$ is also a solution for (5.31), where $F(t_k)$ is an arbitrary function of arguments $t_2, t_3, \ldots$. To determine function (5.34), it was necessary to consider the first three approximations in the above proposed asymptotic construction. Formally, the process of constructing functions w_j with $j \geq 1$ in expansion (5.7) can be extended indefinitely. So, to determine the function $w_1(\xi, t_0, t_1, \ldots)$, it is necessary to consider equation (5.9) at $j = 3$, and the function $w_0(\xi, t_0, t_1, t_2, \ldots)$ will be refined, since its functional dependence on $t_2 = \mu^2 t$ will be determined. Note that the introduction of a function $F(t_2, t_3, \ldots)\mathbf{Y}_m(t_1)$, when considering higher approximations, is necessary for elimination of arising secular terms in inhomogeneous equations with respect to functions w_j for $j \geq 3$. However, it should be borne in mind that for $j > 3$, the correction defined in expansion (5.7) becomes comparable with an error of original equations.

5.1.2 Reconstruction of the asymptotic expansion

It follows from relations (5.26), (5.27) that if $q \approx 1$, then assumption (5.8) for b does not hold. In this case, the required form of localized parametric vibrations has a form different from (5.7). Therefore, the case of $q \approx 1$ deserves separate consideration.

A similar case of reconstruction of an asymptotic solution was considered in [222] under solving the problem on the localized buckling of a non-circular cylindrical shell under a static axial load.

We assume

$$\begin{gathered}
\tilde{\mu} = \mu^{2/3}, \quad q = 1 + \tilde{\mu}^2 \tilde{q}, \quad \tilde{f}_1 = \tilde{\mu}^{-1/2} f_1, \\
T_1(\varphi, t) = f_0(\varphi) + \tilde{\mu}^2 \tilde{f}_1(\varphi) \cos(\Omega t), \\
\Omega = 2\omega_0 + \tilde{\mu}^2 \tilde{\sigma}, \quad \tilde{\sigma} \sim 1 \qquad \tilde{\mu} \to 0, \\
\zeta = \tilde{\mu}^{-1}(\varphi - \varphi_0), \quad t_0 = t, \quad \tilde{t}_1 = \tilde{\mu}^2 t.
\end{gathered} \tag{5.36}$$

An uniformly available asymptotic solution of Eqs. (5.6) satisfying the conditions

$$w, \Phi \to 0 \quad \text{for} \quad |\zeta| \to \infty \tag{5.37}$$

can be found as a series [115]

$$w(\varphi, t) = \sum_{j=0}^{\infty} \tilde{\mu}^j w_j(\zeta, t_0, \tilde{t}_1), \quad \Phi(\varphi, t) = \sum_{j=0}^{\infty} \tilde{\mu}^j \Phi_j(\zeta, t_0, \tilde{t}_1). \tag{5.38}$$

At the same time, assuming further consideration of only the first three approximations and taking into account Remark 5.1, we will seek the functions w_j, Φ_j up to their functional dependence on the variables $t_2, t_3, \ldots$.

As the characteristic dimension, as before, we take $R = R_2(\varphi_0)$. Then $k(\varphi_0) = 1$.

After substituting (5.36), (5.38) into Eqs. (5.6), we again arrive at the sequence of Eqs. (5.9), but now

$$\mathbf{L}_0 z = \frac{\partial^2 z}{\partial t_0^2} + [2 - f_0(\varphi_0)]\, z, \quad \mathbf{L}_1 z = [2k'(\varphi_0) - f_0'(\varphi_0)]\, \zeta z,$$

$$\begin{aligned} \mathbf{L}_2 z = 4\frac{\partial^4 z}{\partial \zeta^4} + 4k'(\varphi_0)\zeta\frac{\partial^2 z}{\partial \zeta^2} + 4k'(\varphi_0)\frac{\partial z}{\partial \zeta} \\ +\frac{1}{2}\left[2k''(\varphi_0) + 2k'^2(\varphi_0) - f_0''(\varphi_0)\right]\zeta^2 z \\ +2\tilde{q}\,[2 - f_0(\varphi_0)]\, z + 2\frac{\partial^2 z}{\partial t_0 \partial \tilde{t}_1} - \tilde{f}_1(\varphi_0)\cos(\Omega t_0) z, \ldots \end{aligned} \tag{5.39}$$

Consider them step-by-step. In the leading approximation (for $j = 0$), one has the homogeneous differential equation $\mathbf{L}_0 w_0 = 0$. Its solution can be written as:

$$w_0(\zeta, t_0, \tilde{t}_1) = w_{0,c}(\zeta, \tilde{t}_1)\cos(\omega_0 t_0) + w_{0,s}(\zeta, \tilde{t}_1)\sin(\omega_0 t_0), \tag{5.40}$$

where

$$\omega_0 = \sqrt{2 - f_0(\varphi_0)} \tag{5.41}$$

is the leading approximation for the natural frequency of a thin cylindrical shell under the axial load

$$T_1^* = \frac{Eh^2}{R\sqrt{12(1-\nu^2)}} f_0(\varphi_0). \tag{5.42}$$

We suppose that $f_0 < f_{\mathrm{cr}} = 2$, where $f_{\mathrm{cr}} = 2$ corresponds to the classical value of the critical buckling force. Otherwise, the shell loses its stability.

In the first-order approximation ($j = 1$), one gets the non-homogeneous differential equation

$$\mathbf{L}_0 w_1 = [2k'(\varphi_0) - f_0'(\varphi_0)]\, \zeta w_0. \tag{5.43}$$

Due to (5.43), the right-hand side of this equation generates secular terms. The condition for their absence is the equation

$$2k'(f_0) - f_0'(f_0) = 0,$$

which serves to determine the weakest generatrix.

In what follows, we consider the special case when

$$k'(\varphi_0) = f_0'(\varphi_0) = 0. \tag{5.44}$$

Then the solution of Eq. (5.43) is

$$w_1(\zeta, t_0, \tilde{t}_1) = w_{1,c}(\zeta, \tilde{t}_1)\cos(\omega_0 t_0) + w_{1,s}(\zeta, \tilde{t}_1)\sin(\omega_0 t_0). \tag{5.45}$$

In the second-order approximation ($j = 2$), with Eqs. (5.40), (5.44), (5.45) taken into account, we arrive at the inhomogeneous equation

$$\begin{gathered}\mathbf{L_0} w_2 = [N_s \sin(\omega_0 t_0) + N_c \cos(\omega_0 t_0)] \\ +\tfrac{1}{2}\tilde{f}_1(\varphi_0)\left[w_{0,s}\sin(3\omega_0 t_0 + \tilde{\sigma}\tilde{t}_1) + w_{0,c}\cos 3\omega_0 t_0 + \tilde{\sigma}\tilde{t}_1)\right]\end{gathered} \tag{5.46}$$

with

$$\begin{aligned} N_s &= 4\frac{\partial^4 w_{0,s}}{\partial\zeta^4} + \frac{1}{2}\left[2k''(\varphi_0) - f_0''(\varphi_0)\right]\zeta^2 w_{0,s} + 2\tilde{q}\left[2 - f_0(\varphi_0)\right] w_{0,s} \\ &\quad -2\omega_0\frac{\partial w_{0,c}}{\partial\tilde{t}_1} + \frac{1}{2}\tilde{f}_1(\varphi_0)\left[w_{0,s}\cos(\tilde{\sigma}\tilde{t}_1) + w_{0,c}\sin(\tilde{\sigma}\tilde{t}_1)\right], \\ N_c &= 4\frac{\partial^4 w_{0,c}}{\partial\zeta^4} + \frac{1}{2}\left[2k''(\varphi_0) - f_0''(\varphi_0)\right]\zeta^2 w_{0,c} + 2\tilde{q}\left[2 - f_0(\varphi_0)\right] w_{0,c} \\ &\quad +2\omega_0\frac{\partial w_{0,s}}{\partial\tilde{t}_1} + \frac{1}{2}\tilde{f}_1(\varphi_0)\left[w_{0,s}\sin(\tilde{\sigma}\tilde{t}_1) - w_{0,c}\cos(\tilde{\sigma}\tilde{t}_1)\right]. \end{aligned} \tag{5.47}$$

The first term on the right-hand side of Eq. (5.46) generates secular terms in the constructed solution. The condition for their absence is the following differential equation

$$4\frac{\partial^4 \mathbf{X}}{\partial\zeta^4} + \left[k''(\varphi_0) - \frac{1}{2}f_0''(\varphi_0)\right]\zeta^2\mathbf{X} + 2\tilde{q}\omega_0^2\mathbf{X} + 2\omega_0\mathbf{E}'\frac{\partial\mathbf{X}}{\partial\tilde{t}_1} + \mathbf{G}\mathbf{X} = 0, \tag{5.48}$$

for the vector $\mathbf{X} = (w_{0,s}, w_{0,c})$, where

$$\mathbf{E}' = \begin{pmatrix} 0 & -1 \\ 1 & 0 \end{pmatrix}, \quad \mathbf{G} = \frac{1}{2}\tilde{f}_1(\varphi_0)\begin{pmatrix} \cos(\tilde{\sigma}\tilde{t}_1) & \sin(\tilde{\sigma}\tilde{t}_1) \\ \sin(\tilde{\sigma}\tilde{t}_1) & -\cos(\tilde{\sigma}\tilde{t}_1) \end{pmatrix}. \tag{5.49}$$

Applying the Fourier transform to Eq. (5.48)

$$\mathbf{X}(\zeta, \tilde{t}_1) = \frac{1}{\sqrt{2\pi}}\int_{-\infty}^{+\infty}\mathbf{X}^F(\eta, \tilde{t}_1)e^{i\eta\zeta}d\eta, \tag{5.50}$$

we get the lower order differential equation

$$\left[\frac{1}{2}f_0''(\varphi_0) - k''(\varphi_0)\right]\frac{\partial^2\mathbf{X}^F}{\partial\eta^2} + 2\omega_0\mathbf{E}'\frac{\partial\mathbf{X}^F}{\partial\tilde{t}_1} + \left[4\eta^4 + 2\tilde{q}\omega_0^2 + \mathbf{G}\right]\mathbf{X}^F = 0 \tag{5.51}$$

with respect to the vector function

$$\mathbf{X}^F(\eta, \tilde{t}_1) = \left(\omega_s^F(\eta, \tilde{t}_1), \omega_c^F(\eta, \tilde{t}_1)\right)^{\mathrm{T}}. \tag{5.52}$$

After changing variables

$$x = \tau\eta, \quad \tau = \sqrt{2}\left[2k''(\varphi_0) - f_0''(\varphi_0)\right]^{-1/6}, \tag{5.53}$$

Eq. (5.51) is readily reduced to the following form:

$$\frac{\partial^2 \mathbf{X}^F}{\partial x^2} - \frac{1}{2}\omega_0 \tau^{-4} \mathbf{E}' \frac{\partial \mathbf{X}^F}{\partial \tilde{t}_1} - \frac{1}{4}\left(4\tau^{-4}x^4 + 2\tilde{q}\omega_0^2 + \mathbf{G}\right)\mathbf{X}^F = 0. \tag{5.54}$$

Its solution can be represented in the form

$$\mathbf{X}^F = y_m(x)\mathbf{Y}_m(\tilde{t}_1), \quad \mathbf{Y}_m = \left(S_m(\tilde{t}_1), C_m(\tilde{t}_1)\right)^{\mathrm{T}}, \tag{5.55}$$

where $y_m (m = 0, 1, 2, \ldots)$ is an eigenfunction of the boundary value problem

$$y''_{xx} + (\lambda - x^4)y = 0, \quad y \to 0 \quad \text{as} \quad x \to \pm\infty, \tag{5.56}$$

and $\mathbf{Y}_m$ is a solution of the homogeneous system of differential equations (5.31) with the periodic matrix (5.32), where σ and t_1 should be replaced by $\tilde{s}$ and $\tilde{t}_1$, respectively, and parameters $a_0, a_{2,m}$ are introduced as

$$a_0 = \frac{\tilde{f}_1(\varphi_0)}{4\omega_0}, \quad a_{2,m} = \frac{\lambda_m}{2\omega_0}\left[2k''(\varphi_0) - f''_0(\varphi_0)\right]^{2/3} + \tilde{q}\omega_0. \tag{5.57}$$

Here $\lambda_m (m = 0, 1, 2, \ldots)$ are eigenvalues of the boundary-value problem (5.56). In particular, $\lambda_0 \approx 1.060$, $\lambda_1 \approx 3.800$, $\lambda_2 \approx 7.456$. Plots of the corresponding eigenfunctions $y_0(x), y_1(x), y_2(x)$ are depicted in Fig. 5.1.

The deflection shape is as follows:

$$w = \sin(\pi n s/l)\left\{\left[S_m(\tilde{\mu}^2 t)\sin(\omega_0 t) + C_m(\tilde{\mu}^2 t)\cos(\omega_0 t)\right] Z_m(\varphi) + O(\tilde{\mu})\right\}, \tag{5.58}$$

where

$$Z_m(\varphi) = \frac{1}{\sqrt{2\pi}} \int_{-\infty}^{+\infty} y_m\{\sqrt{2}\left[2k''(\varphi_0) - f''_0(\varphi_0)\right]^{-1/6}\eta\} \times \exp\left[i\eta\tilde{\mu}^{-1}(\varphi - \varphi_0)\right] d\eta. \tag{5.59}$$

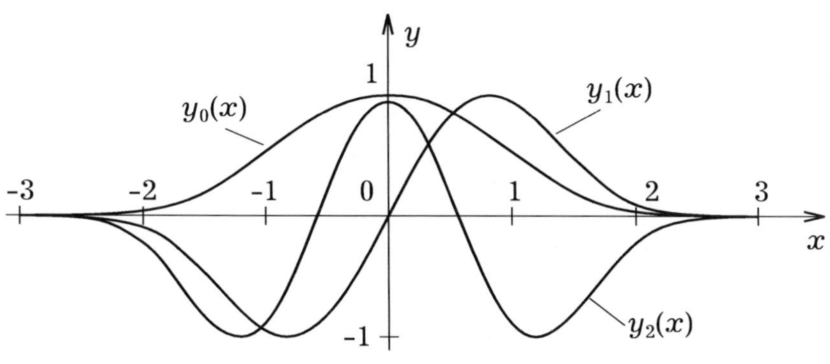

FIGURE 5.1: Eigenfunctions of boundary-value problem (5.56).

The formal asymptotic solutions (5.34), (5.58) describe the excited parametric vibrations of the shell localized in the neighborhood of the weakest generatrix $\varphi = \varphi_0$. The form of parametric vibrations depends both on the number of semi-waves n in the axial direction and on the integer number m (the number of oscillations in an amplitude), which characterizes the variability of the solution in the circumferential direction of the shell. Since solutions (5.34), (5.58) are not periodic with respect to the circumferential coordinate φ and satisfy the attenuation condition of the waves amplitude as $|\varphi| \to \infty$, they should be considered only in the vicinity of the weakest generatrix. Taking into account the weak dependence of amplitudes on time in solutions (5.34), (5.58), the constructed forms of parametric vibrations may be considered as quasi-stationary ones with the frequency ω_0 equal to the leading approximation of the natural frequency of free vibrations localized in the neighborhood of the weakest generatrix $\varphi = \varphi_0$.

5.1.3 Effect of dissipative forces

Consider here the effect of internal friction in the shell material. For this purpose, we introduce the dissipative term $\mu\gamma_v \frac{\partial w}{\partial t}$ into the first equation of system (5.3). Here, $\gamma_v > 0$ is a dimensionless coefficient of viscous friction associated with its dimensioned counterpart γ_v^* by the formula

$$\gamma_v = \gamma_v^* R/(\mu E h t_c). \tag{5.60}$$

The valuation of the dissipative term was made in such a way that its influence appeared itself in the second-order approximation of the asymptotic solution to be constructed below.

Using the procedure for constructing the asymptotic solution (5.5), (5.7) described above, we arrive at the sequence of equations (5.9) in which the term $\gamma_v \frac{\partial}{\partial t_0}$ should be added to the differential operator (5.12). The presence of this term leads to a change in matrix (5.32):

$$\mathbf{A}_m(t_1) = \begin{pmatrix} -a_1 - a_0 \sin\sigma t_1 & -a_{2,m} + a_0 \cos\sigma t_1 \\ a_{2,m} + a_0 \cos\sigma t_1 & -a_1 + a_0 \sin\sigma t_1 \end{pmatrix}, \tag{5.61}$$

where $a_1 = \gamma_v/2$.

Performing the reconstruction of asymptotic solution as it was done above with $q \approx 1$, but taking into account internal friction, we arrive again at the homogeneous system of differential equations (5.31) with the periodic matrix (5.61), in which one needs to assume $a_1 = \tilde{\gamma}_v/2$, where

$$\tilde{\gamma}_v = \gamma_v^* R/(\tilde{\mu}^{1/2} E h t_c). \tag{5.62}$$

5.1.4 Parametric instability domains

Consider the system of differential equations (5.31) with periodic matrix (5.61). Equations (5.31) have a periodic solution $S_m(t_1)$, $C_m(t_1)$ if and only if the multipliers of these equations are equal to one. If the moduli of multipliers are greater than one, then their solutions grow indefinitely as $t_1 \to +\infty$, if less than one, they decrease.

Analysis of equations (5.31) was carried out numerically. We constructed the monodromy matrix and found its eigenvalues, i.e., the multipliers. It was found out that in the absence of dissipative forces ($a_1 = 0$), the plane (σ, a_0) is split by the lines

$$a_0 = \pm(a_{2,m} - \sigma/2) \tag{5.63}$$

onto the regions (see Fig. 5.2), in each of which the nature of the solutions is the same: if the point (σ, a_0) lies in regions I, II, then the solutions increase indefinitely, in regions III the solutions are bounded (aperiodic) functions.

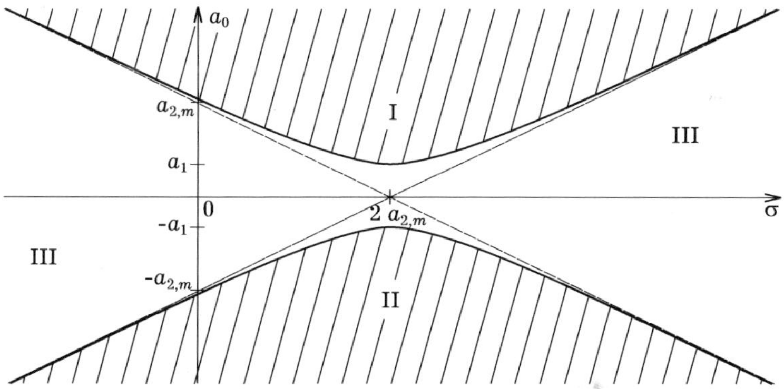

FIGURE 5.2: Main region of parametric instability in the presence of dissipative forces.

From the point of view of the dynamic stability criterion for elastic systems, unlimited increasing the solutions of system (5.31) in regions I, II means a parametric instability of oscillations being excited.

In the presence of structural damping or when taking into account the viscoelastic properties of a material ($a_1 > 0$), the plane (σ, a_0) is divided by lines

$$a_0 = \pm\sqrt{a_1^2 + (a_{2,m} - \sigma/2)^2} \tag{5.64}$$

into three regions (see Fig. 5.2), and the straight lines (5.63) here play the role of asymptotes.

If the point (σ, a_0) lies in region I or II, then the solutions growth without limit (see Fig. 5.3), on the bound of the regions the behavior of the solutions is close to harmonic, and, finally, in region III, the solutions are decreasing functions as shown in Fig. 5.4.

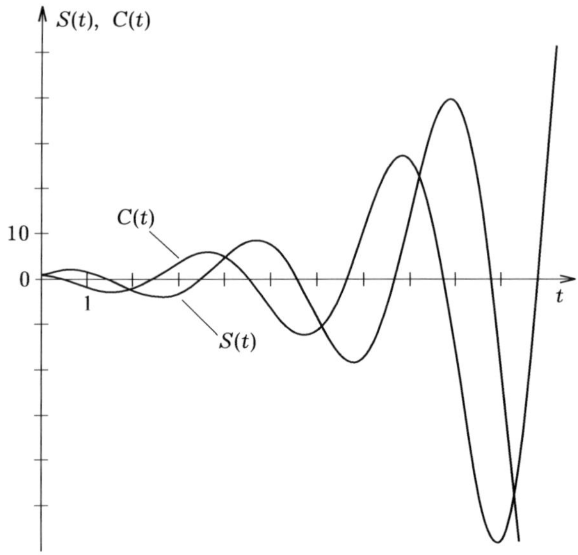

FIGURE 5.3: Unbounded solutions of Eqs. (5.31) for $S(0) = 1$, $C(0) = 1$, $\sigma = 3$, $a_0 = 1$, $a_1 = 0.5$, $a_{2,m} = 2$.

In the case when the axial forces do not depend on time ($a_0 = 0$), Eqs. (5.31) admit an explicit solution. When $q > 1$ and $q < 1$ we have

$$\begin{aligned} S_m(t) &= [c_1 \sin(\mu a_{2,m} t) + c_2 \cos(\mu a_{2,m} t)] \exp(-\mu a_1 t), \\ C_m(t) &= [-c_1 \cos(\mu a_{2,m} t) + c_2 \sin(\mu a_{2,m} t)] \exp(-\mu a_1 t), \end{aligned} \tag{5.65}$$

and if $q \approx 1$, then

$$\begin{aligned} S_m(t) &= \left[c_1 \sin(\mu^{4/3} a_{2,m} t) + c_2 \cos(\mu^{4/3} a_{2,m} t)\right] \exp(-\mu^{4/3} a_1 t), \\ C_m(t) &= \left[-c_1 \cos(\mu^{4/3} a_{2,m} t) + c_2 \sin(\mu^{4/3} a_{2,m} t)\right] \exp(-\mu^{4/3} a_1 t), \end{aligned} \tag{5.66}$$

where c_1, c_2 are arbitrary constants. Then solutions (5.34), (5.65) and (5.59) define the eigenmodes of vibrations for a cylindrical shell non-uniformly compressed in the axial direction with the frequencies

$$\omega = \omega_0 + \mu a_{2,m} + O(\mu^2) \tag{5.67}$$

and

$$\omega = \omega_0 + \mu^{4/3} a_{2,m} + O(\mu^{8/3}), \tag{5.68}$$

for the cases $q \neq 1$ and $q \approx 1$, respectively.

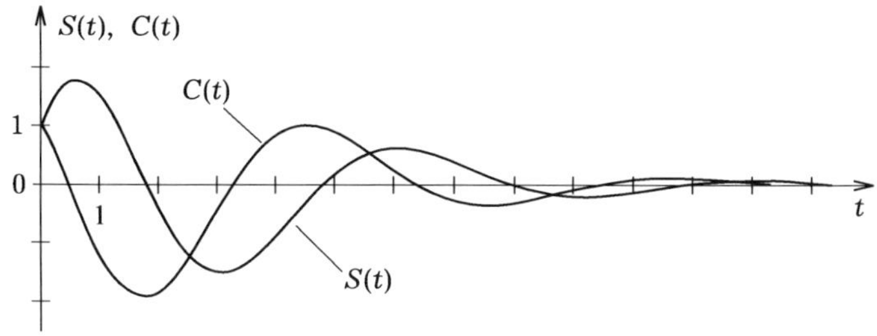

FIGURE 5.4: Decaying solutions of Eqs. (5.31) for $S(0) = 1$, $C(0) = 1$, $\sigma = 2$, $a_0 = 1$, $a_1 = 0.5, a_{2,m} = 2$.

5.1.5 Examples

From the above formulas (5.63), (5.64) one can determine the boundaries of the interval $\Omega^- \leq \Omega \leq \Omega^+$ for the dimensionless excitation frequency $\Omega = \Omega^* t_c$ corresponding to the parametric shell instability:

$$\begin{aligned} \Omega^{\pm} &= 2\omega_0 + \sqrt[3]{\frac{h^2}{12R^2(1-\nu^2)}}\,\sigma^{\pm} \quad \text{for} \quad q \approx 1, \\ \Omega^{\pm} &= 2\omega_0 + \sqrt[4]{\frac{h^2}{12R^2(1-\nu^2)}}\,\sigma^{\pm} \quad \text{for} \quad q < 1 \quad \text{or} \quad q > 1, \end{aligned} \tag{5.69}$$

where $\sigma^{\pm} = 2(a_{2,m} \pm \sqrt{a_0^2 - a_1^2})$. If $q < 1$, then the parameters $\omega_0, a_0, a_{2,m}$ can be found by formulas (5.18), (5.33); at $q \approx 1$, they are calculated by Eqs. (5.41), (5.57), and for $q > 1$, one needs to use relations (5.19), (5.33).

Example 5.1. Consider a circular cylindrical shell (here, $k(\varphi) \equiv 1$) under the action of the combined non-uniform axial load (5.1), where $f_0(\varphi) = 0.5(1+\cos\varphi)$ and f_1 is a constant value. In this case, the weakest generatrix is $\varphi_0 = 0$, along which the axial load (5.1) is maximal. The dependence of dimensionless parameters $q, \Omega^{\pm}, \omega$ on the numbers n and m for $f_1 = 1$ (values on the top row) and $f_1 = 2$ (values on the bottom row) is given in Table 5.1. Here, ω is the eigenfrequency of localized vibrations of the cylindrical shell under the action of the static inhomogeneous force $T_1(\varphi) = f_0(\varphi) = 0.5(1+\cos\varphi)$. The values of ω were determined by formulas (5.67), (5.68). The calculations were carried out for $h/R = 0.01$, $l/R = 0.52$, $\nu = 0.3$ and small values of n and m for which the condition $\sigma \sim 1$ is satisfied. The dimensional frequencies are obtained by dividing the dimensionless counterparts from the table by the characteristic time $t_c = R\sqrt{\rho/E}$.

The table shows that an increase in the amplitude of the periodic component of the axial load results in the expansion of the boundaries of the instability regions.

TABLE 5.1: The boundaries of the main region of parametric instability.

n	q	$m=0$			$m=1$		
		ω	Ω^-	Ω^+	ω	Ω^-	Ω^+
1	0.332	0.351	0.693	0.712	0.389	0.769	0.787
			0.648	0.721		0.760	0.796
2	0.665	0.627	1.327	1.363	0.688	1.358	1.349
			1.308	1.381		1.339	1.413
3	0.997	1.004	1.981	2.036	1.022	2.016	2.071
			1.954	2.064		1.988	2.099
4	1.329	1.550	3.067	3.131	1.579	3.126	3.189
			3.036	3.162		3.094	3.221

Figure 5.5 shows the boundaries of the main region of parametric instability, taking into account the presence of dissipative forces, characterized by the internal friction coefficient γ_v. The calculations were carried out for $n = 1$ and for different values of γ_v and m, the remaining parameters were taken the same as in the previous case. As can be seen, the introduction of viscosity leads to a narrowing of the instability regions.

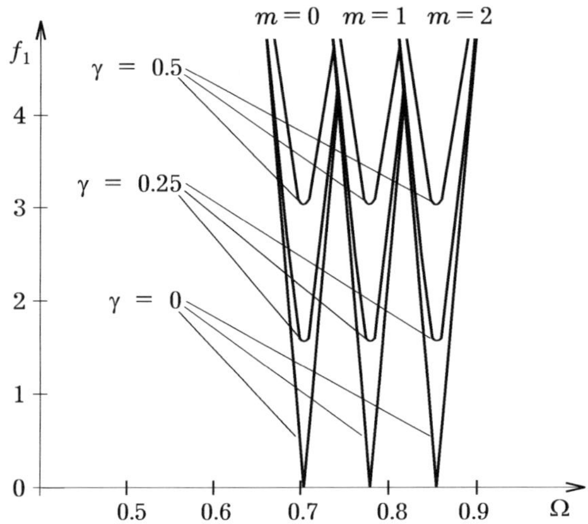

FIGURE 5.5: The boundaries of the main region of parametric instability for different coefficients of internal friction γ_v.

To analyze the reliability of the asymptotic solutions constructed above, we present the results of numerically solving the system of equations (5.6) using the grid method. We will consider two essentially different cases when the applied axial load does not cause parametric resonance and when it leads to its appearance.

First, let $n = 1, \Omega = 0.5, f_1 = 1$. Then, as follows from Table 5.1, the periodic axial load does not lead to parametric resonance. In Fig. 5.6, the asymptotic and numerical solutions are shown for $t = 25$. As can be seen, there is a good agreement between the numerical and analytical solutions, especially in the vicinity of the weakest generatrix $f_0 = 0$. A similar pattern holds for other points in time.

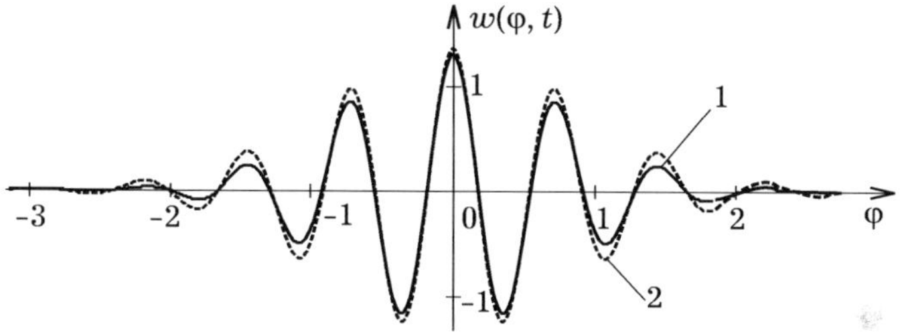

FIGURE 5.6: Plots of the function $w(\varphi, t)$ for $t = 25$: 1 – asymptotic solution, 2 – numerical solution.

Let now the shell be subject to a load corresponding to the previously found main region of parametric instability (see Table 5.1). For this purpose we assume $n = 1, m = 0, \Omega = 0.71$. In this case, there is a pronounced localization of solutions in the neighborhood of the generatrix $\varphi_0 = 0$ with slow increasing the amplitude of oscillations. The plots of the functions $w(\varphi, t)$ for different instants of time t are shown in Fig. 5.7.

Example 5.2. Consider a simply supported elliptic cylindrical shell with semi-axes e_1, e_2 ($e_1 \leqslant e_2$), subject to the axial forces (5.1), where $f_0 = 1$, f_1 are constant values. As a characteristic dimension, the value $R = e_2^2/e_1$ is taken. Then

$$k(\varphi) = (\cos^2\theta + \delta_1^2 \sin^2\theta)^{3/2}, \quad d\theta/d\varphi = k, \quad \delta_1 = e_2/e_1.$$

Here the two generatrices $\varphi_0 = 0, \pi$ ($\theta_0 = 0, \pi$) are the weakest lines, corresponding to the small semi-axes of the ellipse. Consider parametric vibrations near these generatrices in the case when $q < 1$. We obtain

$$\begin{gathered} \omega_0 = q = \frac{h^{1/2} e_1^{1/2} \pi n}{[12(1-\nu^2)]^{1/4} e_2 l} < 1, \\ a_0 = f_1 q/4, \quad a_{2,m} = 4(1+2m)\sqrt{3(e_2^2 - e_1^2) e_1^{-2}(q - q^2)}, \end{gathered} \tag{5.70}$$

where $n = 1, 2, 3, \ldots, m = 0, 1, 2, \ldots$. Substituting Eqs. (5.70) into (5.69) one can find the boundaries of the regions corresponding to parametric resonance for different n and m. Table 5.2 shows these boundaries for $e_1 = 50$ mm, $e_2 = 60$ mm, $h = 0.4$ mm, $l = 4$, $\nu = 0.37$, $f_1 = 1$ and for different n, m. Calculations were performed for small values of n and m for which the conditions $q < 1$, $\sigma \sim 1$ hold.

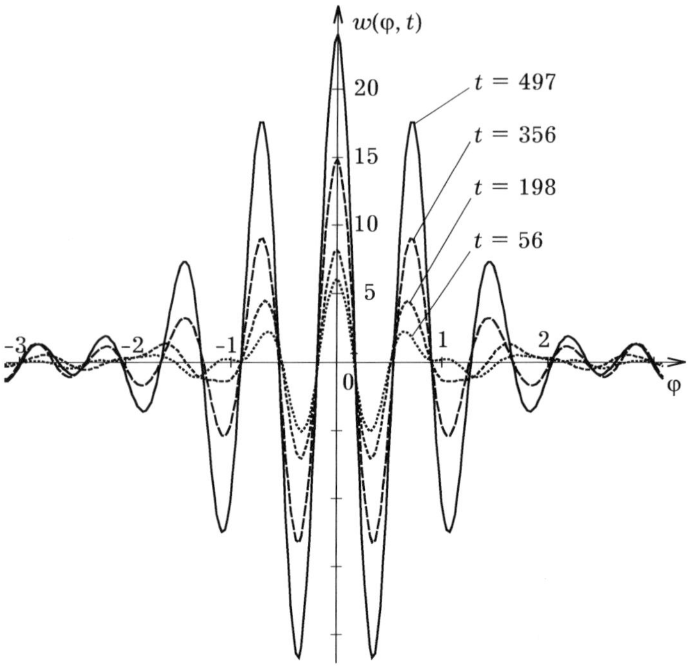

FIGURE 5.7: Plots of the function $w(\varphi, t)$ for different instants of time t.

5.2 Parametric vibrations of laminated cylindrical shells under periodic axial forces: effect of shear

In this section, we consider a thin laminated cylindrical shell of length L consisting of N transversely isotropic layers. Each layer is characterized by thickness h_j, Young's modulus E_j, the shear modulus G_j, Poisson's ratio ν_j, and the density ρ_j, where $j = 1, 2, \ldots, N$ (the numbering of layers begins with the innermost lamina). The middle surface of any fixed layer is taken as the reference surface with the axial and circumferential coordinates α_1 and α_2, respectively. As in the previous section, the shell is non-circular with the radius $R_2(\alpha_2)$ of the reference surface. The shell is under the axial force (5.1) which is non-uniformly distributed along the shell edge and the superposition of static and periodic components.

To predict parametric vibrations of the laminated shell taking into account the shear effect, we shall apply the equivalent single layer shell theory with governing Eqs. (2.109). After introducing the additional terms accounting for

TABLE 5.2: The boundaries of the main regions of parametric instability.

$n\backslash m$	0	1	2
1	$0.130 \leqslant \Omega \leqslant 0.132$	$0.263 \leqslant \Omega \leqslant 0.265$	$0.396 \leqslant \Omega \leqslant 0.398$
2	$0.220 \leqslant \Omega \leqslant 0.223$	$0.405 \leqslant \Omega \leqslant 0.408$	$0.590 \leqslant \Omega \leqslant 0.593$
3	$0.303 \leqslant \Omega \leqslant 0.307$	$0.525 \leqslant \Omega \leqslant 0.529$	$0.747 \leqslant \Omega \leqslant 0.752$

the initial stresses, these equations in the dimensionless form read as:

$$\mu^4\left(1-\mu^3\tau\triangle\right)\triangle^2\widehat{\chi}+\mu^2k(\varphi)\frac{\partial^2\widehat{\Phi}}{\partial s^2}+\mu^2T_1\left(\varphi,t\right)\frac{\partial^2}{\partial s^2}\left(1-\mu^2\kappa\triangle\right)\widehat{\chi}$$
$$+\frac{\partial^2}{\partial t^2}\left(1-\mu^2\kappa\triangle\right)\widehat{\chi}=0, \tag{5.71}$$
$$\mu^2\triangle^2\widehat{\Phi}-k(\varphi)\frac{\partial^2}{\partial s^2}\left(1-\mu^2\kappa\triangle\right)\widehat{\chi}=0,$$

$$\frac{1-\nu}{2}\frac{h^2}{\varpi}\triangle\Theta=\Theta. \tag{5.72}$$

Here the dimensionless curvilinear coordinates s,φ and the time t are introduced as in the previous section, $t_c = R\sqrt{\rho/E}$ is the characteristic time, where E,ρ are the reduced Young modulus and the density for laminates (see Eq. (2.84) in Section 2.10), the required dimensionless displacement and stress functions $\widehat{\chi},\widehat{\Phi}$ are introduced by

$$\chi = R\widehat{\chi}(s,\varphi), \quad \Phi=\mu^2EhR^2\widehat{\Phi}(s,\varphi), \tag{5.73}$$

where

$$\mu^4=\frac{h^2\eta_3}{12R^2\left(1-\nu^2\right)} \tag{5.74}$$

is the small parameter characterizing the shell thickness, and κ,τ are the dimensionless shear parameters which are introduced to assume the following asymptotic estimates:

$$\frac{h_*^2}{\varpi}=\mu^2\kappa, \quad \frac{\theta h_*^2}{\varpi}=\mu^3\tau, \; h_*=\frac{h}{R}, \quad \kappa,\tau\sim 1 \text{ as } \mu\to 0. \tag{5.75}$$

These estimates are valid for sufficiently thin shell containing a "soft" core or several such layers. The parameters ϖ,θ are calculated from Eqs. (2.97) and (2.103), respectively. The last estimation in (5.75) is assumed because of the smallness of a parameter θ; for instance, for a single layer isotropic shell [80], $\theta = 1/85$. Correlations (5.75) imply that the reduced shear modulus $G = \Gamma h$, with Γ calculated by (2.84), is small when comparing with the reduced Young modulus E so that $G \sim h_*E$ as $h_* \to 0$.

In what follows, the symbol $\widehat{\ }$ will be omitted.

Let both edges be simply supported and free of a diaphragm. Then, in terms of the displacement and stress functions, the corresponding conditions read as follows [80]:

$$\chi = \Delta\chi = \Delta^2\chi = \Phi = \Delta\Phi = 0, \quad \frac{\partial\Theta}{\partial\alpha_1} = 0 \quad \text{at} \quad \alpha_1 = 0, L. \tag{5.76}$$

It is seen that Eqs. (5.71) and (5.72) are not coupled, and the boundary condition (5.76) for Θ is independent of the residual conditions. Hence, one may assume $\Theta = 0$.

5.2.1 Asymptotic solution

Let $\varphi = \varphi_0$ be the weakest generatrix in the neighborhood of which the localized parametric vibrations are excited. Following the approach developed in Subsection 5.1.1, the formal asymptotic solution of Eqs. (5.71) is sought again in the form of a series

$$\chi(\varphi, t, \mu) = \sin\left(\tfrac{q_n s}{\mu}\right) \sum_{j=0}^{\infty} \mu^{j/2}\chi_j(\xi, t_0, t_1, ...) \exp\left(\mathrm{i}\left(\mu^{-1/2}p\xi + \tfrac{1}{2}b\xi^2\right)\right),$$
$$q_n = \mu\pi n l^{-1} \sim 1, \quad n = 1, 2, \ldots, \quad \xi = \mu^{-\frac{1}{2}}(\varphi - \varphi_0),$$
$$\mathrm{i} = \sqrt{-1}, \quad t_m = \mu^m t, \quad m = 0, 1, \ldots, \tag{5.77}$$

where $\Im b > 0$, χ_j are polynomials in ξ and t_m is the "slow time" for $m \geq 1$. The function Φ is constructed in the same form. It is also assumed that

$$p,\ q_n,\ |b|,\ |y|,\ \left|\frac{\partial y}{\partial x}\right| \sim 1 \text{ as } \mu \to 0, \tag{5.78}$$

where y is any of functions χ_j, Φ_j and x is any their argument. We fix n and omit this subscript in what follows.

Substituting Eqs. (5.77) into Eqs. (5.71) gives the sequence of equations

$$\sum_{\varsigma=0}^{j} \mathbf{L}_\varsigma \chi_{j-\varsigma} = 0, \quad j = 0, 1, 2, \ldots \tag{5.79}$$

Here,

$$\mathbf{L}_0 z = \frac{\partial^2 z}{\partial t_0^2} + H^2(p, q, \varphi_0)\, z, \tag{5.80}$$

$$H(p, q, \phi_0) = \sqrt{\frac{(p^2 + q^2)^2}{1 + \kappa(p^2 + q^2)} + \frac{k^2(\varphi_0)q^4}{(p^2 + q^2)^2} - f_0(\varphi_0)q^2}, \tag{5.81}$$

and the operators $\mathbf{L}_\varsigma$ for $\varsigma \geq 1$ are introduced by (4.21), where

$$\mathbf{N}z = 2\frac{\partial^2 z}{\partial t_0 \partial t_1} - q^2 f_1(\varphi_0)\cos(\Omega t_0)z + q^6\tau z. \tag{5.82}$$

Function (5.81) contains the shear parameter κ and so it generalize similar formula (5.11) introduced above for a single layer isotropic shell.

Consider Eqs. (5.79) step by step. In the leading approximation ($j = 0$), one has the homogeneous equation $\mathbf{L}_0\chi_0 = 0$ which has the solution

$$\chi_0 = P_{0,c}(\xi, t_1, t_2, ...)\cos(\omega_0 t_0) + P_{0,s}(\xi, t_1, t_2, ...)\sin(\omega_0 t_0), \\ \omega_0 = H(p, q, \varphi_0), \tag{5.83}$$

where $P_{0,c}, P_{0,s}$ are polynomials in ξ with coefficients depending on "slow time".

In the first-order approximation ($j = 1$), one obtains the inhomogeneous equation

$$\frac{\partial^2\chi_1}{\partial t_0^2} + H^2(p, q, \varphi_0)\chi_1 = \frac{\partial (H)^2}{\partial p}\left(\mathrm{i}\frac{\partial \chi_0}{\partial \xi} - b\xi\chi_0\right) - \frac{\partial (H)^2}{\partial \varphi_0}\xi\chi_0. \tag{5.84}$$

which has unlimitedly growing solutions (as $t_0 \to \infty$) called the secular ones. To eliminate them, one needs to assume

$$\frac{\partial (H)^2}{\partial p} = 0, \quad \frac{\partial (H)^2}{\partial \varphi_0} = 0. \tag{5.85}$$

The above equations serve to determine the wave parameter $p = p^\circ$ and the weakest generatrix $\varphi_0 = \varphi_0^\circ$. There are three different cases to consider:

(case I) $q > z_0$;

(case II) $q < z_0$;

(case III) $q = z_0$,

here z_0 is a root of the equation

$$-2k^2(\varphi_0)(1 + \kappa q z)^2 + z^4(2 + \kappa q z) = 0. \tag{5.86}$$

If $q > z_0$ (case I), then

$$p = 0, \quad \omega_0 = \sqrt{k^2(\varphi_0) - f_0(\varphi_0)q^2 + \frac{q^4}{1 + \kappa q^2}}, \tag{5.87}$$

where the weakest generatrix φ_0 is found from the equation

$$2k(\varphi_0)k'(\varphi_0) - f_0'(\varphi_0)q^2 = 0. \tag{5.88}$$

For $q < z_0$ (case II), one arrives at the formulas

$$p = \sqrt{q(z_0 - q)}, \quad \omega_0 = \sqrt{\frac{k^2(\varphi_0)q^2}{z_0^2} - f_0(\varphi_0)q^2 + \frac{q^2 z_0^2}{1 + \kappa q z_0}}, \tag{5.89}$$

and φ_0 is the root of the equation

$$2k(\varphi_0)k'(\varphi_0) - f_0'(\varphi_0)z_0^2 = 0. \tag{5.90}$$

In what follows, we consider the value $R = R_2(\varphi_0)$ as the characteristic dimension of the shell.

For $q = z_0$ (case III), Eqs. (5.87), (5.88) coincide with Eqs. (5.89), (5.90), respectively.

Remark 5.2. In order for expression under radicals (5.87), (5.89) to be positive, we require that the inequality $f_0(\varphi_0) < \min\left\{f_{cr}^{(1)}, f_{cr}^{(2)}\right\}$ be satisfied for any φ_0, where

$$f_{cr}^{(1)} = \min_{\varphi_0} \frac{1}{f_0(\varphi_0)}\left[\frac{k^2(\varphi_0)}{q^2} + \frac{q^2}{1+\kappa q^2}\right] \tag{5.91}$$

for $q > z_0$ and φ_0 defined from Eq. (5.88), and

$$f_{cr}^{(2)} = \min_{\varphi_0} \frac{1}{f_0(\varphi_0)}\left[\frac{k^2(\varphi_0)}{z_0^2} + \frac{z_0^2}{1+\kappa z_0^2}\right] \tag{5.92}$$

for $q \leq z_0$ and φ_0 determined from Eq. (5.90). In other words, it is assumed that the axial force does not reach the critical value $F_{cr} = \mu^2 Eh \min\left\{f_{cr}^{(1)}, f_{cr}^{(2)}\right\}$, which results in the buckling of a thin laminated circular cylindrical shell [137].

Consider Eq. (5.79) in the second-order approximation ($j = 2$) and assume again condition (5.22). Then we arrive at the inhomogeneous equation

$$\begin{aligned}\frac{\partial^2 \chi_2}{\partial t_0^2} + H^2(p,q,\varphi_0)\chi_2 &= -N_c\cos(\omega_0 t_0) - N_s \sin(\omega_0 t_0)\\ &+r\left[P_{0,c}\cos(3\omega_0 t_0 + \sigma t_1) + P_{0,s}\sin(3\omega_0 t_0 + \sigma t_1)\right]\end{aligned} \tag{5.93}$$

with

$$\begin{aligned} N_c &= -\frac{1}{2}\frac{\partial^2 H^2}{\partial p^2}\frac{\partial^2 P_{0,c}}{\partial \xi^2} + a\xi\frac{\partial P_{0,c}}{\partial \xi} + \left[c\xi^2 + \tfrac{1}{2}a + \frac{\tau(p^2+q^2)^3}{1+\kappa(p^2+q^2)}\right]P_{0,c}\\ &+2\omega_0\frac{\partial P_{0,s}}{\partial t_1} + r\left[P_{0,s}\sin(\sigma t_1) - P_{0,c}\cos(\sigma t_1)\right]\\ N_s &= -\frac{1}{2}\frac{\partial^2 H^2}{\partial p^2}\frac{\partial^2 P_{0,s}}{\partial \xi^2} + a\xi\frac{\partial P_{0,s}}{\partial \xi} + \left[c\xi^2 + \tfrac{1}{2}a + \frac{\tau(p^2+q^2)^3}{1+\kappa(p^2+q^2)}\right]P_{0,s}\\ &-2\omega_0\frac{\partial P_{0,c}}{\partial t_1} + r\left[P_{0,s}\cos(\sigma t_1) - P_{0,c}\sin(\sigma t_1)\right],\end{aligned} \tag{5.94}$$

$$a = -\mathrm{i}\left(b\frac{\partial^2 H^2}{\partial p^2} + \frac{\partial^2 H^2}{\partial p\partial\varphi_0}\right), \quad c = \frac{1}{2}\left(b^2\frac{\partial^2 H^2}{\partial p^2} + 2b\frac{\partial^2 H^2}{\partial p\partial\varphi_0} + \frac{\partial^2 H^2}{\partial\varphi_0^2}\right),$$

$$r = \tfrac{1}{2}q^2 f_1(\varphi_0).$$

The first two components on the right-hand side of Eq. (5.93) generate the secular solutions. The conditions for their absence lead to the vector differential equation

$$-\frac{1}{2}\frac{\partial^2 H}{\partial p^2}\frac{\partial^2 \mathbf{X}}{\partial \xi^2} + a\xi\frac{\partial \mathbf{X}}{\partial \xi} + \left[c\xi^2 + \frac{1}{2}a + \frac{\tau\left(p^2+q^2\right)^3}{1+\kappa\left(p^2+q^2\right)}\right]\mathbf{X}$$
$$+2\omega_0\left(\begin{smallmatrix}0 & -1\\ 1 & 0\end{smallmatrix}\right)\frac{\partial \mathbf{X}}{\partial t_1} + r(\varphi_0)\begin{pmatrix}\cos\sigma t_1 & \sin\sigma t_1\\ \sin\sigma t_1 & -\cos\sigma t_1\end{pmatrix}\mathbf{X} = 0 \tag{5.95}$$

which differs from Eq. (5.25) due to the presence of the term depending on the shear parameters κ, τ.

For the vector equation (5.95) to have a solution in the form of polynomials in ξ, we assume $c = 0$. Hence,

$$b = \left\{\frac{\partial^2(H^2)}{\partial p^2}\right\}^{-1}\left\{-\frac{\partial^2(H^2)}{\partial p\partial\varphi_0} \pm \mathrm{i}\sqrt{\frac{\partial^2(H^2)}{\partial p^2}\frac{\partial^2(H^2)}{\partial\varphi_0^2} - \left[\frac{\partial^2(H^2)}{\partial p\partial\varphi_0}\right]^2}\right\}. \tag{5.96}$$

It is seen that $\Im b > 0$ if the inequalities

$$\frac{\partial^2(H^2)}{\partial p^2} > 0, \quad \frac{\partial^2(H^2)}{\partial p^2}\frac{\partial^2(H^2)}{\partial\varphi_0^2} - \left[\frac{\partial^2(H^2)}{\partial p\partial\varphi_0}\right]^2 > 0 \tag{5.97}$$

hold simultaneously.

If $q > z_0$, then inequalities (5.97) imply

$$2[(k'(\varphi_0)]^2 + 2k(\varphi_0)k''(\varphi_0) - f_0''(\varphi_0)q^2 > 0, \tag{5.98}$$

and for $q < z_0$, one has

$$\left\{2[k'(\varphi_0)]^2 + 2k(\varphi_0)k''(\phi_0) - f_0''(\varphi_0)z^2\right\}\left[8 + 9\kappa qz + 3(\kappa qz)^2\right] >$$
$$4\left[f_0'(\varphi_0)\right]^2(1+\kappa qz)^3. \tag{5.99}$$

Below, we consider the special case when $k \equiv 1$ (a circular cylinder) and $f_0(\varphi)$ is a function of φ. From Eq. (5.88), (5.90) and inequalities (5.98), (5.99) we get the conditions

$$f_0'(\varphi_0) = 0, \quad f_0''(\varphi_0) < 0, \tag{5.100}$$

which mean that the weakest generatrix $\varphi = \varphi_0$ is that which is more compressed by a static component of the axial force. The required parameter b becomes

$$b = \mathrm{i}\sqrt{\frac{-q^4(1+\kappa q^2)^2 f_0''(\varphi_0)}{2q^4(2+\kappa q^2) - 4(1+\kappa q^2)^2}} \tag{5.101}$$

for $q > z_0$ (case I), and

$$b = \mathrm{i}\sqrt{\frac{-q(1+\kappa q^2)^3 f_0''(\varphi_0)}{4(z_0-q)[8+9\kappa q z_0+3(\kappa q z_0)^2]}} \tag{5.102}$$

for $q < z_0$ (case II).

It can be seen that $\lim_{q\to z_0}|b| = +\infty$ for both cases, (I) and (II), and requirement (5.78) for b does not hold if the root q is close to z_0. Thus, the case (III), where q is close to z_0, requires an additional consideration.

We will not consider higher approximations because Eqs. (5.71) are not sufficiently accurate. To construct higher approximations, one needs to apply the complete system of governing equations written in terms of displacements u_i, w, ψ_i.

From Eq. (5.96), it follows that the vector equation (5.95) admits a solution in the form (5.30), where the vector $\mathbf{Y}_m = (S_m(t_1), C_m(t_1))^{\mathrm{T}}$ is found from the homogeneous system of differential equations (5.31) with the periodic matrix

$$\mathbf{A}_m(t_1) = \begin{pmatrix} -a_0\sin\sigma t_1 & -a_{2,m}-a_\tau+a_0\cos\sigma t_1 \\ a_{2,m}+a_\tau+a_0\cos\sigma t_1 & a_0\sin\sigma t_1 \end{pmatrix}, \tag{5.103}$$

where

$$a_\tau = \frac{\tau(p^2+q^2)^3}{2\omega_0[1+\kappa(p^2+q^2)]}. \tag{5.104}$$

5.2.2 Reconstruction of asymptotic solution

To construct a solution at $q \approx z_0$ (case III), we assume the following relations:

$$\begin{aligned} &\widetilde{\mu} = \mu^{2/3}, \quad q = q_* + \widetilde{\mu}^2\widetilde{q}, \quad \widetilde{f_1} = \widetilde{\mu}^{-1/2} f_1, \\ &T_1(\varphi,t) = f_0(\varphi) + \widetilde{\mu}^2 \widetilde{f_1}(\varphi)\cos(\Omega t), \\ &\Omega = 2\omega_0 + \widetilde{\mu}^2\widetilde{\sigma}, \quad \widetilde{\sigma}\sim 1 \quad \text{as} \quad \widetilde{\mu}\to 0, \\ &\zeta = \widetilde{\mu}^{-1}(\varphi-\varphi_0), \quad t_0 = t, \quad \widetilde{t_1} = \widetilde{\mu}^2 t, \end{aligned} \tag{5.105}$$

where $q_* = z_0$.

A solution of Eqs. (5.71), satisfying the conditions

$$\chi,\ \Phi \to 0 \quad \text{as} \quad |\zeta|\to\infty, \tag{5.106}$$

can be constructed in the following form (we recall that the symbol $\widehat{\ }$ was omitted above):

$$\chi(\varphi,t) = \sum_{j=0}^{\infty}\widetilde{\mu}^j\chi_j\left(\zeta,t_0,\widetilde{t_1}\right), \quad \Phi(\varphi,t) = \sum_{j=0}^{\infty}\widetilde{\mu}^j\Phi_j\left(\zeta,t_0,\widetilde{t_1}\right). \tag{5.107}$$

As above, we chose the characteristic dimension $R = R_2(\varphi_0)$. Then $k(\varphi_0) = 1$. The substitution of Eqs. (5.105), (5.107) into the governing equations (5.71) produces the system of differential equations (5.79), where

$$
\begin{aligned}
&L_0 z = \tfrac{\partial^2 z}{\partial t_0^2} + \left[1 - f_0\left(\varphi_0\right) q_\star^2 + \tfrac{q_\star^4}{1+\kappa q_\star^2}\right] z,\\
&L_1 z = \left[2k'\left(\varphi_0\right) - f_0'\left(\varphi_0\right) q_\star^2\right] \zeta z,\\
&L_2 z = \tfrac{1}{1+\kappa q_\star^2} \left(1 + \tfrac{\kappa}{q_\star^2} + \tfrac{3}{q_\star^4}\right) \tfrac{\partial^4 z}{\partial \zeta^4} + \tfrac{1}{1+\kappa q_\star^2} \left\{ \tfrac{4k'(\varphi_0)\zeta}{q_\star^2}\right.\\
&\left. + \zeta\kappa \left[f_0'\left(\varphi_0\right) q_\star^2 + 2k'\left(\varphi_0\right)\right]\right\} \tfrac{\partial^2 z}{\partial \zeta^2}\\
&+ \tfrac{1}{2}\left[2k''\left(\varphi_0\right) + 2k'^2\left(\varphi_0\right) - f_0''\left(\varphi_0\right) q_\star^2\right] \zeta^2 z\\
&+ 2q_\star \widetilde{q} \left[\tfrac{q_\star^2(2+\kappa q_\star^2)}{(1+\kappa q_\star^2)^2} - f_0\left(\varphi_0\right)\right] z + 2\tfrac{\partial^2 z}{\partial t_0 \partial \widetilde{t_1}} - \widetilde{f_1}\left(\varphi_0\right) q_\star^2 \cos\left(\Omega t_0\right) z,
\end{aligned}
\tag{5.108}
$$

When $j = 0$, one gets the homogeneous differential equation $L_0 \chi_0 = 0$. Its solution can be written as follows:

$$
\chi_0\left(\zeta, t_0, \widetilde{t_1}\right) = \chi_{0,c}\left(\zeta, \widetilde{t_1}\right) \cos\left(\omega_0 t_0\right) + \chi_{0,s}\left(\zeta, \widetilde{t_1}\right) \sin\left(\omega_0 t_0\right), \tag{5.109}
$$

$$
\omega_0^2 = 1 - f_0\left(\varphi_0\right) q_\star^2 + \frac{q_\star^4}{1 + \kappa q_\star^2}. \tag{5.110}
$$

For $j = 1$, one arrives at the inhomogeneous differential equation

$$
L_0 \chi_1 = -\left(2k'\left(\varphi_0\right) - f_0'\left(\varphi_0\right) q_\star^2\right) \zeta \chi_0. \tag{5.111}
$$

The right-hand side of this equation generates secular terms. The following equation is the condition of their absence

$$
2k'\left(\varphi_0\right) - f_0'\left(\varphi_0\right) q_\star^2 = 0. \tag{5.112}
$$

In what follows, we consider the particular case, when

$$
k'\left(\varphi_0\right) = f_0'\left(\varphi_0\right) = 0. \tag{5.113}
$$

Then the solution of Eq. (5.79) at $j = 1$ can be written as

$$
\chi_1\left(\zeta, t_0, \widetilde{t_1}\right) = \chi_{1,c}\left(\zeta, \widetilde{t_1}\right) \cos\left(\omega_0 t_0\right) + \chi_{1,s}\left(\zeta, \widetilde{t_1}\right) \sin\left(\omega_0 t_0\right). \tag{5.114}
$$

For $j = 2$, we arrive again at the inhomogeneous differential equation

$$
\begin{aligned}
&L_0 \chi_2 = -\left[N_s \sin\left(\omega_0 t_0\right) + N_c \cos\left(\omega_0 t_0\right)\right]\\
&+ \tfrac{1}{2} q_\star^2 \widetilde{f_1}\left(\varphi_0\right) \left[\chi_{0,s} \sin\left(3\omega_0 t_0 + \widetilde{\sigma} \widetilde{t_1}\right) + \chi_{0,c} \cos\left(3\omega_0 t_0 + \widetilde{\sigma} \widetilde{t_1}\right)\right] = 0
\end{aligned}
\tag{5.115}
$$

with

$$N_s = \frac{1}{1+\kappa q_\star^2}\left(1+\frac{\kappa}{q_\star^2}+\frac{3}{q_\star^4}\right)\frac{\partial^4\chi_{0,s}}{\partial\zeta^4}+\frac{1}{2}\left[2k''(\varphi_0)-f_0''(\varphi_0)\,q_\star^2\right]\zeta^2\chi_{0,s}$$
$$+2q_\star\widetilde{q}\left[\frac{2}{q_\star^2}-f_0(\varphi_0)\right]\chi_{0,s}-2\omega_0\frac{\partial\chi_{0,c}}{\partial\widetilde{t_1}}$$
$$+\frac{1}{2}\widetilde{f_1}(\phi_0)\,q_\star^2\left[\chi_{0,s}\cos\left(\widetilde{\sigma}\widetilde{t_1}\right)+\chi_{0,c}\sin\left(\widetilde{\sigma}\widetilde{t_1}\right)\right]$$
$$N_c = \frac{1}{1+\kappa q_\star^2}\left(1+\frac{\kappa}{q_\star^2}+\frac{3}{q_\star^4}\right)\frac{\partial^4\chi_{0,c}}{\partial\zeta^4}+\frac{1}{2}\left[2k''(\varphi_0)-f_0''(\varphi_0)\,q_\star^2\right]\zeta^2\chi_{0,c}$$
$$+2q_\star\widetilde{q}\left[\frac{2}{q_\star^2}-f_0(\varphi_0)\right]\chi_{0,c}+2\omega_0\frac{\partial\chi_{0,s}}{\partial\widetilde{t_1}}$$
$$+\frac{1}{2}\widetilde{f_1}(\varphi_0)\,q_\star^2\left[\chi_{0,s}\sin\left(\widetilde{\sigma}\widetilde{t_1}\right)-\chi_{0,c}\cos\left(\widetilde{\sigma}\widetilde{t_1}\right)\right]. \tag{5.116}$$

To eliminate the secular solutions generated by the fist term on the right-hand side of (5.115), we assume the conditions $N_c = 0, N_s = 0$ which result in the following vector differential equation

$$A\frac{\partial^4\mathbf{X}}{\partial\zeta^4}+\frac{1}{2}\left[2k''(\varphi_0)-f_0''(\varphi_0)\,q_\star^2\right]\zeta^2\mathbf{X}+B\mathbf{X}+2\omega_0\mathbf{E}'\frac{\partial\mathbf{X}}{\partial\widetilde{t_1}}+\mathbf{G}\mathbf{X}=0, \tag{5.117}$$

with respect to the vector $\mathbf{X} = (\chi_{0,s},\chi_{0,c})^T$, where

$$A = \frac{1}{1+\kappa q_\star^2}\left(1+\frac{\kappa}{q_\star^2}+\frac{3}{q_\star^4}\right), \quad B = 2q_\star\widetilde{q}\left[\frac{2}{q_\star^2}-f_0(\varphi_0)\right],$$
$$\mathbf{G} = \frac{1}{2}\widetilde{f_1}(\varphi_0)\,q_\star^2\begin{pmatrix}\cos\widetilde{\sigma}\widetilde{t_1} & \sin\widetilde{\sigma}\widetilde{t_1}\\ \sin\widetilde{\sigma}\widetilde{t_1} & -\cos\widetilde{\sigma}\widetilde{t_1}\end{pmatrix}, \tag{5.118}$$

and the matrix $\mathbf{E}'$ is introduced by (5.49).

Applying the Fourier transform (5.50) to Eq. (5.117), we arrive at the differential equation

$$\left[\frac{1}{2}f_0''(\varphi_0)\,q_\star^2-k''(\varphi_0)\right]\frac{\partial^2\mathbf{X}^F}{\partial\eta^2}+2\omega_0\mathbf{E}'\frac{\partial\mathbf{X}^F}{\partial\widetilde{t_1}}+\left(A\eta^4+B+\mathbf{G}\right)\mathbf{X}^F=0 \tag{5.119}$$

with respect to the vector-function

$$\mathbf{X}^F\left(\eta,\widetilde{t_1}\right) = \left(\chi_s^F\left(\eta,\widetilde{t_1}\right),\chi_c^F\left(\eta,\widetilde{t_1}\right)\right)^{\mathrm{T}}. \tag{5.120}$$

It should be noticed that for $\kappa = 0$, Eq. (5.119) coincides with Eq. (5.48) derived above for a single layer isotropic shell without taking into account a shear.

We introduce the new variable

$$x = \varsigma\eta, \quad \varsigma = \sqrt[6]{A}\left(k''(\varphi_0)-\frac{1}{2}f_0''(\varphi_0)\,q_\star^2\right)^{-1/6}. \tag{5.121}$$

Now Eq. (5.119) can be rewritten as follows:

$$\frac{\partial^2 \mathbf{X}^F}{\partial x^2} - \frac{2\omega_0 \varsigma^4}{A} E' \frac{\partial \mathbf{X}^F}{\partial \widetilde{t_1}} - \frac{\varsigma^4}{A} \left(A\varsigma^{-4} x^4 + B + \mathbf{G} \right) \mathbf{X}^F = 0. \tag{5.122}$$

A solution of above equation can be constructed in the following form

$$\mathbf{X}^F = y_m(x) Y_m(\widetilde{t_1}), \quad \mathbf{Y}_m = \left(S_m(\widetilde{t_1}), C_m(\widetilde{t_1}) \right)^T, \tag{5.123}$$

where $y_m\,(m = 0, 1, 2, ...)$ is an eigenfunction of the boundary-value problem (5.56), and $\mathbf{Y}_m$ is a solution of Eqs. (5.31) with periodic matrix (5.103), where σ and t_1 should be replaced by $\widetilde{\sigma}$ and $\widetilde{t_1}$, respectively, and the parameters $a_0, a_1, a_{2,m}$ are introduced as follows:

$$a_0 = \frac{\widetilde{f_1}(\varphi_0)\, q_\star^2}{4\omega_0}, \quad a_{2,m} = \frac{\lambda_m A^{1/3}}{2\omega_0} \left[k''(\varphi_0) - \frac{1}{2} f_0''(\varphi_0)\, q_\star^2 \right]^{2/3} + \frac{B}{2\omega_0}. \tag{5.124}$$

Finally, an approximate solution of Eqs. (5.71) can written in the form (5.58):

$$\chi = \sin\left(\frac{\pi n s}{l}\right) \left\{ \left[S_m(\widetilde{\mu}^2 t) \sin(\omega_0 t) + C_m(\widetilde{\mu}^2 t) \cos(\omega_0 t) \right] Z_m(\varphi) + O(\widetilde{\mu}) \right\}, \tag{5.125}$$

where

$$Z_m(\varphi) = \frac{1}{\sqrt{2\pi}} \int_{-\infty}^{+\infty} y_m \left\{ \sqrt[6]{A} \left[k''(\varphi_0) - \frac{1}{2} f_0''(\varphi_0)\, q_\star^2 \right]^{-1/6} \eta \right\} e^{i\eta\widetilde{\mu}^{-1}(\varphi - \varphi_0)} d\eta \tag{5.126}$$

5.2.3 Main region of instability

It is seen that matrices (5.32) and (5.103) coincide with an accuracy up to constant a_τ. It is obvious that the main regions of parametric instability for a layered shell, taking into account the effect of a shift, are obtained by shifting all regions in Fig. 5.2 to the right by the value $2a_\tau$. In other words, the σa_0 - plane is divided by the two lines

$$a_0 = \pm \left(a_{2,m} + a_\tau - \frac{\sigma}{2} \right) \tag{5.127}$$

into the following two domains:

$$\begin{aligned} \mathcal{D}^- &= \left\{ (\sigma, a_0) : \left| a_{2,m} + a_\tau - \tfrac{\sigma}{2} \right| < |a_0| \right\}, \\ \mathcal{D}^+ &= \left\{ (\sigma, a_0) : \left| a_{2,m} + a_\tau - \tfrac{\sigma}{2} \right| > |a_0| \right\}. \end{aligned} \tag{5.128}$$

On lines (5.127), solutions Eqs. (5.77), (5.125) are bounded and close to harmonic functions; if a point $M(\sigma, a_0) \in \mathcal{D}^+$, then solutions are decreasing functions, and in the domain $\mathcal{D}^-$ they grow unboundadly. Thus, the domain $\mathcal{D}^-$

corresponds to the parametrically unstable vibrations localized in the neighborhood of the weakest generatrix $\varphi = \varphi_0$. The farther a point $M(\sigma, a_0) \in \mathcal{D}^-$ from lines (5.127) is situated, the faster the amplitude of the resonance parametric vibrations increases.

5.3 Parametric vibrations of cylindrical and conical shells under pulsing pressure

In this section, the problems of parametric instability of the dynamic momentless stress state of thin conical and cylindrical shells are considered. They are characterized by the fact that waves are elongated along asymptotic lines and localized in the neighborhood of the weakest one of them. The additional stress state arising at parametric resonance is semi-momentless. In contrast to the problems considered in the previous section, it is assumed that of all the in-plane forces arising in a shell under dynamic loading, the forces in the circumferential direction are decisive. The method is applicable to conical and cylindrical shells of medium length. The original 2D problem is reduced to a sequence of 1D boundary-value problems on the weakest generatrix. For cylindrical shells under certain particular assumptions, the asymptotic solution is obtained explicitly.

5.3.1 Parametric vibrations of conical shells

Consider a thin non-circular conical shell of thickness h. On the reference surface, we introduce an orthogonal system of curvilinear coordinates s, φ, where $s = R^{-1}s'$, s' is the distance to the cone vertex, φ is the coordinate on the directrix, which is chosen in such a way that the first quadratic form of the surface had the form $d\sigma^2 = R^2\left(ds^2 + s^2 d\varphi^2\right)$ with R being the characteristic dimension of the shell. The curvature radii are $R_1 = \infty$, $R_2 = Rsk^{-1}(\varphi)$. The shell is assumed to be closed in the direction of φ and bounded by two edges,

$$s_1(\varphi) \leq s \leq s_2(\varphi), \quad 0 \leq \varphi \leq \varphi_1, \tag{5.129}$$

where φ_1 is the length of the curve formed when the cone and the sphere of unit radius with the centre at the vertex of this cone intersect.

Let the shell be subject to external slowly varying dynamic forces that lead to the emergence of the periodic membrane stress resultants

$$T_i(s, \varphi, t) = \varepsilon^6 E_0 h_0 R^{-1}[T_i^\circ(s, \varphi) + \varepsilon T_i^c(s, \varphi)\cos\Omega t], \quad i = 1, 2, 3. \tag{5.130}$$

Here $\varepsilon^8 = h_0^2/[12R^2(1 - \nu_0^2)]$ is the small parameter, h_0, E_0, ν_0 are the characteristic values of variable thickness $h(s, \varphi)$, Young's modulus $E(s, \varphi)$ and

Poisson's ratio $\nu(s,\varphi)$, respectively, Ω is the dimensionless frequency of excitation, t is time related to the characteristic time $t_c = R\sqrt{\rho_0/E_0}$, and ρ_0 is the material density. It is assumed also that $T_i^\circ, T_i^c = O(1)$.

The dynamic response of the shell can be described by Eqs. (2.48). Taking into account the presence of the initial membrane stress resultants (5.130), they should be modified. In the dimensionless form they read as

$$\varepsilon^4\Delta\left(d\Delta w\right) + \varepsilon^2\Delta_{\mathrm{T}} w + \Delta_k\Phi + \gamma\frac{\partial^2 w}{\partial t^2} = 0,$$
$$\varepsilon^4\Delta\left(g^{-1}\Delta\Phi\right) - \Delta_k w = 0, \tag{5.131}$$

where w, Φ are the normal displacement and stress function related to R and $\varepsilon^4 R^2 E_0 h_0$, respectively. The differential operators are introduced by

$$\Delta z = \frac{1}{s^2}\frac{\partial^2 z}{\partial\varphi^2} + \frac{1}{s}\frac{\partial z}{\partial s} + \frac{\partial^2 z}{\partial s^2}, \quad \Delta_k z = \frac{k(\varphi)}{s}\frac{\partial^2 z}{\partial s^2},$$
$$\Delta_{\mathrm{T}} z = \frac{1}{s}\left[\frac{1}{s}\frac{\partial}{\partial\varphi}\left(T_2\frac{\partial z}{\partial\varphi}\right) + \frac{\partial}{\partial s}\left(T_3\frac{\partial z}{\partial\varphi}\right) + \frac{\partial}{\partial\varphi}\left(T_3\frac{\partial z}{\partial s}\right) + \frac{\partial}{\partial s}\left(sT_1\frac{\partial z}{\partial s}\right)\right], \tag{5.132}$$

and the dimensionless variable coefficients are as follows:

$$d = \frac{Eh^3\left(1-\nu_0^2\right)}{E_0 h_0^3\left(1-\nu^2\right)}, \quad g = \frac{Eh}{E_0 h_0}, \quad \gamma = \frac{\rho h}{\rho_0 h_0}. \tag{5.133}$$

The functions $k(\varphi), s_i(\varphi), E(s,\varphi), \nu(s,\varphi), \rho(s,\varphi), T_i^\circ(s,\varphi), T_i^c(s,\varphi)$ are assumed to be infinitely differentiable.

Different variants of boundary conditions were discussed in Chapter 4. Here, we consider only simply supported edges. So, to study the basic dynamic stress-strain state with an error up to values of order ε^2, one needs to satisfy conditions

$$w = \Phi = 0. \tag{5.134}$$

Let at first $T_i^c = 0$ for $i = 1, 2, 3$. We assume that the shell loaded by only static forces has the natural modes of low-frequency vibrations localized in the neighborhood of some line $\varphi = \varphi_0$ being the weakest. This localization may be induced by both the shell geometry and the non-uniform character of loading. We study the dynamic stability of these eigenmodes under the presence of the periodic components in (5.130).

Presupposing the localized character of excited parametric vibrations, we seek a solution of Eqs. (5.131) in the form:

$$w(s,\varphi,t,\varepsilon) = w_* \exp\left[\mathrm{i}\left(\varepsilon^{-1/2}p\xi + \tfrac{1}{2}b\xi^2\right)\right],$$
$$w_* = \sum_{j=0}^{\infty}\varepsilon^{j/2}w_j(s,\xi,t_0,t_1,\ldots), \quad \xi = \varepsilon^{-1/2}(\varphi-\varphi_0), \tag{5.135}$$
$$\mathrm{i} = \sqrt{-1}, \quad \Im b > 0, \quad t_m = \varepsilon^m t, \quad m = 0, 1, 2, \ldots,$$

where $w_j(s, \xi, t_0, t_1, ...)$ are polynomials in ξ with coefficients depending on s, t_m. The function Φ is sought in the same form (5.135), where w_j are replaced by Φ_j.

It is assumed that the parameters p, b and functions w_j, Φ_j satisfy estimates (5.8) at some initial interval of time.

To define the required p, b, w_j, Φ_j, we substitute expansion (5.135) into Eqs. (5.131) and equate the coefficients with the same powers of $\varepsilon^{1/2}$. As a result, we arrive at the sequence of differential equations (5.9). However, now

$$\mathbf{L}_0 z = \mathbf{D} z + s\,\gamma(s, \varphi_0)\frac{\partial^2 z}{\partial t_0^2}, \tag{5.136}$$

where

$$\mathbf{D} z = \frac{k^2(\varphi_0)}{p^4}\frac{\partial^2}{\partial s^2}\left[s^3 g(s, \varphi_0)\frac{\partial^2 z}{\partial s^2}\right] + \frac{p^4 d(s, \varphi_0)}{s^3} z - \frac{p^2}{s} T_2^\circ(s, \varphi_0) z, \tag{5.137}$$

and operators $\mathbf{L}_j$ for $j \geq 1$ are introduced by Eqs. (4.21) in which

$$\mathbf{N} = 2s\frac{\partial^2}{\partial t_0 \partial t_1} - \frac{p^2}{s} T_2^c(s, \varphi_0)\cos\Omega t + ip\left[\frac{\partial T_3^\circ(s, \varphi_0)}{\partial s} + 2T_3^\circ(s, \varphi_0)\frac{\partial}{\partial s}\right]. \tag{5.138}$$

The substitution of (5.135) into the boundary conditions (5.134) produces the sequence of boundary conditions for w_j:

$$\sum_{m=0}^{j} \mathbf{\Gamma}_m w_{j-m} = \sum_{m=0}^{j} \mathbf{\Gamma}_m \frac{\partial^2 w_{j-m}}{\partial s^2} = 0, \quad s = s_1(\varphi_0), s = s_2(\varphi_0), \tag{5.139}$$

where operators $\mathbf{\Gamma}_m$ are defined as follows:

$$\mathbf{\Gamma}_0 = 1, \quad \mathbf{\Gamma}_1 = \xi s'(\varphi_0)\frac{\partial}{\partial s}, \quad \mathbf{\Gamma}_2 = \frac{\xi^2}{2}\left(s'^2(\varphi_0)\frac{\partial^2}{\partial s^2} + s''(\varphi_0)\frac{\partial}{\partial s}\right), \ldots \tag{5.140}$$

Consider the boundary-value problems (5.9), (5.139) step-by-step for $j = 0, 1, 2, ...$

In the leading order approximation, one has the homogeneous boundary-value problem

$$\begin{aligned} &\frac{k^2(\varphi_0)}{p^4}\frac{\partial^2}{\partial s^2}\left[s^3 g(s, \varphi_0)\frac{\partial^2 w_0}{\partial s^2}\right] + \frac{p^4 d(s, \varphi_0)}{s^3} w_0 \\ &-\frac{p^2}{s} T_2^\circ(s, \varphi_0) w_0 + s\,\gamma(s, \varphi_0)\frac{\partial^2 w_0}{\partial t_0^2} = 0, \\ &w_0 = \frac{\partial^2 w_0}{\partial s^2} = 0, \qquad s = s_1(\varphi_0),\ s_2(\varphi_0). \end{aligned} \tag{5.141}$$

Its solution may be presented in the form:

$$w_0(s,\xi,t_0,t_1,...) = y(s,\varphi_0)P_0(\xi,t_0,t_1,...),$$
$$P_0(\xi,t_0,t_1,...) = P_{0,c}(\xi,t_1,t_2,..)\cos\omega_0 t_0 + P_{0,s}(\xi,t_1,t_2,..)\sin\omega_0 t_0, \tag{5.142}$$

where $P_{0,c}(\xi,t_1,t_2,..)$, $P_{0,s}(\xi,t_1,t_2,..)$ are polynomials in ξ with coefficients depending on $t_m (m = 1,2,\ldots)$, ω_0 is the zero order approximation for the natural frequency of the shell vibrations, and $y(s,\varphi_0)$ is an eigenfunction of the homogeneous boundary-value problem

$$\mathbf{D}y - s\gamma(s,\varphi_0)\lambda_0 y = 0,$$
$$y = \frac{\partial^2 y}{\partial s^2} = 0 \quad \text{at} \quad s = s_1(\varphi_0),\ s_2(\varphi_0), \tag{5.143}$$

where $\lambda_0 = \omega_0^2$.

We note that λ_0 is the function of the parameters p,φ_0. As in the previous section, we will study parametric vibrations as a form of quasi-stationary vibrations with frequency ω_0° close to the lowest eigenfrequency vibrations localized near the weakest generatrix $\varphi = \varphi_0$. Let

$$\lambda_0^\circ = \min_{p,\varphi_0} \lambda_0(p,\varphi_0) = \lambda_0(p^\circ,\varphi_0^\circ), \tag{5.144}$$

where p°,φ_0° are determined from the equations

$$\frac{\partial\lambda_0}{\partial\varphi_0} = \frac{\partial\lambda_0}{\partial p} = 0. \tag{5.145}$$

To calculate the derivatives in (5.145), we differentiate the boundary-value problem (5.143) with respect to p and φ_0:

$$\mathbf{D}y_p - s\gamma(s,\varphi_0)\lambda_0 y_p + \frac{\partial\mathbf{D}}{\partial p}y - s\,\gamma(s,\varphi_0)\frac{\partial\lambda_0}{\partial p}y = 0,$$
$$y_p = \frac{\partial^2 y_p}{\partial s^2} = 0 \quad \text{for} \quad s = s_j(\varphi_0), \tag{5.146}$$

$$\mathbf{D}y_\varphi - s\gamma(s,\varphi_0)\lambda_0 y_\varphi + \frac{\partial\mathbf{D}}{\partial\varphi_0}y - s\,\gamma(s,\varphi_0)\frac{\partial\lambda_0}{\partial\varphi_0}y = 0,$$
$$y_\varphi + s'(\varphi_0)\frac{\partial y}{\partial s} = \frac{\partial^2 y_\varphi}{\partial s^2} + s'(\varphi_0)\frac{\partial^3 y}{\partial s^3} = 0 \quad \text{for} \quad s = s_j(\varphi_0). \tag{5.147}$$

Problems (5.146), (5.147) are the inhomogeneous boundary-value problems "on the spectrum." Conditions for the existence of solutions of these problems read

$$\int_{s_1}^{s_2} y\frac{\partial\mathbf{D}}{\partial p}y\,ds = 0,$$
$$\int_{s_1}^{s_2} y\frac{\partial\mathbf{D}}{\partial\varphi_0}y\,ds + \frac{2k^2(\varphi_0)s^3 s'}{p^4}\frac{\partial y}{\partial s}\frac{\partial^3 y}{\partial s^3}\Big|_{s_1}^{s_2} = 0. \tag{5.148}$$

We note that Eqs. (5.148) are equivalent to conditions (5.145). Let the characteristic dimension of the shell reference surface be equal to $R = R_2(s, \varphi_0^\circ)/s$, then $k(\varphi_0^\circ) = 1$. In what follows, the superscribe $^\circ$ in λ_0°, p°, φ_0° will be omitted.

For $j = 1$, one has the inhomogeneous boundary-value problem (5.9), (5.139):

$$\mathbf{L}_0 w_1 + \mathbf{L}_1 w_0 = 0,$$
$$\mathbf{\Gamma}_0 w_1 + \mathbf{\Gamma}_1 w_0 = \mathbf{\Gamma}_0 \frac{\partial^2 w_1}{\partial s^2} + \mathbf{\Gamma}_1 \frac{\partial^2 w_0}{\partial s^2} = 0 \qquad s = s_i(\varphi_0). \tag{5.149}$$

Note that the resolvability conditions for the boundary-value problem (5.149) coincide with equalities (5.145) as well. The solution of problem (5.149) is sought in the form

$$w_1(s, \xi, t_m) = \left[b\xi y_p + \xi y_\varphi - \mathrm{i} y_p \frac{\partial}{\partial \xi} \right] P_0(\xi, t_m) + y P_1(\xi, t_m),$$
$$P_1(\xi, t_m) = P_{1,c}(\xi, t_1, t_2, ..) \cos \omega_0 t_0 + P_{1,s}(\xi, t_1, t_2, ..) \sin \omega_0 t_0, \tag{5.150}$$

where $m = 0, 1, 2, ...,$ $P_{1,c}(\xi, t_1, t_2, ...)$, $P_{1,s}(\xi, t_1, t_2, ..)$ are polynomials in ξ with coefficients depending on $t_1, t_2, ...$, and $y_p(s, \varphi_0)$, $y_\varphi(s, \varphi_0)$ are the eigenfunctions of the boundary-value problems (5.146) and (5.147), respectively.

As in Section 5.3, consider the case when $\Omega \approx 2\omega_0$. We perform the detuning frequency of the excited vibrations

$$\Omega = 2\omega_0 + \varepsilon\sigma, \quad \sigma \sim 1 \quad \text{at} \quad \varepsilon \to 0. \tag{5.151}$$

Then, taking into account (5.148), (5.150), the inhomogeneous boundary-value problem (5.9), (5.139), arising in the second-order approximation ($j = 2$), can be written down in the form:

$$\mathbf{L}_0 w_2 = N^{(1)}(s, \varphi_0, \xi, t_0, t_1, ...) + N^{(2)}(s, \varphi_0, \xi, t_0, t_1, ...), \tag{5.152}$$

$$w_2 + [\mathbf{G} P_{0,c} + \Gamma_1 y(s) P_{1,c}] \cos \omega_0 t_0$$
$$+ [\mathbf{G} P_{0,s} + \Gamma_1 y(s) P_{1,s}] \sin \omega_0 t_0 = 0 \quad \text{for} \quad s = s_i(\varphi_0),$$
$$\frac{\partial^2 w_2}{\partial s^2} + \frac{\partial^2}{\partial s^2} [\mathbf{G} P_{0,c} + \Gamma_1 y(s) P_{1,c}] \cos \omega_0 t_0 \tag{5.153}$$
$$+ \frac{\partial^2}{\partial s^2} [\mathbf{G} P_{0,s} + \Gamma_1 y(s) P_{1,s}] \sin \omega_0 t_0 = 0 \quad \text{for} \quad s = s_i(\varphi_0),$$

where

$$
\begin{aligned}
N^{(1)}(s,\varphi_0,\xi,t_0,t_1,...) &= \frac{p^2 y(s,\varphi_0)T_2^c(s,\varphi_0)}{2s}[P_{0,c}\cos(3\omega_0 t_0+\sigma t_0) \\
&\quad + P_{0,s}\sin(3\omega_0 t_0+\sigma t_0)], \\
N^{(2)}(s,\varphi_0,\xi,t_0,t_1,...) &= -\left[\mathbf{M}P_{0,c}+\mathbf{L}_1 y(s)P_{1,c}\right]\cos\omega_0 t_0 \\
&\quad -\left[\mathbf{M}P_{0,s}+\mathbf{L}_1 y(s)P_{1,s}\right]\sin\omega_0 t_0 \\
&\quad -y(s)\left[2\omega_0\frac{\partial P_{0,s}}{\partial t_1}+\frac{p^2T_2^c(\varphi_0)}{2s}(P_{0,s}\sin\sigma t_1-P_{0,c}\cos\sigma t_1)\right]\cos\omega_0 t_0 \\
&\quad +y(s)\left[2\omega_0\frac{\partial P_{0,c}}{\partial t_1}-\frac{p^2T_2^c(\varphi_0)}{2s}(P_{0,s}\cos\sigma t_1+P_{0,c}\sin\sigma t_1)\right]\sin\omega_0 t_0.
\end{aligned}
\tag{5.154}
$$

Here $\mathbf{M},\mathbf{G}$ are the differential operators which are introduced as follows:

$$
\begin{aligned}
&\mathbf{M} = \xi^2(b^2\mathbf{M}_p + b\mathbf{M}_{p\varphi}+\mathbf{M}_\varphi) \\
&-\mathrm{i}\xi(2b\mathbf{M}_p+\mathbf{M}_{p\varphi})\frac{\partial}{\partial\xi}-\mathrm{i}(b\mathbf{M}_p+\mathbf{M}_{p\varphi})-\mathbf{M}_p\frac{\partial^2}{\partial\xi^2}, \\
&\mathbf{M}_p = \frac{1}{2}\frac{\partial^2\mathbf{D}}{\partial p^2}y+\frac{\partial\mathbf{D}}{\partial p}y_p, \quad \mathbf{M}_\varphi = \frac{1}{2}\frac{\partial^2\mathbf{D}}{\partial\varphi_0^2}y+\frac{\partial\mathbf{D}}{\partial\varphi_0}y_\varphi, \\
&\mathbf{M}_{p\varphi} = \frac{\partial^2\mathbf{D}}{\partial\varphi_0\partial p}y+\frac{\partial\mathbf{D}}{\partial\varphi_0}y_p+\frac{\partial\mathbf{D}}{\partial p}y_\varphi, \\
&\mathbf{G} = \xi s'\left(b\xi\frac{\partial y_p}{\partial s}+\xi\frac{\partial y_\varphi}{\partial s}-\mathrm{i}\frac{\partial y_p}{\partial s}\frac{\partial}{\partial\xi}\right)+\Gamma_2 y\,.
\end{aligned}
\tag{5.155}
$$

The solution of problem (5.152), (5.153) can be represented as a sum of two functions

$$
w_2(s,\xi,t_0,t_1,...) = w_2^{(1)}(s,\xi,t_0,t_1,...)+w_2^{(2)}(s,\xi,t_0,t_1,...), \tag{5.156}
$$

where $w_2^{(1)}$ is a general solution of the inhomogeneous boundary-value problem

$$
\begin{aligned}
&\mathbf{L}_0 w_2 = N^{(1)}(s,\varphi_0,\xi,t_0,t_1,...), \\
&w_2 = \frac{\partial^2 w_2}{\partial s^2} = 0 \quad \text{for} \quad s=s_i(\varphi_0),
\end{aligned}
\tag{5.157}
$$

and $w_2^{(2)}$ is a partial solution of the inhomogeneous differential equation

$$\mathbf{L}_0 w_2 = N^{(2)}(s, \varphi_0, \xi, t_0, t_1, ...) \tag{5.158}$$

with the boundary conditions (5.153).

It is obvious that the solution to problem (5.157) exists and is a periodic in time function. Consider the inhomogeneous problem (5.158), (5.153). Proceeding from the Fredholm alternative, it is easy to show that its solution has the form

$$w_2^{(2)} = P_{2,c}(s, \xi, t_1, t_2, ...) \cos \omega_0 t_0 + P_{2,s}(s, \xi, t_1, t_2, ...) \sin \omega_0 t_0, \tag{5.159}$$

if the vector-function $\mathbf{X} = (P_{0,s}, P_{0,c})^T$ satisfies the differential equation

$$\begin{gathered} -\frac{1}{2}\frac{\partial^2 \lambda_0}{\partial p^2}\frac{\partial^2 \mathbf{X}}{\partial \xi^2} + a\xi\frac{\partial \mathbf{X}}{\partial \xi} + \left(c\xi^2 + \frac{1}{2}a + \eta\right)\mathbf{X} \\ +2\omega_0 \begin{pmatrix} 0 & -1 \\ 1 & 0 \end{pmatrix} \frac{\partial \mathbf{X}}{\partial t_1} + r \begin{pmatrix} \cos \sigma t_1 & \sin \sigma t_1 \\ \sin \sigma t_1 & -\cos \sigma t_1 \end{pmatrix} \mathbf{X} = 0 \end{gathered} \tag{5.160}$$

with

$$a = -\mathrm{i}\left(b\frac{\partial^2 \lambda_0}{\partial p^2} + \frac{\partial^2 \lambda_0}{\partial \varphi_0 \partial p}\right), \quad c = \frac{1}{2}\left(b^2\frac{\partial^2 \lambda_0}{\partial p^2} + 2b\frac{\partial^2 \lambda_0}{\partial \varphi_0 \partial p} + \frac{\partial^2 \lambda_0}{\partial \varphi_0^2}\right),$$

$$r = \frac{p^2}{2z}\int_{s_1}^{s_2} \frac{T_2^c y^2}{s} ds, \quad z = \int_{s_1}^{s_2} s\gamma(s, \varphi_0) y^2 ds,$$

$$\eta = \frac{\mathrm{i}}{2z}\left[\int_{s_1}^{s_2} y\left(\frac{\partial \mathbf{D}}{\partial \varphi_0} y_p - \frac{\partial \mathbf{D}}{\partial p} y_\varphi\right) ds + \frac{s^3 s'}{p^4}\left(\frac{\partial y_p}{\partial s}\frac{\partial^3 y}{\partial s^3} + \frac{\partial^3 y_p}{\partial s^3}\frac{\partial y}{\partial s}\right)\Bigg|_{s_1}^{s_2}\right]. \tag{5.161}$$

To calculate the derivatives

$$\frac{\partial^2 \lambda_0}{\partial p^2}, \frac{\partial^2 \lambda_0}{\partial \varphi_0 \partial p}, \frac{\partial^2 \lambda_0}{\partial \varphi_0^2},$$

we differentiate problems (5.146), (5.147) with respect to p and φ_0 once more. In this way, we arrive again at an inhomogeneous boundary value-problems with respect to the functions $y_{pp}, y_{p\varphi}, y_{\varphi\varphi}$. From the solvability conditions of these problems (which are not written out here), we obtain the following relations:

$$\frac{\partial^2 \lambda_0}{\partial p^2} = \frac{1}{z}\int_{s_1}^{s_2} y\left(2\frac{\partial \mathbf{D}}{\partial p}y_p + \frac{\partial^2 \mathbf{D}}{\partial p^2}y\right)ds,$$

$$\begin{aligned}\frac{\partial^2 \lambda_0}{\partial \varphi_0^2} &= \frac{1}{z}\int_{s_1}^{s_2} y\left(2\frac{\partial \mathbf{D}}{\partial \varphi_0}y_\varphi + \frac{\partial^2 \mathbf{D}}{\partial \varphi_0^2}y\right)ds \\ &+ \frac{s^3 s'}{zp^4}\left[\frac{\partial^3 y}{\partial s^3}\left(2s'\frac{\partial y_\varphi}{\partial s} + s'^2\frac{\partial^2 y}{\partial s^2} + s''\frac{\partial y}{\partial s}\right)\right. \\ &+ \left.\left.\frac{\partial y}{\partial s}\left(2s'\frac{\partial^3 y_\varphi}{\partial s^3} + s'^2\frac{\partial^4 y}{\partial s^4} + s''\frac{\partial^3 y}{\partial s^3}\right)\right]\right|_{s_1}^{s_2},\end{aligned} \tag{5.162}$$

$$\begin{aligned}\frac{\partial^2 \lambda_0}{\partial \varphi_0 \partial p} &= \frac{1}{z}\int_{s_1}^{s_2} y\left(\frac{\partial^2 \mathbf{D}}{\partial \varphi_0 \partial p}y + \frac{\partial \mathbf{D}}{\partial \varphi_0}y_p + \frac{\partial \mathbf{D}}{\partial p}y_\varphi\right)ds \\ &+ \frac{s^3 s'}{zp^4}\left.\left(\frac{\partial^3 y}{\partial s^3}\frac{\partial y_p}{\partial s} + \frac{\partial y}{\partial s}\frac{\partial^3 y_p}{\partial s^3}\right)\right|_{s_1}^{s_2}.\end{aligned}$$

In order for the system of equations (5.160) to have solutions in the form of polynomials in ξ, it is necessary to assume $c = 0$. From here, we determine the unique parameter b such that $\Im b > 0$.

The solution of the coupled equations (5.160) can be written in the explicit form (5.30)

$$\mathbf{X} = \mathcal{H}_{\mathbf{m}}(\theta\xi)\mathbf{Y}_{\mathbf{m}}, \tag{5.163}$$

where $\mathcal{H}_m(x)$ is the Hermite polynomial of degree m, and

$$\theta = \left[\frac{\partial^2 \lambda_0}{\partial p^2}\frac{\partial^2 \lambda_0}{\partial \varphi_0^2} - \left(\frac{\partial^2 \lambda_0}{\partial \varphi_0 \partial p}\right)^2\right]^{1/4}\left(\frac{\partial^2 \lambda_0}{\partial p^2}\right)^{-1/2}. \tag{5.164}$$

Substituting (5.164) into (5.160) yields the homogeneous system of differential equations (5.31) with the periodic matrix (5.32), where

$$a_0 = \frac{r}{2\omega_0}, \quad a_1 = 0, \quad a_{2,m} = \frac{2m+1}{4\omega_0}a + \frac{\eta}{2\omega_0}. \tag{5.165}$$

Finally, the shape of the conical shell deflection under parametric vibrations localized near the weakest generatrix $\varphi = \varphi_0$ has the form

$$\begin{aligned}w &= y(s,\varphi_0)\exp\left\{\mathrm{i}\varepsilon^{-1}\left[p(\varphi-\varphi_0) + \frac{1}{2}b(\varphi-\varphi_0)^2\right]\right\} \\ &\times \left\{\mathcal{H}_m\left[\varepsilon^{-1/2}(\varphi-\varphi_0)\theta\right]\left[S_m(\varepsilon t)\sin(\omega_0 t) + C_m(\varepsilon t)\cos(\omega_0 t)\right] + O\left(\varepsilon^{1/2}\right)\right\}.\end{aligned} \tag{5.166}$$

As in Section 5.2, solution (5.166) does not take into account the dependence on $t_2 = \varepsilon^2 t, t_3 = \varepsilon^3 t, \ldots$. To determine this dependence, it is necessary to continue the procedure of constructing the functions w_j in the higher-order approximations by means of eliminating the arising secular terms (see Remark 5.1).

5.3.2 Parametric vibrations of a cylindrical shell under pulsing pressure

Consider a medium-length cylindrical shell of thickness h, whose the shape and dimensions are determined as

$$s_1(\varphi) \le s \le s_2(\varphi), \quad d\sigma^2 = R^2(ds^2 + d\varphi^2), \quad R_2 = Rk^{-1}(\varphi),$$

where, R is the characteristic dimension of a shell. In particular, for the circular cylindrical shell, R is its radius, and $k \equiv 1$.

All the formulas derived in the previous subsection may be readily used to study the localized parametric vibrations of the cylindrical shell. To make this transition, the factor s^m appearing explicitly in the differential operators (5.136)-(5.138) should be substituted by unity.

We consider here the case when the shell is under the action of a pulsing pressure q_n^* which is non-uniform in the circumferential direction:

$$q_n^* = \varepsilon^6 EhR^{-1} q, \quad q = q_0(\varphi) + \varepsilon q_c(\varphi) \cos \Omega t. \tag{5.167}$$

Here q is the dimensionless counterpart of q_n^*, ε is the small parameter introduced above, and Ω and t are the dimensionless frequency of excitation and time, respectively.

If q_0, q_c are the slowly varying functions of φ, and the frequency Ω/t_* belongs to the lowest part of the eigenfrequency spectrum of the shell (here, t_* is the characteristic time introduced in Subsection 5.3.1), then pressure (5.167) results in the momentless dynamic stress state which is specified by the in-plane hoop stress resultant

$$T_2 = T_2^\circ(\varphi) + \varepsilon T_2^c(\varphi) \cos \omega t, \tag{5.168}$$

where

$$T_2^\circ(\varphi) = q^\circ(\varphi) k^{-1}(\varphi), \quad T_2^c(\varphi) = q^c(\varphi) k^{-1}(\varphi) \tag{5.169}$$

are the dimensionless amplitudes of its static and periodic components.

Now the boundary-value problem (5.141) reads

$$\begin{gathered} \frac{k^2(\varphi_0)}{p^4} \frac{\partial^4 w_0}{\partial s^4} + p^4 w_0 - p^2 T_2^\circ(\varphi_0) w_0 + \frac{\partial^2 w_0}{\partial t_0^2} = 0, \\ w_0 = \frac{\partial^2 w_0}{\partial s^2} = 0 \quad \text{for} \quad s = s_1(\varphi_0), \, s_2(\varphi_0). \end{gathered} \tag{5.170}$$

This problem has a solution in the form of (5.142), where in contrast to a conical shell, the eigenvalue ω_0 and the associated eigenfunction $y(s, \varphi_0)$ can be written out explicitly

$$\omega_0 = \lambda_0^{1/2} = \sqrt{p^4 - p^2 T_2^\circ(\varphi_0) + \frac{\Lambda_n^4 k^2(\varphi_0)}{p^4 l^4(\varphi_0)}}, \qquad (5.171)$$

$$y(s, \varphi_0) = \sin \frac{\Lambda_n [s - s_1(\varphi_0)]}{l(\varphi_0)}, \quad l(\varphi_0) = s_2(\varphi_0) - s_1(\varphi_0),$$

where $\Lambda_n = \pi n \, (n = 1, 2, ...)$. For other variants of boundary conditions, formula (5.171) for ω_0 differs only in the value of Λ_n. So, for the boundary conditions

$$w_0 = \frac{\partial w_0}{\partial s} = 0 \quad \text{for} \quad s = s_1(\varphi_0), \, s_2(\varphi_0), \qquad (5.172)$$

and

$$w_0 = \frac{\partial^2 w_0}{\partial s^2} = 0 \quad \text{for} \quad s = s_1(\varphi_0),$$
$$w_0 = \frac{\partial w_0}{\partial s} = 0 \quad \text{for} \quad s = s_2(\varphi_0), \qquad (5.173)$$

one has $\Lambda_1 \approx 4,730$ and $\Lambda_1 \approx 3,927$, respectively, and the eigenfunction $y(s)$ is a solution of the equation

$$y^{IV} - \frac{\Lambda_n^4}{l(\varphi_0)^4} y = 0 \qquad (5.174)$$

with the required variant of boundary conditions at $s = s_1(\varphi_0), s_2(\varphi_0)$.

Now, Eqs. (5.145) for determining parameters p°, φ_0° have the form:

$$-\frac{4\Lambda_n^4 k^2(\varphi_0)}{p^5 l^4(\varphi_0)} + 4p^3 - 2pT_2^\circ(\varphi_0) = 0,$$

$$l(\varphi_0)\left[\frac{\Lambda_n^4 k(\varphi_0) k'(\varphi_0)}{p^4 l^4(\varphi_0)} - \frac{1}{2} p^2 T_2^{\circ\prime}(\varphi_0)\right] \qquad (5.175)$$

$$-\frac{2\Lambda_n^4 k^2(\varphi_0)}{p^4 l^4(\varphi_0)} \left[s_2^3(\varphi_0) s_2'(\varphi_0) - s_1^3(\varphi_0) s_1'(\varphi_0)\right] = 0.$$

In Eqs. (5.175) and below, the prime $'$ denotes the differentiation with respect to φ_0, and the superscript $^\circ$ is omitted.

It is seen from (5.171), that the static component $q(\varphi_0)$ of the external pressure (5.167) ought to be less than the critical value

$$q = \frac{4\Lambda_1 \sqrt[4]{3k^2(\varphi_0)}}{3l(\varphi_0)} \qquad (5.176)$$

resulting in the buckling of a circular cylindrical shell of radius $R/k(\varphi_0)$ and the constant length $Rl(\varphi_0)$ of generatrix [226]. Note that the dimension counterpart of (5.176) for a simply supported cylinder ($\Lambda_1 = \pi$) reads

$$Q = \frac{\pi\sqrt{6}Eh}{9L(1-\nu^2)^{3/4}}\left(\frac{h}{R}\right)^{3/2}. \tag{5.177}$$

Equation (5.177) is the well-known Southwell-Papkovich formula for the critical buckling pressure [77].

In the following, we consider the particular case with

$$k'(\varphi_0) = s_1'(\varphi_0) = s_2'(\varphi_0) = T_2^{\circ\prime}(\varphi_0) = 0. \tag{5.178}$$

Now Eq. (5.175) for determining the weakest generatrix $\varphi = \varphi_0^\circ$ is satisfied. In this case, correlations (5.162) can be written in the explicit form:

$$\begin{gathered}\frac{\partial^2\lambda_0}{\partial p^2} = \frac{20\Lambda_n^4}{p^6 l^4} + 12p^2 - 2T_2^\circ, \\ \frac{\partial^2\lambda_0}{\partial \varphi_0^2} = \frac{2\Lambda_n^4 k}{p^4 l^2}\left(\frac{k}{l^2}\right)'' - p^2 T_2^{\circ\prime\prime}, \qquad \frac{\partial^2\lambda_0}{\partial\varphi_0\partial p} = 0.\end{gathered} \tag{5.179}$$

Also, we readily determine

$$r = \frac{1}{2}p^2T_2^\circ, \quad \eta = 0. \tag{5.180}$$

In Eqs. (5.179), (5.180), all functions of the argument φ are calculated at $\varphi = \varphi_0^\circ$. As the next step, using Eqs. (5.165), we can find the coefficients $a_0, a_{2,m}$ of the coupled equations (5.31) with respect to the vector function $\mathbf{Y}_m = (S_m(t_1), C_m(t_1))^{\mathrm{T}}$ and the boundaries for the main region of localized parametric instability as well.

5.3.3 Examples

The boundaries of the interval $\Omega^- \le \Omega \le \Omega^+$ for dimensionless excitation frequency $\Omega = \Omega^* t_c$ corresponding to the parametric instability of the shell, are determined by the formulas:

$$\Omega^\pm = 2\omega_0 + \sqrt[8]{\frac{h^2}{12R^2(1-\nu^2)}}\,\sigma^\pm, \quad \sigma^\pm = 2(a_{2,m} \pm \sqrt{a_0^2 - a_1^2}), \tag{5.181}$$

where parameters $\omega_0, a_0, a_{2,m}$ are found from Eqs. (5.171), (5.165).

Consider some examples on parametric instability of a cylindrical shell.

Example 5.3. Let a circular cylindrical shell ($k \equiv 1$) bounded by the plane edges $s_1 \equiv 0$, $s_2 \equiv l$ be in the momentless dynamic stress state characterized by the hoop stress resultant

$$T_2(\varphi, t) = \frac{1}{2}(1 + \cos\varphi) + 2\varepsilon\cos\Omega t \tag{5.182}$$

the static component of which is non-uniformly distributed in the circumferential direction.

In this case, the weakest generatrix is the line $\varphi_0 = 0$ on which the static component of force (5.182) is maximal. In the vicinity of this generatrix, at some initial interval of time, one has the localization of excited parametric vibrations.

Figure 5.8 shows the boundaries $\Omega^{\pm}$ corresponding to the main regions of parametric instability versus the shell length for three variants of boundary conditions. The calculations were performed for $h/R = 0.01$, $\nu = 0.3$. Regions S-S and C-C correspond to the cases when both edges either simply supported or clamped, respectively, and C-S does to the combined boundary conditions. It may be seen that increasing the shell length results in some expanding of the parametric instability regions, and differences caused by the way of ends restraint become less essential that coincides with the conclusions made in paper [170].

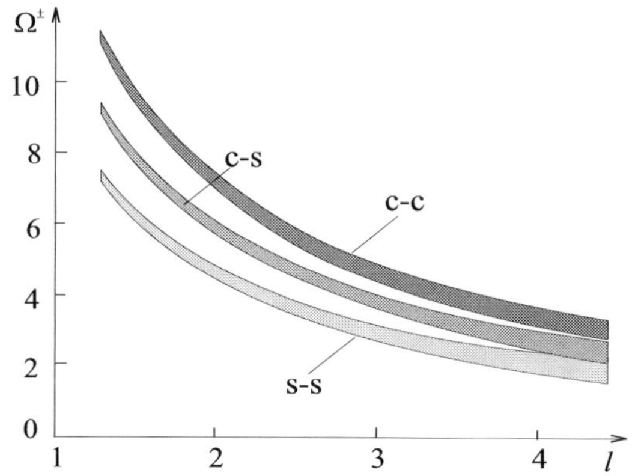

FIGURE 5.8: Interval boundaries for the excitation frequency Ω corresponding to the parametric resonance versus the shell length for different variants of boundary conditions.

Example 5.4. Consider a circular cylindrical shell, in which one edge is straight ($s_2 \equiv l$), and the second one is slanted:

$$s_1(\varphi) = -(1 + \cos\varphi)\tan\theta, \tag{5.183}$$

where θ is the cut angle.

The shell is under pulsing external pressure which generates in the middle surface the hoop stress resultant

$$T_2(t) = 2\varepsilon\cos\Omega t. \tag{5.184}$$

In this case, the longest generatrix $\varphi_0 = 0$ is the weakest one. In Figure 5.9, the influence of the inclination angle θ on the boundaries of the main region of parametric instability for various variants of boundary conditions is demonstrated. The figure shows that with increasing the inclination angle of the

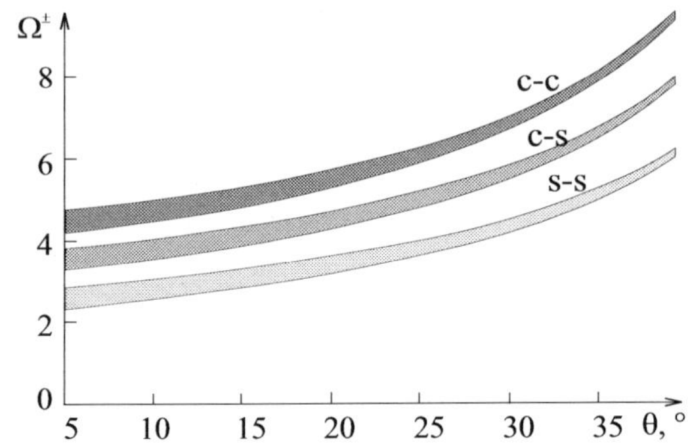

FIGURE 5.9: Interval boundaries for the excitation frequency Ω corresponding to the parametric resonance versus the inclination angle θ,° for different variants of boundary conditions.

shell edge, the boundaries of the main regions of local parametric instability narrow, and the difference between these regions for shells with different boundary conditions becomes more noticeable.

5.4 Parametric vibrations of nearly cylindrical shells

Here, we consider localized parametric vibrations of a shell which is close in a shape to the base non-circular cylindrical surface and is affected by the periodic membrane forces $T_i\,(i = 1, 2, 3)$.

We introduce the curvilinear coordinates s, φ so that the first quadratic form of the base surface is $R^2(ds^2 + d\varphi^2)$, where R is the characteristic dimension of the base surface. The geometry of the shell middle surface which is close to the base one is described by Eq. (4.61). Then the coefficients of the first quadratic form of the shell middle surface will be determined by formulas (4.62), where $F(s, \varphi) \sim 1$ is a function describing the deviation shape of the shell surface from the base one. The order of the amplitude of the deviation δ of the middle surface from the base one is taken equal to (4.63).

To describe the parametric vibrations of the shell under the action of periodic membrane stress resultants, we use Eqs. (5.131) as the governing ones, in which it is assumed that $d = g = \gamma = 1$:

$$\begin{gathered}\varepsilon^4\Delta^2 w + \varepsilon^2\Delta_{\rm T} w + \Delta_k\Phi + \frac{\partial^2 w}{\partial t^2} = 0,\\ \varepsilon^4\Delta^2\Phi - \Delta_k w = 0.\end{gathered} \tag{5.185}$$

The operators $\Delta, \Delta_{\rm T}, \Delta_k$ in the accepted coordinate system are given in [222]. Setting the goal to construct the leading approximation for the mode of localized parametric vibrations as well as to find the main region of instability, we shall omit the terms in these operators that do not appear in the asymptotic construction used below. So we can assume

$$\Delta z = \frac{\partial^2 z}{\partial s^2} + \frac{\partial^2 z}{\partial \varphi^2}, \quad \Delta_R z = \varepsilon^2 \frac{\partial}{\partial \varphi}\left[k_1(s,\varphi)\frac{\partial z}{\partial \varphi}\right] + k(\varphi)\frac{\partial^2 z}{\partial s^2},$$
$$\Delta_{\rm T} z = \left[\frac{\partial}{\partial \varphi}\left(T_2\frac{\partial z}{\partial \varphi}\right) + \frac{\partial}{\partial s}\left(T_3\frac{\partial z}{\partial \varphi}\right) + \frac{\partial}{\partial \varphi}\left(T_3\frac{\partial z}{\partial s}\right) + \frac{\partial}{\partial s}\left(T_1\frac{\partial z}{\partial s}\right)\right], \tag{5.186}$$

where $k(\varphi)$ is the curvature of the base cylindrical surface, and $k_1(s,\varphi) = -\partial^2 F/\partial s^2$.

Let the membrane stress resultants be given by (5.130). The shell edges $s = s_1(\varphi)$, $s = s_2(\varphi)$, which are not necessarily plane curves, are assumed to be simply supported.

A solution of Eqs. (5.185) is sought in the form (5.135). The procedure for finding unknown parameters and functions in ansatz (5.131) remains the same. Substituting (5.135) into (5.185) leads to the sequence of differential equations (5.9), where

$$\mathbf{L}_0 = \frac{\partial^2}{\partial t_0^2} + \mathbf{D}, \tag{5.187}$$

$$\begin{aligned}\mathbf{D} = {} & \frac{k^2(\varphi_0)}{p^4}\frac{\partial^4}{\partial s^4} - \frac{2k(\varphi_0)}{p^2}\frac{\partial}{\partial s}\left[k_1(s,\varphi_0)\frac{\partial}{\partial s}\right]\\ & + p^4 - p^2 T_2^\circ(s,\varphi_0) + k_1^2(s,\varphi_0) - \frac{1}{p^2}\frac{\partial^2 k_1(s,\varphi_0)}{\partial s^2},\end{aligned} \tag{5.188}$$

and the operators $\mathbf{L}_j$ for $j \geq 1$ are introduced by formulas (4.21). However, now the operator $\mathbf{N}$ appearing in $\mathbf{L}_2$ is defined as

$$\begin{aligned}\mathbf{N} = {} & 2\frac{\partial^2}{\partial t_0 \partial t_1} - p^2 T_2^c(s,\varphi_0)\cos\Omega t + {\rm i}p\left[\frac{\partial T_3^\circ(s,\varphi_0)}{\partial s} + 2T_3^\circ(s,\varphi_0)\frac{\partial}{\partial s}\right]\\ & - \frac{4{\rm i}k(\varphi_0)}{p^3}\left[\frac{\partial^3 k_1(s,\varphi_0)}{\partial^2 s\,\partial\varphi_0} + 2\frac{\partial^2 k_1(s,\varphi_0)}{\partial s\,\partial\varphi_0}\frac{\partial}{\partial s}\right]\\ & + \frac{2{\rm i}\,k'(\varphi_0)}{p^3}\left[\frac{\partial k_1(s,\varphi_0)}{\partial s}\frac{\partial}{\partial s} + \frac{\partial^2 k_1(s,\varphi_0)}{\partial s^2}\right].\end{aligned} \tag{5.189}$$

We write down the boundary-value problem arising in the leading approximation:

$$\frac{k^2(\varphi_0)}{p^4}\frac{\partial^4 w_0}{\partial s^4} - \frac{2k(\varphi_0)}{p^2}\frac{\partial}{\partial s}\left[k_1(s,\varphi_0)\frac{\partial w_0}{\partial s}\right]$$
$$+\left[p^4 - p^2 T_2^\circ(s,\varphi_0) + k_1^2(s,\varphi_0) - \frac{1}{p^2}\frac{\partial^2 k_1(s,\varphi_0)}{\partial s^2}\right] w_0 + \frac{\partial^2 w_0}{\partial t_0^2} = 0,$$
$$w_0 = \frac{\partial^2 w_0}{\partial s^2} = 0 \quad \text{for} \quad s = s_1(\varphi_0),\, s_2(\varphi_0). \tag{5.190}$$

Its solution can be found in the form of (5.142), where $\omega_0 = \lambda_0^{1/2}$ is the zero approximation for the natural frequency of the shell, and $y(s,\varphi_0)$ is the eigenfunction of the homogeneous boundary-value problem

$$\mathbf{D}y - \lambda_0 y = 0, \quad y = \frac{\partial^2 y}{\partial s^2} = 0 \quad \text{for} \quad s = s_1(\varphi_0),\, s_2(\varphi_0). \tag{5.191}$$

Constructing the next two approximations in the case when $\Omega \approx 2\omega_0$ can be implemented using the formulas obtained in Subsection 5.3.1. For this purpose, in all relations it is necessary to replace the factor s^m appearing explicitly by unity, and also take into account Eqs. (5.188), (5.189).

If the functions T_2°, k_1 do not depend on s, then the boundary-value problem (5.191) has a solution in the explicit form. Consider a particular case when the deviation shape of the shell surface from the base cylinder (see Eqs. (4.62) is defined by the function

$$F(s,\varphi) = \left(1 - \frac{s^2}{l_1^2}\right) f(\varphi), \quad -l_1 \le s_1(\varphi) \le s \le s_2(\varphi) \le l_1. \tag{5.192}$$

Now $k_1 = 2f(\varphi)/l_1^2$. Hence, one obtains

$$y = \sin\frac{\pi n\,[s - s_1(\varphi_0)]}{l(\varphi_0)}, \quad l(\varphi_0) = s_2(\varphi_0) - s_1(\varphi_0),$$
$$\lambda_0 = \left[\frac{\pi^2 n^2 k(\varphi_0)}{p^2 l^2(\varphi_0)} + \frac{2f(\varphi_0)}{l_1^2}\right]^2 + p^4 - T_2^\circ(\varphi_0)p^2, \tag{5.193}$$

where n is a natural number. Substituting formula (5.193) for λ_0 into conditions (5.145), one arrives at the following two equations for determining the parameters p°, φ_0°:

$$\frac{\pi^4 n^4}{p^4}\frac{d}{d\varphi_0}\left(\frac{k^2}{l^4}\right) + \frac{4\pi^2 n^2}{l_1^2 p^2}\frac{d}{d\varphi_0}\left(\frac{kf}{l^2}\right) - p\frac{dT_2^\circ}{d\varphi_0} + \frac{8}{l_1^4}\frac{df}{d\varphi_0} = 0,$$
$$2p^8 - T_2^\circ p^6 - \frac{4\pi^2 n^2 kf}{l_1^2 l^2}p^2 - \frac{2\pi^4 n^4 k^2}{l^4} = 0. \tag{5.194}$$

Let us study examples of the influence of the shell shape and degree of its deviation from the base cylindrical surface on the main region of local parametric instability.

Example 5.5. Consider the shell of revolution with a parabolic flexure (see Fig. 5.10), for which the distance from the base circular cylinder (here $k = 1$) to the middle surface is determined as

$$r(s) = \varepsilon^2 F(s), \quad F(s) = \kappa\left(1 - \frac{s^2}{l_1^2}\right). \tag{5.195}$$

For $\kappa > 0$, we have the barrel-shaped shell, and for $\kappa < 0$, one has the shell

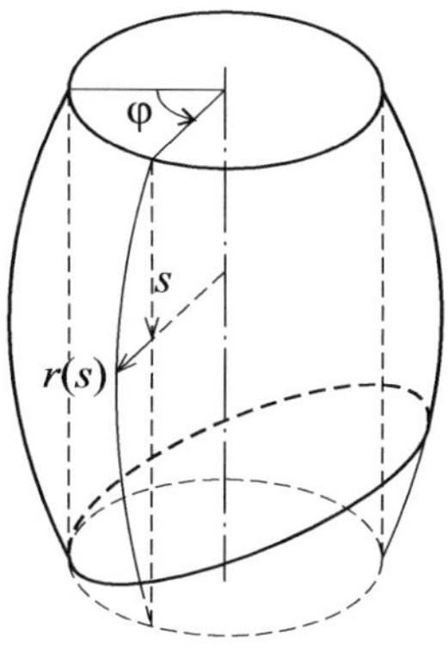

FIGURE 5.10: Shell of revolution close to cylindrical surface with a parabolic bend.

of the negative Gaussian curvature. It is assumed that one of the shell edges is straight and the second one is oblique:

$$s_1(\varphi) = -l_1 + (1 - \cos\varphi)\tan\theta, \quad s_2 = l_1, \tag{5.196}$$

where α is the slope angle. Here the maximum length of the generatrix of the base cylinder is $l = 2l_1$ and it is attained when $\varphi = 0$.

Let the shell be under the normal periodic pressure

$$q_n = 1 + \varepsilon \cos \Omega t. \tag{5.197}$$

Then

$$T_2^\circ = T_2^c = 1 + O(\varepsilon^2), \quad T_1 = T_3 = O(\varepsilon^2). \tag{5.198}$$

The first equation in (5.194) shows that the weakest generatrix is the longest one, $\varphi_0 = 0$. In the vicinity of this generatrix, it is observed the localization of parametric vibrations with forms determined by expansion (5.135).

Formulas (5.181) for the boundaries of the main region of parametric instability allow us to determine the boundaries of the interval $\Omega^- \leq \Omega \leq \Omega^+$ for the dimensionless excitation frequency $\Omega = \Omega^* t_c$ corresponding to the dynamic instability of the shell. Figure 5.11 shows the boundaries of the main region of parametric instability versus a parameter κ characterizing the depth of deviation of the shell surface from the base cylinder for $\theta = 30°$ and different values of the dimensionless length l. Similar curves illustrating the effect of an angle θ on the boundaries $\Omega^\pm$ at the fixed $\kappa = -0.5$ and different l are drawn in Fig. 5.12 In both figures, the dotted lines correspond to values of the double

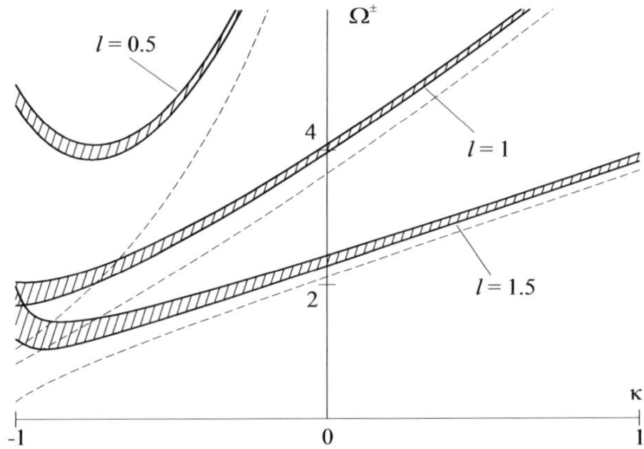

FIGURE 5.11: Boundaries of the main region of parametric instability for the shell of revolution versus a parameter κ for the fixed angle θ and different values of l.

natural frequency of vibrations of the shell ($2\omega_0$). It can be seen from figures that for $\kappa < 0$ the boundaries of dynamic instability expand. It should be noted that for concave shells ($\kappa < 0$), an error of the obtained formulas increases together with a parameter $|\kappa|$, since for such shells assumption (5.198) about the smallness of stresses T_1, T_3 compared to T_2 is not fulfilled .

Example 5.6. Let us now consider the parametric vibrations of a nearly cylindrical shell with a slightly curved axis (see Fig. 4.8) under the action of the periodic pressure (5.197). It is assumed that $k = 1$, l_1 are constant values, and the form of deviation of the shell surface from the base cylindrical one is

$$F(s, \varphi) = \left(1 - \frac{s^2}{l_1^2}\right) f(\varphi), \quad f(\varphi) = -\kappa \cos(\varphi) \tag{5.199}$$

with $\kappa > 0$, $-l_1 \leq s \leq l_1$.

If the deviation of the shell middle surface from the base cylinder is of the order ε^2, then (5.198) hold. As follows from Eqs. (5.194), here the weakest

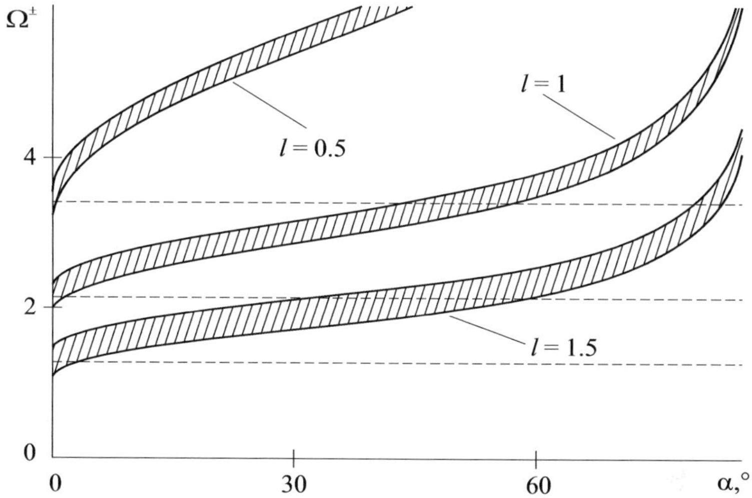

FIGURE 5.12: Boundaries of the main region of parametric instability for the shell of revolution versus an angle θ for the fixed $\kappa = -0.5$ and different values of l.

generatrix is the line $\varphi = 0$, where the shell surface is most curved with the negative Gaussian curvature.

Figure 5.13 shows the boundaries of the main region of parametric instability versus a parameter κ for different values of l.

5.5 Localized axisymmetric vibrations of long cylindrical shells resting on elastic foundation

In this section, we study localized parametric vibrations of a very long cylindrical shell of radius R, whose thickness $h(x)$, Young's modulus $E(x)$, Poisson's ratio $\nu(x)$ and material density $\rho(x)$ are functions of the axial coordinate $x = Rs$. The shell rests on a non-uniform elastic foundation or contains an elastic filler with a variable spring "constant" $c_f(x)$ and is pre-stressed by membrane forces T_1°, T_2° which are the periodic functions of time.

To study axisymmetric vibrations, we shall apply the Flügge equations in the form (3.123). Introduce the dimensionless functions $d(s), g(s), \gamma(s), c(s)$ in accordance with (3.125) and dimensionless periodic stress resultants

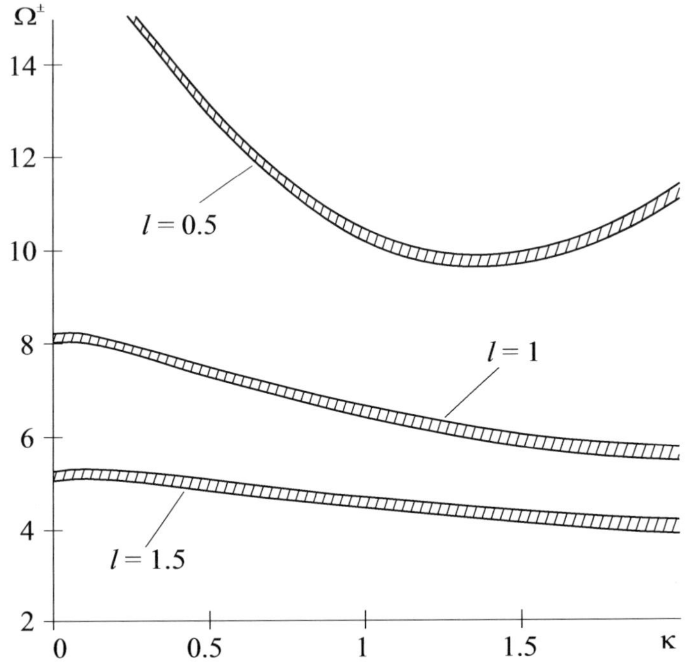

FIGURE 5.13: Boundaries of the main region of parametric instability for the nearly cylindrical shell with a slightly curved axis versus a parameter $\kappa = -0.5$ and different values of l.

$f_1(s,t), f_2(s,t)$:

$$T_1^\circ = \mu_1^2 E_0 h_0 f_1(s,t), \quad T_2^\circ = E_0 h_0 f_2(s,t),$$
$$f_1(s,t) = f_{10}(s) + \mu_1 f_{1c}(s)\cos\Omega t, \tag{5.200}$$
$$f_2(s,t) = f_{20}(s) + \mu_1 f_{2c}(s)\cos\Omega t,$$

where $s = x/R$ is the dimensionless axial coordinate, $t = t^* t_c^{-1}$ is the dimensionless time, $t_c = R\sqrt{\rho_0/E_0}$ is the characteristic time, h_0, ρ_0, E_0 are the characteristic values of the thickness, material density and Young's modulus, respectively, $\mu_1^4 = h_0^2/(12R^2)$ is the small parameter, and Ω is the frequency of pulsation of the membrane forces.

If $f_{1c} = f_{2c} = 0$, and the remaining parameters are variable, then we arrive at the problem on free localized vibrations considered in Section 3.10. Let $s = s_0$ be the weakest parallel in the neighborhood of which it is observed the localization of free vibrations with eigenfrequency ω_0. We will show here, that if $f_{1c} \neq 0, f_{2c} \neq 0$ and $\Omega \approx 2\omega_0$, then the constructed above natural modes (3.131) might be dynamically unstable.

To examine bending vibrations, we assume in (3.128) the following estimates for displacements:

$$u_1 = \mu_1 R u, \quad u_3 = R w, \quad \text{where} \quad u, w \sim 1 \quad \text{at} \quad \mu_1 \to 0. \tag{5.201}$$

A solution of Eqs. (3.128) satisfying conditions (3.130) is sought in the form of asymptotic expansion

$$w(s, t, \mu_1) = \sum_{j=0}^{\infty} \mu_1^{j/2} w_j(\xi, t_0, t_1) \exp\left\{ \mathrm{i}\left(\mu_1^{-1/2} p\xi + \frac{1}{2} b\xi^2 \right) \right\}, \quad \operatorname{Im} b > 0, \tag{5.202}$$

where $\xi = \mu_1^{-1/2}(s - s_0)$, $t_0 = t$, $t_1 = \mu_1 t$, and w_j are polynomials in ξ with coefficients depending on the "slow" time t_1. The longitudinal displacement u is assumed in the same form (5.202).

The substitution of (5.201), (5.202) into Eqs. (3.128) results in the sequence of differential equations (5.9). The operator $\mathbf{L}_0$ of the leading approximation has the from (5.10), where

$$H(p, s_0) = \sqrt{\frac{d(s_0)p^4 + [1 - \nu^2(s_0)]g(s_0) + c(s_0) - f_{20}(s_0) - f_{10}(s_0)p^2}{\gamma(s_0)}}. \tag{5.203}$$

The operators $\mathbf{L}_j$ for $j \geq 1$ are defined in accordance with (4.21) with

$$\mathbf{N}z = 2\frac{\partial^2 z}{\partial t_0 \partial t_1} - \frac{1}{\gamma(s_0)}\{[f_{2c}(s_0) + f_{1c}(s_0)p^2] \cos \Omega t + 2d'(s_0)p^3\}z. \tag{5.204}$$

In the leading approximation, one has the homogeneous differential equation

$$\frac{\partial^2 w_0}{\partial t_0^2} + H^2(p, s_0) w_0 = 0. \tag{5.205}$$

Its solution reads

$$\begin{gathered} w_0 = P_{0,c}(\xi, t_1) \cos \omega_0 t_0 + P_{0,s}(\xi, t_1) \sin \omega_0 t_0, \\ \omega_0 = H(p, s_0), \end{gathered} \tag{5.206}$$

where $P_{0,c}, P_{0,s}$ are the unknown polynomials which will be defined below.

For $j = 1$, Eq. (5.9) is inhomogeneous. With relations (5.206) taken into account, the right-hand side of this equation generates secular solutions with respect to time t_0. The conditions for the absence of these solutions lead to the relation (3.136) for the wave number

$$p^\circ = \sqrt{\frac{f_{10}}{2d(s_0^\circ)}}, \tag{5.207}$$

and the equation

$$\gamma\left[\frac{d'}{4d^2}f_{10}^2+c'-f_{20}'-\frac{f_{10}f_{10}'}{2d}-(\nu^2 g)'\right] \\ -\gamma'\left[(1-\nu^2)g+c-f_{20}-\frac{f_{10}^2}{4d}\right]=0. \tag{5.208}$$

for seeking the weakest parallel $s=s_0$. The prime in Eq. (5.208) and below means the differentiation of a function with respect to s. Let $\omega_0^\circ=H(p^\circ,s_0^\circ)$. In what follows, the superscript $^\circ$ is omitted.

As before, consider the case of $\Omega\approx 2\omega_0$ which is of most interest and introduce the detuning frequency of the periodic forces, $\Omega=2\omega_0+\mu\sigma$. Then, taking into account (5.207), (5.208), equation (5.9) for $j=2$ reads

$$\mathbf{L}_0 w_2=-(N_c\cos\omega_0 t_0+N_s\sin\omega_0 t_0) \\ +r[P_{0,c}\cos(3\omega_0 t_0+\sigma t_1)+P_{0,s}\sin(3\omega_0 t_0+\sigma t_1)], \tag{5.209}$$

$$N_c=-\frac{1}{2}\frac{\partial^2 H}{\partial p^2}\frac{\partial^2 P_{0,c}}{\partial\xi^2}+a\xi\frac{\partial P_{0,c}}{\partial\xi}+\left(c\xi^2+\frac{1}{2}a+\eta\right)P_{0,c} \\ +2\omega_0\frac{\partial P_{0,s}}{\partial t_1}+r(P_{0,s}\sin\sigma t_1-P_{0,c}\cos\sigma t_1), \\ N_s=-\frac{1}{2}\frac{\partial^2 H}{\partial p^2}\frac{\partial^2 P_{0,s}}{\partial\xi^2}+a\xi\frac{\partial P_{0,s}}{\partial\xi}+\left(c\xi^2+\frac{1}{2}a+\eta\right)P_{0,s} \\ +2\omega_0\frac{\partial P_{0,c}}{\partial t_1}+r(P_{0,s}\cos\sigma t_1-P_{0,c}\sin\sigma t_1), \\ a=-i\left(b\frac{\partial^2 H}{\partial p^2}+\frac{\partial^2 H}{\partial\varphi_0\partial p}\right),\ c=\frac{1}{2}\left(b^2\frac{\partial^2 H}{\partial p^2}+2b\frac{\partial^2 H}{\partial\varphi_0\partial p}+\frac{\partial^2 H}{\partial\varphi_0^2}\right), \\ r=\frac{1}{2\gamma(s_0)}\left[f_{2c}(s_0)+\frac{f_{1c}(s_0)f_{10}(s_0)}{2d(s_0)}\right],\quad \eta=\frac{2d'(s_0)p^3}{\gamma(s_0)}. \tag{5.210}$$

The first two terms on the right-hand side of equation (5.209) generate secular partial solutions. To eliminate them, one needs to put $N_c=0$, $N_s=0$. Hence, one obtains the system of differential equations (5.25) with respect to the vector $\mathbf{X}=(P_{0,s},\,P_{0,c})^{\mathrm{T}}$, in which the coefficient $r(s_0)$ is calculated by the formula (5.210). Under the condition $c=0$, Eq. (5.25) has a solution in the polynomial form (5.30), where the vector function $\mathbf{Y}_m=(S_m(t_1),\,C_m(t_1))^{\mathrm{T}}$ is again found from the homogeneous system of differential equations (5.31) with a periodic matrix (5.32). Here,

$$a_0=\frac{r(s_0)}{2\omega_0},\quad a_1=0,\quad a_{2,m}=\frac{(2m+1)a+\eta}{2\omega_0}. \tag{5.211}$$

Formulas (5.211) allow one to determine the boundaries for dimensionless frequency of excitation corresponding to the parametric resonance:

$$\Omega^- \le \Omega \le \Omega^+,$$
$$\Omega^\pm = 2\omega_0 + \mu_1\sigma^\pm, \quad \sigma^\pm = 2\,(a_{2,m} \pm a_0)\,. \tag{5.212}$$

Example 5.7. We shall return to the shell considered in Example 3.7. Let E, ρ, ν, c, f_{10}, f_{20}, f_{2c} be constant, $f_{1c} = 0$, and the thickness

$$h(s) = h_0\left(1 + \frac{1}{2}\kappa s^2 + O(s^3)\right)$$

be variable. In this case, only the hoop stress resultant is periodic, and the axial one is stationary. A similar case takes place, for example, in a long pipe lying in a ground, with a flowing fluid under pressure having a weak non-stationary component. We remind that the weakest line is here the parallel $s = s_0 = 0$, and the condition $\Im b > 0$ is equivalent to the inequality $\kappa\left[f_{20} - c + (1-\nu^2)f_{10}^2\right] > 0$. In this example, one obtains

$$b = \frac{\mathrm{i}}{2}\sqrt{\frac{\kappa\left[t_2 - c + (1-\nu^2)t_1^2\right]}{t_1}}, \quad \omega_0 = 1 - f_{20} + c - \frac{1-\nu^2}{4}f_{10}^2,$$
$$a_0 = \frac{f_{2c}}{4\omega_0}, \quad a_{2,m} = (1+2m)\sqrt{\kappa\left[f_{20} - c + (1-\nu^2)f_{10}^2\right]f_{10}}\,. \tag{5.213}$$

As in the problem on free localized vibrations considered in Example 3.7, the localization of parametric vibrations takes place in a vicinity of parallel where the thickness is minimal. If the condition $c > f_{20} + (1-\nu^2)f_{10}^2$ holds, then the parametric resonance may occur and in the neighborhood of the parallel where the thickness is maximum.

Concluding this chapter, we note that in [160], a problem was studied on localized thermo-parametric vibrations of a non-circular cylindrical shell being in a pulsating temperature field. Unlike the algorithm of constructing localized solutions like (5.7), (5.135), in the mentioned paper the form of parametric vibrations localized in the vicinity of the weakest generatrix had been constructed using the multiple-scale method by both time and the circumferential coordinate.

Chapter 6

Wave Packets in Medium-length Cylindrical Shells

This chapter deals with non-stationary localized vibrations propagating in the circumferential direction in medium-length cylindrical shells.

A non-circular thin elastic cylinder with not necessarily plane edges subject to different variants of boundary conditions is considered. The coupled equations of the technical shell theory in the framework of the Kirchhoff-Love model are used to govern dynamic behavior of the shell. The peculiarity of the initial-boundary-value problem is to set the initial distribution of transversal displacements and velocities of points on the middle surface, which are localized near some generatrix of this surface. To study the non-stationary localized dynamics of shells, an asymptotic method being the generalization of the asymptotic approach proposed in Chapter 4 is developed. A solution of the problem is constructed in the form of superposition of *wave packets* (WPs) travelling in the circumferential direction. The asymptotic method allows splitting the initial-boundary-value problem into the sequence of 1D boundary-value problems being considered on the moving generatrix which is the center of running WP. As an alternative approach, the Maslov-WKB method [131] is utilized to study localized waves in a cylindrical shell with the generatrix of constant length and to compare the asymptotic solutions constructed by different methods.

The detailed analysis of WPs behavior versus geometrical parameters of a shell is performed. The effects of reflection of WPs from some generatrix (a rigid area on the shell surface), being accompanied by focusing of WPs and increasing amplitudes are revealed.

6.1 Wave packets in a non-circular cylindrical shell

Consider a non-circular cylindrical shell of constant thickness h. Young's modulus, Poisson's ratio and the mass density are also assumed to be constant. We introduce on the middle surface of the shell an orthogonal coordinate system $s = x/R, \varphi = y/R$, where x and y are the lengths of the segment and arc on the shell surface in the axial and circumferential directions, respectively, and R is the characteristic dimension of the shell. The variable radius of curvature is $R_2 = R/k(\varphi)$. Let the shell be bounded by two not necessarily plane edges and not necessarily closed (in the case of a panel) in the direction of φ:

$$s_1(\varphi) \le s \le s_2(\varphi), \quad \varphi_1 \le \varphi \le \varphi_2.$$

6.1.1 Setting the problem

Assuming the shell to be thin, we introduce the estimates for derivatives with respect to the coordinates x, y and time t^*:

$$\frac{\partial F}{\partial x} \sim R^{-1} h_*^{-\varsigma_1} F, \quad \frac{\partial F}{\partial y} \sim R^{-1} h_*^{-\varsigma_2} F, \quad \frac{\partial F}{\partial t^*} \sim \omega_c h_*^{-\varsigma_t} F, \tag{6.1}$$

where F is any of unknown functions, $\omega_c = \sqrt{E/(\rho R^2)}$ is the characteristic frequency, ς_1, ς_2 and ς_t are indices of variability and dynamism of the stress-strain state (SSS) of the shell [63, 64].

We shall study bending wave forms of motion of the shell for which

$$\varsigma_1 = 0, \quad \varsigma_2 = 1/4, \quad \varsigma_t = -1/2. \tag{6.2}$$

For the above dynamic SSS, it may be used equations of the technical shell theory (2.41). In the dimensionless form, they read [64, 233]

$$\begin{aligned} \varepsilon^4 \Delta^2 W + k(\varphi) \frac{\partial^2 \Phi}{\partial s^2} + \varepsilon^2 \frac{\partial^2 W}{\partial t^2} &= 0, \\ \varepsilon^4 \Delta^2 \Phi - k(\varphi) \frac{\partial W}{\partial s} &= 0. \end{aligned} \tag{6.3}$$

Here ε is the small parameter introduced by (4.13), $t = \varepsilon^3 t^* \omega_c$ is dimensionless time, $W^* = \varepsilon^{-4} R W$, $\Phi^* = R^2 h E \Phi$ are the normal deflection and the stress function, respectively, and W, Φ are their dimensionless counterparts.

On the edges $s = s_1(\varphi)$, $s = s_2(\varphi)$, we consider one of the two groups of boundary conditions corresponding to a simple support and/or rigid clamping. When one of these conditions holds on one of the boundaries, one can consider the group of free edge on the second boundary. As in the problems on low-frequency free or parametric vibrations, we shall study the basic SSS satisfying the main boundary conditions. Up to values of the order of ε^2, these conditions are

$$W = \Phi = 0, \quad W = \frac{\partial W}{\partial s} = 0, \quad \Phi = \frac{\partial \Phi}{\partial s} = 0 \tag{6.4}$$

for the group of simple support, rigid clamping and free edge, respectively.

Consider the following Cauchy problem:

$$\begin{aligned} W|_{t=0} &= W_0(s, \varphi, \varepsilon) \exp\left[\mathrm{i}\varepsilon^{-1} S_0(\varphi)\right], \\ \dot{W}|_{t=0} &= \mathrm{i}\varepsilon^{-1} V_0(s, \varphi, \varepsilon) \exp\left[\mathrm{i}\varepsilon^{-1} S_0(\varphi)\right], \end{aligned} \tag{6.5}$$

where

$$S_0(\varphi) = a_0 \varphi + \frac{1}{2} b_0 \varphi^2, \quad \mathrm{i} = \sqrt{-1}, \quad a_0 \neq 0, \quad \Im b_0 > 0, \tag{6.6}$$

a_0 is a real number, and W_0, V_0 are complex-valued functions infinitely differentiable with respect to φ and s, and satisfying conditions

$$\frac{\partial W_0}{\partial s}, \frac{\partial V_0}{\partial s} \sim 1 \quad \text{as} \quad \varepsilon \to 0. \tag{6.7}$$

The dot in (6.5) and below means differentiation with respect to dimensionless time t. Additional assumptions for the functions W_0, V_0 will be made below.

The real and imaginary parts of functions (6.5) define on the shell surface the initial WPs with the center on the generatrix $\varphi = 0$. With distance from this generatrix, the amplitude of waves in WPs decreases due to inequality (6.6). The odd and even forms of the initial WP for constant values of W_0, V_0 and $\Re b_0 = 0$ are shown in Fig. 6.1. Functions like (6.5) can be used to approximate the response of a shell to a localized short-time force impact [211] or concentrated static forces [127]

In what follows, the value $R = R_2(0)$ is assumed as the characteristic dimension of the shell.

For the first time, the initial-boundary-value problem (6.3)-(6.6) in the given formulation was solved in paper [150].

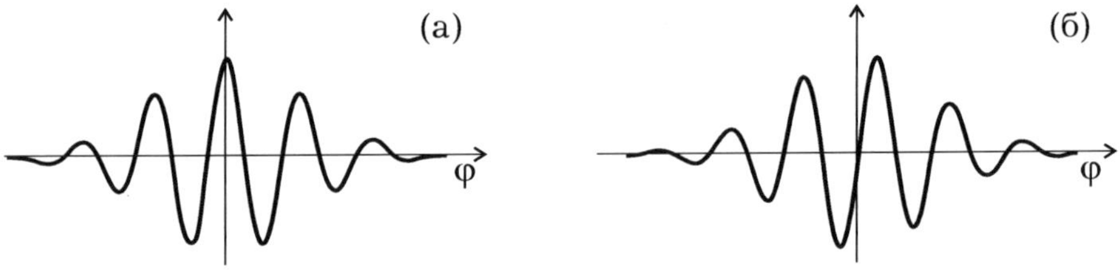

FIGURE 6.1: The odd and even forms of the initial WP.

6.1.2 Splitting the initial WP

We consider the ordinary differential equation

$$\frac{\partial^4 z}{\partial s^4} - \lambda z = 0 \tag{6.8}$$

with one of the three variants of boundary conditions :

$$\begin{aligned} z = 0, \quad & \frac{\partial^2 z}{\partial s^2} = 0, \\ z = 0, \quad & \frac{\partial z}{\partial s} = 0, \\ \frac{\partial^2 z}{\partial s^2} = 0, \quad & \frac{\partial^3 z}{\partial s^3} = 0 \end{aligned} \tag{6.9}$$

for $s = s_1(\varphi)$, $s = s_2(\varphi)$. This boundary-value problem is related to the one on free bending vibrations of the Bernoulli-Euler beam of the length $l = s_2(\varphi) - s_1(\varphi)$. Let $z_1(s,\varphi), z_2(s,\varphi), \ldots$ be an infinite normalized system of eigenfunctions of this boundary-value problem, and $\lambda_1(\varphi), \lambda_2(\varphi), \ldots$ is the corresponding sequence of eigenvalues. We note that z_n and λ_n are parametrically dependent of φ.

The functions $z_n(s,\varphi)(n = 1, 2, \ldots)$ form a complete orthogonal system [116]. Let the amplitudes W_0, V_0 of the initial WP satisfy one of the above variants of boundary conditions (6.9). Then for any $\varphi_1 \leq \varphi \leq \varphi_2$ functions W_0, V_0 can be expanded by the system of eigenfunctions $z_n(s)$ into uniformly converging series on the interval $s_1(\varphi) \leq s \leq s_2(\varphi)$:

$$\begin{aligned} W_0 &= \sum_{n=1}^{\infty} w_n^\circ(\varphi,\varepsilon) z_n(s,\varphi), \quad w_n^\circ = \int_{s_1}^{s_2} W_0(s,\varphi,\varepsilon) z_n(s,\varphi) ds, \\ V_0 &= \sum_{n=1}^{\infty} v_n^\circ(\varphi,\varepsilon) z_n(s,\varphi), \quad v_n^\circ = \int_{s_1}^{s_2} V_0(s,\varphi,\varepsilon) z_n(s,\varphi) ds. \end{aligned} \tag{6.10}$$

Thus, the initial WP (6.5) can be treated as the superposition of set of the nth initial WPs corresponding to eigenvalues λ_n $(n = 1, 2, \ldots)$.

Remark 6.1. Due to estimates (6.7), the functions W_0, V_0 have a small variability with respect to the coordinate s. Hence $|w_n^\circ|, |v_n^\circ|$ are small for large values of n and for practical calculations, one can limit ourselves to a finite number of terms in expansions (6.10).

Let w_n°, v_n° admit the following expansions:

$$\begin{aligned} w_n^\circ &= \sum_{j=0}^{\infty} \varepsilon^{j/2} w_{n,j}^\circ(\zeta), \quad w_{n,j}^\circ = \sum_{k=0}^{M_{n,j}} c_{n,jk}^\circ \zeta^k, \\ v_n^\circ &= \sum_{j=0}^{\infty} \varepsilon^{j/2} v_{n,j}^\circ(\zeta), \quad v_{n,j}^\circ = \sum_{k=0}^{M_{n,j}} d_{n,jk}^\circ \zeta^k, \end{aligned} \tag{6.11}$$

where $\zeta = \varepsilon^{-1/2}\varphi$, and $w_{n,j}^\circ, v_{n,j}^\circ$ are polynomials of degrees $M_{n,j}$ with complex coefficients $c_{n,jk}^\circ, d_{n,jk}^\circ$ such that $|c_{n,jk}^\circ|, |d_{n,jk}^\circ| \sim 1$.

Taking into account the linearity of the original equations, the solution of the initial-boundary-value problem (6.3)-(6.5) is sought in the form of series

$$W = \sum_{n=1}^{\infty} w_n(s, \varphi, t, \varepsilon), \quad \Phi = \sum_{n=1}^{\infty} f_n(s, \varphi, t, \varepsilon), \tag{6.12}$$

where w_n, f_n are unknown functions localized at any point of time in the neighborhood of some generatrix $\varphi = q_n(t)$. Here, $q_n(t)$ is a twice differentiable function, such that

$$q_n(0) = 0. \tag{6.13}$$

A pair of functions w_n, f_n will be called the nth WP with the center on the line $\varphi = q_n(t)$. To study behavior of the nth WP, we fix a number n and omit the index n in all functions, while implying that the latter correspond to the nth WP. For example, instead of $\lambda_n, z_n, w_n, q_n, \xi_n, w_n^\circ, w_{n,j}^\circ, c_{n,jk}^\circ$, we will write $\lambda, z, w, q, \xi, w^\circ, w_j^\circ, c_{jk}^\circ$, respectively.

6.1.3 The initial-boundary-value problem for a fixed WP

We introduce a new local coordinate system associated with the center $\varphi = q(t)$ of travelling the nth WP:

$$\varphi = q(t) + \varepsilon^{1/2}\,\xi. \tag{6.14}$$

Assuming s, ξ, t to be independent variables, one arrives at the coupled equations

$$\begin{aligned} &\varepsilon^2 \frac{\partial^4 w}{\partial \xi^4} + 2\varepsilon^3 \frac{\partial^4 w}{\partial \xi^2\,\partial s^2} + \varepsilon^4 \frac{\partial^4 w}{\partial s^4} + k(\varphi)\frac{\partial^2 f}{\partial s^2} \\ &+ \varepsilon^2 \frac{\partial^2 w}{\partial t^2} - 2\varepsilon^{3/2}\dot{q}\,\frac{\partial^2 w}{\partial \xi\,\partial t} + \varepsilon \dot{q}^2 \frac{\partial^2 w}{\partial \xi^2} - \varepsilon^{3/2}\,\ddot{q}\,\frac{\partial w}{\partial \xi} = 0, \\ &\varepsilon^2 \frac{\partial^4 f}{\partial \xi^4} + 2\varepsilon^3 \frac{\partial^4 f}{\partial \xi^2\,\partial s^2} + \varepsilon^4 \frac{\partial^4 f}{\partial s^4} - k(\varphi)\frac{\partial^2 w}{\partial s^2} = 0, \end{aligned} \tag{6.15}$$

governing the localized dynamics of the shell in the neighborhood of moving generatrix $\varphi = q(t)$. We note that now values of the functions $k(\varphi), s_i(\varphi)$ are calculated at $\varphi = q(t) + \varepsilon^{1/2}\,\xi$.

The initial conditions for the nth WP read

$$w|_{t=0} = \sum_{j=0}^{\infty} \varepsilon^{j/2} w_j^{\circ}(\zeta) z(s,\varphi) \exp\left[\mathrm{i}\varepsilon^{-1} S_0(\varphi)\right],$$
$$\dot{w}|_{t=0} = \mathrm{i}\varepsilon^{-1} \sum_{j=0}^{\infty} \varepsilon^{j/2} v_j^{\circ}(\zeta) z(s,\varphi) \exp\left[\mathrm{i}\varepsilon^{-1} S_0(\varphi)\right]. \tag{6.16}$$

The main boundary conditions

$$w = f = 0, \quad w = \frac{\partial w}{\partial s} = 0, \quad f = \frac{\partial f}{\partial s} = 0 \tag{6.17}$$

for simply supported, clamped and free edges, respectively, are applied for $s = s_i\left[q(t) + \varepsilon^{1/2}\xi\right], \quad i = 1, 2.$

6.1.4 Reduction of the original 2D problem to the sequence of 1D problems on moving generatrix

The formal asymptotic solution of the 2D problem (6.15)-(6.17) is sought in the form of the following expansions

$$w = \sum_{j=0}^{\infty} \varepsilon^{j/2} w_j(s,\xi,t) \exp\left[\mathrm{i}\varepsilon^{-1} S(\xi,t,\varepsilon)\right],$$
$$f = \sum_{j=0}^{\infty} \varepsilon^{j/2} f_j(s,\xi,t) \exp\left[\mathrm{i}\varepsilon^{-1} S(\xi,t,\varepsilon)\right], \tag{6.18}$$
$$S = \int_0^t \omega(\tau)d\tau + \varepsilon^{1/2} p(t)\xi + \tfrac{1}{2}\varepsilon b(t)\xi^2,$$

where

$$\Im b(t) > 0 \quad \text{for any} \quad 0 \le t \le t' < +\infty, \tag{6.19}$$

$$\left\{\omega,\ p,\ b,\ \dot{\omega},\ \dot{p},\ \dot{b},\ \dot{q},\ w_j,\ f_j,\ \frac{\partial w_j}{\partial x},\ \frac{\partial f_j}{\partial x}\right\} = O(1) \quad \text{as} \quad \varepsilon \to 0. \tag{6.20}$$

Here, ω, p, b are twice differentiable functions of t, the unknowns w_j, f_j are polynomials in ξ with complex-valued coefficients sufficiently differentiable with respect to s and t. In (6.20), x denotes one of the independent variables s, ξ, t. Conditions (6.20) are necessary for the constructed solution to be asymptotic, furthermore, they guarantee the fulfilment of above assumptions (6.1), (6.2) regarding the indices of variability and dynamism of the main dynamic stress-strain state of the shell.

All the functions appearing in (6.18) have the define mechanical sense: $|\omega(t)|$ is the instantaneous frequency of vibrations, $p(t)$ determines the variability of waves in the φ-direction, the function $\Im b(t)$ under condition (6.19) characterizes the decay rate of the wave amplitudes far away from the center $\varphi = q(t)$, and w_j, f_j are the amplitude functions.

The asymptotic expansion (6.18) is significantly different from the solutions constructed in Chapters 4 and 5. Indeed, formulas (4.18) and (5.77) determine the stationary and quasi-stationary WPs, respectively, for low-frequency free and parametric vibrations of a shell near the weak generatrix. In both cases, the parameters q, ω, p, b (in other notations) were considered constant, and amplitudes were assumed to be decreasing or increasing functions of time. Formulas (6.18) define a non-stationary WP because now all the quantities q, ω, p, b, w_j depend on time.

We proceed to the definition of the unknown functions in ansatz (6.18). At first, the functions $k(\varphi)$, $s_i(\varphi)$, $z(s, \varphi)$ should be expanded into the Taylor series in powers of $\varepsilon^{1/2}\xi$ in the neighborhood of the moving generatrix $\varphi = q(t)$. For instance,

$$k(\varphi) = k[q(t)] + \varepsilon^{1/2}k'[q(t)]\xi + \frac{1}{2}\varepsilon k''[q(t)]\xi^2 + \dots \tag{6.21}$$

We substitute (6.18), (6.21) into Eqs. (6.15). Equating the coefficients by the same powers of $\varepsilon^{1/2}$ and eliminating f_j, one arrives at the sequence of differential equations

$$\sum_{m=0}^{j} \mathbf{L}_m w_{j-m} = 0, \quad j = 0, 1, 2, \dots \tag{6.22}$$

with

$$\mathbf{L}_0 = \frac{k^2[q(t)]}{p^4(t)}\frac{\partial^4}{\partial s^4} + p^4(t) - [\omega(t) - \dot{q}(t)\,p(t)]^2,$$

$$\mathbf{L}_1 = (b\mathbf{L}_p + \mathbf{L}_q + \dot{p}\mathbf{L}_\omega)\,\xi - \mathrm{i}\mathbf{L}_p\frac{\partial}{\partial \xi},$$

$$\begin{aligned}\mathbf{L}_2 = {} & \frac{1}{2}\Big(b^2\mathbf{L}_{pp} + 2b\mathbf{L}_{pq} + \mathbf{L}_{qq} + \dot{p}^2\mathbf{L}_{\omega\omega} \\ & +2\dot{p}\mathbf{L}_{\mathbf{pq}} + 2\dot{p}b\mathbf{L}_{\omega p} + \dot{b}\mathbf{L}_\omega\Big)\,\xi^2 \\ & -\frac{1}{2}\mathbf{L}_{pp}\frac{\partial^2}{\partial \xi^2} - \mathrm{i}\,(b\mathbf{L}_{pp} + \mathbf{L}_{pq} + \dot{p}\mathbf{L}_{\omega p})\,\xi\frac{\partial}{\partial \xi} - \mathrm{i}\mathbf{L}_\omega\frac{\partial}{\partial t} \\ & -\mathrm{i}\left(\frac{1}{2}b\mathbf{L}_{pp} + \frac{1}{2}\dot{\omega}\mathbf{L}_{\omega\omega} + \dot{p}\mathbf{L}_{\omega p} + \mathbf{N}\right), \dots,\end{aligned} \tag{6.23}$$

$$\mathbf{N} = \ddot{q}p + \frac{1}{2}\mathbf{L}_{pq}. \tag{6.24}$$

The indices p, q, ω in (6.23) and below mean differentiation with respect to the corresponding variables. The operators $\mathbf{L}_m$ for $m \geq 3$ are not written out in an expanded form, because they are very cumbersome. However, when introducing the ordered operator [177]

$$\mathbf{G} = \overset{1}{\mathbf{A}}_1 \overset{3}{\mathbf{X}} + \overset{1}{\mathbf{A}}_2 \overset{2}{\mathbf{D}} \tag{6.25}$$

where

$$\mathbf{A}_1 = b\frac{\partial}{\partial p} + \frac{\partial}{\partial q} + \dot{p}\frac{\partial}{\partial \omega}, \quad \mathbf{A}_2 = \frac{\partial}{\partial p}, \quad \mathbf{X} = \xi, \quad \mathbf{D} = -\mathrm{i}\frac{\partial}{\partial \xi}, \tag{6.26}$$

and figures 1,2,3 indicate the order of action of the corresponding operators, then the following recurrence relations take place:

$$\begin{aligned} &\mathbf{L}_1 = \mathbf{G}\mathbf{L}_0, \quad \mathbf{L}_2 = \frac{1}{2}\mathbf{G}^2\mathbf{L}_0 + \mathbf{K}_2, \\ &\mathbf{L}_3 = \frac{1}{6}\mathbf{G}^3\mathbf{L}_0 + \mathbf{G}\mathbf{K}_2 + \mathbf{K}_3, \ldots, \end{aligned} \tag{6.27}$$

$$\begin{aligned} \mathbf{L}_m = \frac{1}{m!}\mathbf{G}^m\mathbf{L}_0 + \Bigg[&\frac{1}{(m-2)!}\mathbf{G}^{m-2}\mathbf{K}_2 \\ &+ \frac{1}{(m-3)!}\mathbf{G}^{m-3}\mathbf{K}_3 + \ldots + \mathbf{K}_m\Bigg], \quad m = 2, 3, \ldots \end{aligned} \tag{6.28}$$

$$\begin{aligned} \mathbf{K}_2 = &-\frac{\mathrm{i}}{2}\left(b\mathbf{L}_p p + 2\dot{p}\mathbf{L}_{\omega p} + \dot{\omega}\mathbf{L}_{\omega\omega}\right) - \mathbf{L}_\omega\left(\mathrm{i}\frac{\partial}{\partial t} - \frac{1}{2}\dot{b}\,\xi^2\right) \\ &-\mathrm{i}\left(\frac{1}{2}\mathbf{L}_{pq} + p\ddot{q}\right), \quad \mathbf{K}_3 = -\frac{\mathrm{i}}{2}\ddot{p}\,\mathbf{L}_{\omega\omega}\,\xi, \ldots \end{aligned} \tag{6.29}$$

It should be noted that the order of action of the operators $\mathbf{A}_1$, $\mathbf{A}_2$, $\mathbf{X}$, $\mathbf{D}$ introduced by (6.26) is preserved in the operators $\mathbf{G}^m$ $(m \geq 2)$. For instance,

$$\mathbf{G}^2 = (\overset{1}{\mathbf{A}}_1)^2(\overset{3}{\mathbf{X}})^2 + 2(\overset{1}{\mathbf{A}}_1\overset{1}{\mathbf{A}}_2)\overset{3}{\mathbf{X}}\overset{2}{\mathbf{D}} + (\overset{1}{\mathbf{A}}_2)^2(\overset{2}{\mathbf{D}})^2. \tag{6.30}$$

The functions f_j are found step-by-step from inhomogeneous equations and are expressed in terms of w_j. In particular,

$$\begin{aligned} f_0 = \frac{k(q)}{p^4}\frac{\partial^2 w_0}{\partial s^2}, \quad f_1 = &\frac{k(q)}{p^4}\frac{\partial^2 w_1}{\partial s^2} + \frac{4\mathrm{i}k(q)}{p^5}\frac{\partial^3 w_0}{\partial s^2 \partial \xi} \\ &-\frac{\xi}{p^4}\left[\frac{k(q)b}{p} + k'(q)\right]\frac{\partial^2 w_0}{\partial s^2}. \end{aligned} \tag{6.31}$$

Substituting (6.18) into (6.17) gives a sequence of boundary conditions for the functions w_j, f_j at $s = s_i[q(t)]$. In what follows, for definiteness, we consider the case of simply supported ends for which

$$w_0 = 0, \quad \frac{\partial^2 w_0}{\partial s^2} = 0, \tag{6.32}$$

$$w_1 + \xi\, s_i^{'} \frac{\partial w_0}{\partial s} = 0, \quad \frac{\partial^2 w_1}{\partial s^2} + \xi\, s_i^{'} \frac{\partial^3 w_0}{\partial s^3} = 0, \tag{6.33}$$

$$w_2 + \xi\, s_i^{'} \frac{\partial w_1}{\partial s} + \frac{1}{2}\xi^2 \left(s_i^{''} \frac{\partial w_0}{\partial s} + s_i^{'2} \frac{\partial^2 w_0}{\partial s^2} \right) = 0,$$
$$\frac{\partial^2 w_2}{\partial s^2} + \xi\, s_i^{'} \frac{\partial^3 w_1}{\partial s^3} + \frac{1}{2}\xi^2 \left(s_i^{''} \frac{\partial^3 w_0}{\partial s^3} + s_i^{'2} \frac{\partial^4 w_0}{\partial s^4} \right) - \frac{4\mathrm{i}s_{\mathrm{i}}^{'}}{p} \frac{\partial^3 w_0}{\partial s^3} = 0, \ldots \tag{6.34}$$

at $s = s_i[q(t)], \quad i = 1, 2.$

Equations (6.32)-(6.34) guarantee the fulfilment of boundary conditions (6.17) only in a small neighborhood of the generatrix $\varphi = q(t)$. To satisfy the boundary conditions at the whole interval $\varphi_1 \le \varphi \le \varphi_2$ does not make sense, because far from the center of travelling WP, it takes place the estimate $|w| = O\left(\varepsilon^{\infty}\right)$.

6.1.5 Solution of the sequence of 1D problems

Consider the sequence of 1D boundary-value problems (6.22), (6.32)-(6.34).

The zero-order approximation. In the leading approximation ($j = 0$), one has the homogeneous boundary-value problem

$$\mathbf{L}_0\, w_0 \equiv \frac{k^2[q(t)]}{p^4(t)} \frac{\partial^4 w_0}{\partial s^4} + \{p^4(t) - [\omega(t) - \dot q(t)\, p(t)]^2\}\, w_0 = 0, \tag{6.35}$$

$$w_0 = 0, \quad \frac{\partial^2 w_0}{\partial s^2} = 0 \quad \text{at} \quad s = s_i[q(t)]. \tag{6.36}$$

Its solution may be written in the form of

$$w_0 = P_0(\xi, t)\, z[s, q(t)]. \tag{6.37}$$

Here $P_0(\xi, t)$ is a polynomial in ξ with coefficients smoothly depending on t, and

$$z_n = z[s, q(t)] = \sin \frac{\pi n \{s - s_1[q(t)]\}}{l[q(t)]} \tag{6.38}$$

for the simply supported edges, where $l[q(t)] = s_2[q(t)] - s_1[q(t)]$.

The substitution of (6.37) into Eq. (6.35) results in the formula for the instantaneous frequency of vibrations

$$\omega(t) = \dot{q}p(t) \mp H[p(t), q(t)], \tag{6.39}$$

where

$$H(p,q) = \sqrt{p^4 + \frac{\lambda(q)\,k^2(q)}{p^4}} \tag{6.40}$$

is the Hamilton function, and $\lambda[q(t)]$ is an eigenvalue of the boundary-value problem for Eq. (6.8) with one of variants of boundary conditions (6.9) at the edges $s = s_i[q(t)]$. In particular, for the simply supported edges

$$\lambda_n = \lambda[q(t)] = \frac{\pi^4 n^4}{l^4[q(t)]}. \tag{6.41}$$

Non-uniqueness in the definition of ω is associated with the presence of two branches of solutions corresponding to the functions H and $-H$. In what follows, for definiteness in formula (6.39), it has taken the minus sign which corresponds to the function H. In formula (6.39), the first term determines the frequency of passing the waves crests relatively to an "unmoving observer"on the shell surface, and the Hamiltonian H for a fixed t is the instantaneous frequency of oscillations of points on the middle surface within the running WP.

The first-order approximation. In the first-order approximation ($j = 1$), one obtains the inhomogeneous boundary-value problem (6.22), (6.33) with respect to w_1. Its solution is sought in the form of

$$w_1 = P_1(\xi, t)\, z[s, q(t)] + w_1^p(s, \xi, t), \tag{6.42}$$

where P_1 is a polynomial of the argument ξ, and w_1^p is some partial solution of the inhomogeneous equation (6.22) for $j = 1$. Taking into account the self-adjointness of the boundary-value problem (6.35), (6.36), the condition for the existence of a solution w_1 reads

$$\int_{s_1[q(t)]}^{s_2[q(t)]} z\,(\mathbf{L}_0 w_1 + \mathbf{L}_1 P_0 z)\, ds = 0. \tag{6.43}$$

To calculate the second integrand in (6.43), one needs to define the operators $\mathbf{L}_p, \mathbf{L}_q$ (see correlations (6.23)). For this purpose, we differentiate the boundary-value problem (6.35), (6.36) with respect to p and q:

$$\begin{gathered}
\mathbf{L}_0 w_p + \mathbf{L}_p w_0 + 2H\,(\dot{q} - H_p)\, w_0 = 0, \\
w_p = 0, \quad \frac{\partial^2 w_p}{\partial s^2} = 0 \quad \text{for} \quad s = s_i[q(t)]; \\
\mathbf{L}_0 w_q + \mathbf{L}_q w_0 - 2H H_q w_0 = 0, \\
w_q + s_i' \frac{\partial w_0}{\partial s} = 0, \quad \frac{\partial^2 w_q}{\partial s^2} + s_i' \frac{\partial^3 w_0}{\partial s^3} = 0 \quad \text{for} \quad s = s_i[q(t)].
\end{gathered} \tag{6.44}$$

Considering Eq. (6.35), we find that

$$\mathbf{L}_p w_0 = -2H\,(\dot{q} - H_p)\,P_0\,z, \quad \mathbf{L}_q w_0 = 2HH_q P_0\,z - \mathbf{L}_0 P_0\,z_q. \tag{6.45}$$

Substituting (6.45) into (6.43) and taking into account the boundary conditions (6.44) for w_q (here, $w_p = 0$), we arrive at the differential equation

$$b\,(\dot{q} - H_p)\,\xi\,P_0 + (\dot{p} + H_q)\,\xi\,P_0 - \mathrm{i}\,(\dot{q} - H_p)\,\frac{\partial P_0}{\partial \xi} = 0 \tag{6.46}$$

with respect to P_0.

By the assumption, $\Im b(t) > 0$ for any $0 \le t \le t' < +\infty$, and the coefficients of Eq. (6.46) appearing in brackets are real. Hence, Eq. (6.46) has a solution in the form of a polynomial in ξ if and only if the functions $p(t), q(t)$ satisfy the Hamilton system:

$$\dot{q} = H_p, \quad \dot{p} = -H_q. \tag{6.47}$$

Comparing (6.6) and (6.18), and taking into account (6.13), one obtains the initial conditions for $p(t)$ and $q(t)$:

$$p(0) = a_0, \quad q(0) = 0. \tag{6.48}$$

Let $p(t), q(t)$ be a solution of the Cauchy problem (6.47), (6.48). We note in passing that

$$H[p(t), q(t)] = H_0 \tag{6.49}$$

is the integral of problem (6.47), (6.48), where $H_0 = H\,(a_0, 0)$.

In view of (6.47), we get from Eqs. (6.45) the identity $\mathbf{L}_p w_p \equiv 0$, and the operator $\mathbf{L}_1$ in (23) is simplified:

$$\mathbf{L}_1 = (\mathbf{L_q} + \dot{p}\,\mathbf{L}_\omega)\,\xi. \tag{6.50}$$

Hence,

$$w_1^{(p)} = \xi\,P_0\,z_q. \tag{6.51}$$

In this approximation, the polynomials P_0, P_1 remain undefined.

The second-order approximation. Consider Eq. (6.22) for $j = 2$ with the boundary conditions (6.34). The compatibility condition of this inhomogeneous boundary-value problem reads

$$\int\limits_{s_1[q(t)]}^{s_2[q(t)]} z\,[\mathbf{L}_0 w_2 + \mathbf{L}_1\,(P_1 z + \xi P_0 z_q) + \mathbf{L}_2 P_0 z]\,ds = 0. \tag{6.52}$$

To define the operators $\mathbf{L}_{pp}$, $\mathbf{L}_{pq}$, $\mathbf{L}_{qq}$ appearing in (6.52), we differentiate the boundary-value problems (6.44) again with respect to p and q:

$$\begin{gathered}
\mathbf{L}_0 w_p p + 2\mathbf{L}_p w_p + \mathbf{L}_{pp} w_0 - \mathbf{L}_\omega H_{pp} w_0 = 0, \\
w_{pp} = 0, \quad \frac{\partial^2 w_{pp}}{\partial s^2} = 0 \quad \text{for} \quad s = s_i[q(t)]; \\
\mathbf{L}_0 w_{pq} + \mathbf{L}_p w_q + \mathbf{L}_q w_p - \mathbf{L}_\omega H_q w_p \\
+ \mathbf{L}_{pq} w_0 - \mathbf{L}_{\omega p} H_q w_0 - \mathbf{L}_\omega H_{pq} w_0 = 0, \\
w_{pq} + s_i^{'} \frac{\partial w_p}{\partial s} = 0, \quad \frac{\partial^2 w_{pq}}{\partial s^2} + s_i^{'} \frac{\partial^3 w_p}{\partial s^3} = 0 \quad \text{for} \quad s = s_i[q(t)]; \\
\mathbf{L}_0 w_{qq} + 2\mathbf{L}_q w_q - 2\mathbf{L}_\omega H_q w_q + \mathbf{L}_{qq} w_0 \\
- 2\mathbf{L}_{\omega q} H_q w_0 + \mathbf{L}_{\omega\omega} H_q^2 w_0 - \mathbf{L}_\omega H_{qq} w_0 = 0, \\
w_{qq} + +2 s_i^{'} \frac{\partial w_q}{\partial s} + s_i^{''} \frac{\partial w_0}{\partial s} + s_i^{'2} \frac{\partial^2 w_0}{\partial s^2} = 0, \\
\frac{\partial^2 w_{qq}}{\partial s^2} + 2 s_i^{'} \frac{\partial^3 w_q}{\partial s^3} + s_i^{''} \frac{\partial^3 w_0}{\partial s^3} + s_i^{'2} \frac{\partial^4 w_0}{\partial s^4} = 0 \quad \text{for} \quad s = s_i[q(t)].
\end{gathered} \tag{6.53}$$

We substitute the operators $\mathbf{L}_{pp}$, $\mathbf{L}_{pq}$, $\mathbf{L}_{qq}$ found from (6.53) into the compatibility condition (6.52). Taking into account the boundary conditions for w_{qq} as well as holding identities $\mathbf{L}_p \equiv 0$, $w_p \equiv 0$, $w_{pq} \equiv 0$, $w_{pp} \equiv 0$, one obtains the differential equation

$$\left(\xi^2 \mathbf{D}_b - 2\mathbf{D}_{\xi t}\right) P_0 = 0 \tag{6.54}$$

with respect to P_0. Here,

$$\begin{gathered}
\mathbf{D}_b = \dot{b} + H_{pp} b^2 + 2H_{pq} b + H_{qq}, \\
\mathbf{D}_{\xi t} = h_0 \frac{\partial^2}{\partial \xi^2} + h_1 \xi \frac{\partial}{\partial \xi} + h_2 \frac{\partial}{\partial t} + h_3
\end{gathered} \tag{6.55}$$

are the differential operators with

$$\begin{gathered}
h_0(t) = \frac{1}{2} H_{pp}, \quad h_1(t) = \mathrm{i}\left(b H_{pp} + H_{pq}\right), \quad h_2 = \mathrm{i}, \\
h_3 = \frac{\mathrm{i}}{2H} \{ b H H_{pp} - \dot{\omega} - 2 H_q H_p + \ddot{q} p + \frac{1}{\eta} \int\limits_{s_1[q(t)]}^{s_2[q(t)]} \mathbf{L}_\omega \dot{z} z ds + \mathbf{\Gamma} \}, \\
\eta(t) = \int\limits_{s_1[q(t)]}^{s_2[q(t)]} z^2 ds, \quad \mathbf{\Gamma} = -\frac{4k[q(t)] k^{'}[q(t)] \lambda[q(t)]}{p^5(t)}.
\end{gathered} \tag{6.56}$$

Equation (6.54) has a solution in the polynomial form in ξ if and only if $\mathbf{D}_b \equiv 0$. Hence, we obtain the Riccati equation

$$\dot{b} + H_{pp}b^2 + 2H_{pq}b + H_{qq} = 0. \tag{6.57}$$

Comparing (6.6) and (6.18) once more, we find the initial condition for $b(t)$:

$$b(0) = b_0. \tag{6.58}$$

Let us return to equation (6.54). Taking into account the Riccati equation, we arrive at the differential equation

$$\mathbf{D}_{\xi t} P_0 = 0, \tag{6.59}$$

which will be called the amplitudes equation.

Higher-order approximations. To find the corrections $\varepsilon^{j/2} w_j$ in (6.18) for $j \geq 1$, it is required to consider the correspondent boundary-value problems in the $(j+2)$-order approximations. For instance, to find the function w_1, one needs to consider the inhomogeneous Eq. (6.22) for $j = 3$ with the correspondent inhomogeneous boundary conditions . Conditions for existence of solutions of boundary-value problems in the $(j+1)$-order approximations lead to the sequence of inhomogeneous differential equations with respect to polynomials $P_j(\xi, t)$:

$$\mathbf{D}_{\xi t} P_j = P_j^*, \quad j = 1, 2, \ldots, \tag{6.60}$$

where P_j^* are some polynomials in ξ depending on the polynomials $P_{j-1}, \ldots, P_0$ which were found at the previous steps.

It should be however noted that the procedure for constructing the functions w_j with $j \geq 4$ loses its meaning for two reasons. First, at the shell edges, the correction adjusted by the boundary-value problem at the sixth step is of the order of $O\left(\varepsilon^2\right) \exp\left(\mathrm{i}\varepsilon^{-1} S\right)$ which is comparable with the error of the main boundary conditions (6.17). Secondly, when writing the governing Eqs. (6.3), we have rejected some terms which must be taken into account when $j \geq 4$.

The function

$$w = \left[w_0(s, \xi, t) + O\left(\varepsilon^{1/2}\right)\right] \exp\{\mathrm{i}\varepsilon^{-1} S(\xi, t, \varepsilon\}, \tag{6.61}$$

found from the first three approximations is the leading term in the asymptotic expansion of the required solution (6.18).

6.1.6 Integration of the amplitude equation

Consider the amplitudes Eq. (6.59)

$$h_0 \frac{\partial^2 P_0}{\partial \xi^2} + h_1 \xi \frac{\partial P_0}{\partial \xi} + h_2 \frac{\partial P_0}{\partial t} + h_3 P_0 = 0. \tag{6.62}$$

Its solution can be sought in the form of polynomial

$$P_0(\xi, t) = \sum_{k=0}^{M} A_k(t)\xi^k \tag{6.63}$$

of order M. Substituting (6.63) into Eq. (6.62) and equating coefficients at the same powers of ξ, one obtains the sequence of differential equations with respect to $A_k(t)$:

$$\begin{gathered} h_2\dot{A}_M + (Mh_1 + h_3)\,A_M = 0, \\ h_2\dot{A}_{M-1} + [(M-1)h_1 + h_3]\,A_{M-1} = 0, \ldots, \\ h_2\dot{A}_{M-r} + [(M-r)h_1 + h_3]\,A_{M-r} \\ + (M-r+2)(M-r+1)h_0 A_{M-r+2} = 0, \end{gathered} \tag{6.64}$$

where $r = 2, 3, \ldots, M$. Sequential integration of Eqs. (6.64) leads to the recurrence relations for the coefficients

$$\begin{gathered} A_M(t) = d_M\Psi_M(t), \quad A_{M-1}(t) = d_{M-1}\Psi_{M-1}(t), \quad A_{M-r}(t) = \\ = \Psi_{M-r}(t)\left[d_{M-r} - (M-r+2)(M-r+1)\int\limits_0^t \frac{h_0(t)A_{M-r+2}(t)}{h_2(t)\Psi_{M-r}(t)}dt\right], \\ \Psi_m(t) = \exp\left\{-\int\limits_0^t \frac{mh_1(t) + h_3(t)}{h_2(t)}dt\right\}, \end{gathered} \tag{6.65}$$

where $r = 2, 3, \ldots, M$; $m = 0, 1, \ldots, M$; d_m are arbitrary complex numbers to be determined from the initial conditions.

The solution of Eq. (6.62) can be expressed in terms of the Hermite polynomials. Such a representation is convenient for comparing expansion (6.18) in a particular case with the solution describing stationary free vibrations of a cylindrical shell near a weak generatrix (see Chapter 4).

We introduce a new independent variable $x = \varrho(t)\xi$. Now Eq. (6.62) can be rewritten as

$$\frac{\partial^2 P_0}{\partial x^2} + \frac{1}{\varrho^2}\left(\frac{h_1}{h_0} + \frac{h_2}{h_0}\frac{\dot{\varrho}}{\varrho}\right)x\frac{\partial P_0}{\partial x} + \frac{h_2}{h_0\varrho^2}\frac{\partial P_0}{\partial t} + \frac{h_3}{h_0\varrho^2}P_0 = 0. \tag{6.66}$$

The function $\varrho(t)$ is found from the nonlinear equation

$$\frac{1}{\varrho^2}\left(\frac{h_1}{h_0} + \frac{h_2}{h_0}\frac{\dot{\varrho}}{\varrho}\right) = -2. \tag{6.67}$$

This equation has two the solution branches:

$$\varrho(t) = \pm\frac{\exp\left[-\int(h_1/h_2)dt\right]}{\sqrt{\tilde{c} + 4\int(h_0/h_2)\exp\left[-2\int(h_1/h_2)dt\right]dt}} \tag{6.68}$$

where $\tilde{c}$ is an arbitrary constant. The choice of the sign as well as the constant $\tilde{c}$ in (6.67) does not matter. For definiteness, we set $\tilde{c} = 0$ and consider the positive branch of the solution, which corresponds to the plus sign in (6.67).

Then, taking into account (6.67), the solution of Eq. (6.66) can be found in the form

$$P_0 = \chi(t)X(x). \tag{6.69}$$

The substitution of (6.69) into Eq. (6.66) results in the equation

$$\frac{1}{X}\left(\frac{d^2X}{dx^2} - 2x\frac{dX}{dx}\right) = -\frac{1}{\chi}\left(\frac{h_2}{h_0\varrho^2}\frac{d\chi}{dt} + \frac{h_3}{h_0\varrho^2}\chi\right). \tag{6.70}$$

Hence,

$$\begin{aligned} &\frac{d^2X}{dx^2} - 2x\frac{dX}{dx} - c^*X = 0, \\ &\frac{h_2}{h_0\varrho^2}\frac{d\chi}{dt} + \frac{h_3}{h_0\varrho^2}\chi - c^*\chi = 0, \end{aligned} \tag{6.71}$$

where c^* is an arbitrary constant. The first of (6.71) has the solution in the form of the Hermit polynomial

$$X = \mathcal{H}_m(x) \tag{6.72}$$

of order m if $c^* = -2m$. Solving the second equation, we find the partial solution

$$\chi = \chi_m(t) = \frac{\{4\int(h_0/h_2)\exp\left[-2\int(h_1/h_2)dt\right]dt\}^{m/2}}{\exp\left[\int(h_3/h_2)dt\right]} \tag{6.73}$$

Thus, the function $P_0 = \chi(t)\mathcal{H}_m(x)$ satisfies Eq. (6.62). Obviously, the polynomial of order M,

$$P_0(\xi, t; c_m) = \sum_0^M c_m\chi_m(t)\mathcal{H}_m[\varrho(t)\xi], \tag{6.74}$$

where c_m are arbitrary constants, is also the solution of Eq. (6.62). The constants c_m are to be found from the initial conditions.

6.1.7 Definition of constants of integration

Let $p^\pm$, $q^\pm$, $\omega^\pm$, $b^\pm$, $P_j^\pm$, $w_j^\pm$, $f_j^\pm$ be the positive and negative branches of the found above functions corresponding to the Hamiltonians H and $-H$. We also introduce new notations for the local coordinates associated with the positive and negative WPs:

$$\xi^\pm = \varepsilon^{-1/2}\left[\varphi - q^\pm(t)\right]. \tag{6.75}$$

Then $P_0^\pm$ are polynomials in $\xi^\pm$ containing undefined constants $d_m^\pm$ (in the case if the polynomials are constructed by formulas (6.65)).

We compose the functions

$$w = w^+ + w^-, \quad f = f^+ + f^-, \tag{6.76}$$

where

$$\begin{aligned} w^\pm &= \left[w_0^\pm + O\left(\varepsilon^{1/2}\right)\right] \exp\left(\mathrm{i}\varepsilon^{-1} S^\pm\right), \\ f^\pm &= \left[f_0^\pm + O\left(\varepsilon^{1/2}\right)\right] \exp\left(\mathrm{i}\varepsilon^{-1} S^\pm\right), \\ w_0^\pm &= P_0^\pm\left(\xi^\pm, t; d_m^\pm\right) z\left[s, q^\pm(t)\right], \\ f_0^\pm &= P_0^\pm\left(\xi^\pm, t; d_m^\pm\right) \frac{k\left[q^\pm(t)\right]}{\left[p^\pm(t)\right]^4} \frac{\partial^2 z}{\partial s^2}\left[s, q^\pm(t)\right], \\ S^\pm &= \int_0^t \omega^\pm(\tau) d\tau + \varepsilon^{1/2} p^\pm(t)\xi^\pm + \frac{1}{2}\varepsilon b^\pm(t){\xi^\pm}^2. \end{aligned} \tag{6.77}$$

By virtue of the performed constructions, functions (6.76), (6.77) satisfy the governing Eqs. (6.15) in the leading approximation. To determine arbitrary constants $d_m^\pm$ $(m = 0, 1, 2, \ldots, M)$ in $P_0^\pm$, we substitute (6.77) into the initial conditions (6.16), where the function $z(s, \varphi)$ is represented by the series

$$z(s, \varphi) = z(s, 0) + \varepsilon^{1/2}\zeta \frac{\partial z}{\partial \varphi}(s, 0) + \frac{1}{2}\varepsilon\zeta^2 \frac{\partial^2 z}{\partial \varphi^2}(s, 0) + \ldots \tag{6.78}$$

We also take into account the identities $\xi^\pm \equiv \zeta$, $z\left(s, q^\pm\right) \equiv z(s, 0)$ holding at $t = 0$. As a result, we obtain the system of two equations

$$P_0^\pm\left(\zeta, 0; d_m^\pm\right) = \frac{1}{2}\left[w_0^\circ(\zeta) \mp \frac{v_0^\circ(\zeta)}{H_0}\right], \tag{6.79}$$

where $P_0^\pm$ are calculated by (6.63), (6.65), and w_0°, v_0° are polynomials of the order M_0 introduced above in accordance with (6.11). Equating in (6.79) coefficients with the same powers of ζ, one obtains

$$d_m^\pm = \frac{1}{2}\left(c_{0,m}^\circ \mp \frac{d_{0,m}^\circ}{H_0}\right), \quad m = 0, 1, 2, \ldots, M_0. \tag{6.80}$$

If the polynomials $P_0^\pm\left(\xi^\pm, t; c_m^\pm\right)$ are expressed in terms of the Hermite polynomials accordingly to (6.74), then indefinite constants $c_m^\pm$, $m = 0, 1, \ldots, M_0$ contained in $P_0^\pm$ are found as follows:

$$c_m^\pm = \frac{1}{2^{m+1} m! \sqrt{\pi}\chi_m(0)} \int_{-\infty}^{+\infty} e^{-\zeta^2} \mathcal{H}_m[\varrho(0)\zeta] \left[w_0^\circ(\zeta) \mp \frac{v_0^\circ(\zeta)}{H_0}\right] d\zeta. \tag{6.81}$$

6.2 Analysis of the constructed solutions

If the functions $q^{\pm}$ are identically non-zero for $t > 0$, then the initial n-th WP defined in accordance with (6.16), splits into a pair of WPs with the centers $\varphi = q^{\pm}(t)$ moving in the opposite directions with the group velocities

$$v_g^{\pm} = \dot{q}^{\pm}(t) = \pm \frac{2\{[p^{\pm}(t)]^8 - \lambda\,[q^{\pm}(t)]\,k^2\,[q^{\pm}(t)]\}}{[p^{\pm}(t)]^5\,H_0}. \tag{6.82}$$

The order number of each of these WPs is equal to that of the eigenvalue $\lambda = \lambda_n$. The width of each packet is a magnitude of the order of $\varepsilon^{1/2}/\Im b^{\pm}(t)$.

Supplying all functions in (6.76), (6.77) with the index n and performing summation in accordance with (6.12), we obtain a solution of the original problem (6.3)—(6.5) in the form of superposition of travelling the $n^{\pm}$th packets.

Solution (6.76), (6.77) is valid in the asymptotic sense at some interval of time $0 \le t \le t'$, where

$$\operatorname{Im} b^{\pm}(t) > 0, \tag{6.83}$$

$$p^{\pm}(t) \sim 1 \tag{6.84}$$

and estimates (6.20) hold.

We prove that inequality (6.83) is fulfilled if $\Im b_0 > 0$. Let $B(t), C(t)$ be a solution of the following system in variations:

$$\begin{gathered} \dot{B} = -H_{pq}B - H_{qq}C, \quad \dot{C} = H_{pp}B + H_{pq}C, \\ B(0) = b_0, \quad C(0) = 1. \end{gathered} \tag{6.85}$$

Consider the function

$$b = B(t)/C(t). \tag{6.86}$$

A direct substitution of (6.86) into the Riccati equation (6.57) turns it into the identity. Thus, function (6.86) is the solution of the Cauchy problem (6.57), (6.58).

It should be noted that the inequality $|b(t)| < +\infty$ holds at any finite interval of time $t \in [0, t']$. The proof of this statement is given in [16]. Furthermore,

$$\operatorname{Im} b(t) = -\frac{\mathrm{i}\,[B(t)\bar{C}(t) - \bar{B}(t)C(t)]}{2|C(t)|^2} = -\frac{ib_0}{|C(t)|^2}, \tag{6.87}$$

where $\bar{B}, \bar{C}$ are the complex-valued functions, which are complex conjugate to B and C. Hence, if $\operatorname{Im} b_0 > 0$, then

$$0 < \operatorname{Im} b(t) < +\infty \tag{6.88}$$

for any $t \in [0, t']$. This means that at any finite interval of time, the width of travelling WP is a finite nonzero value.

Condition (6.84) is introduced to provide the boundedness of the function $f_0^{\pm}$ and is more strong than the correspondent estimate (6.20) for $p^{\pm}(t)$. The fulfilment of conditions (6.20), (6.84) depends on the correlation of parameters included in the problem. Partially, this issue will be discussed below.

Remark 6.2. In the case of a cylindrical shell closed in the circumferential direction, solution (6.76), (6.77) does not possess the property of periodicity by the coordinate φ. Under constructing (6.76), (6.77), the region $-\pi < \varphi \leq \pi$ was substituted for the infinite strip $-\infty < \varphi < +\infty$. However, this circumstance does not prevent us from using formulas (6.76), (6.77) to predict the dynamic behavior of a closed cylindrical shell for times when $|q^{\pm}(t)| > \pi$. For instance, let $q^{+} \to \pi$ as $t \to t_{\pi}$ where $t \in [0, t')$. In that case, to calculate the amplitudes of waves for the n^{+}-th WP in the region $-\pi < \varphi < 0$ at $t_{\pi} \leq t < t'$, one needs to assume $\xi^{+} = q^{+}(t) - 2\pi - \varphi$ in the constructed solution (6.76), (6.77).

Remark 6.3. If $\varphi_1 < \varphi < \varphi_2$ ($\varphi_1 > -\pi$, $\varphi_2 < \pi$), that is a shell is not closed in the circumferential direction, then solution (6.76), (6.77) is valid if the WP centers $\varphi = q^{\pm}$ are sufficiently far from the straight edge so that its influence may be disregarded.

Let us study solution (6.76), (6.77) versus the functions $k(\varphi), s_i(\varphi)$ and parameters included into the initial conditions. We introduce the notation $g = \lambda(\varphi)k^2(\varphi)$ and consider different cases.

6.2.1 Stationary wave packet (localized eigenmode)

Let

$$g'(0) = 0, \quad g''(0) > 0. \tag{6.89}$$

This case is of a particular interest, since the center of the initial WP is on the generatrix $\varphi = 0$ which is the weakest one for a shell. The forms of free low-frequency vibrations localized in the vicinity of this line are approximated by function (4.18):

$$w = \left[\mathcal{H}_m(\varrho^{\circ}\zeta) z(s, 0) + O\left(\varepsilon^{1/2}\right)\right] \exp\left[\mathrm{i}\varepsilon^{-1}\left(\omega^{\circ} t + p^{\circ}\varphi + \frac{1}{2} b^{\circ}\varphi^2\right)\right]. \tag{6.90}$$

Here, $\varrho^{\circ} = (H_{qq}/H_{pp})^{1/4}$, and

$$\omega^{\circ} = \omega_0^{\circ} + \varepsilon\omega_1^{(m)} \tag{6.91}$$

is the dimensionless natural frequency of the shell with

$$\omega_0^{\circ} = H\left(p^{\circ}, q^{\circ}\right), \quad \omega_1^{(m)} = \left(m + \frac{1}{2}\right) \sqrt{H_{pp} H_{qq}}\Big|_{\substack{p=p^{\circ} \\ q=q^{\circ}}}, \qquad m = 0, 1, 2, \dots, \tag{6.92}$$

The parameters $p^{\circ} = g^{1/8}(0)$, $q^{\circ} = 0$ $b^{\circ} = \mathrm{i}\sqrt{H_{qq}/H_{pp}}$ are the solutions of the equations

$$H_p = 0, \quad H_q = 0 \tag{6.93}$$

and

$$H_{pp}b^2 + 2H_{pq}b + H_{qq} = 0, \tag{6.94}$$

respectively (here, $H_{pq}(p^\circ, 0) \equiv 0$).

The eigenmode (6.90) corresponds to the number n of semi-waves along the generatrix and depends on degree m of the Hermite polynomial.

We note that (6.93), (6.94) are degenerate analogues of the Hamiltonian system and the Riccati equation (if $\dot{p} \equiv 0$, $\dot{q} \equiv 0$, $\dot{b} \equiv 0$). Let $a_0 = p^\circ$, $b_0 = b^\circ$, where a_0, b_0 are parameters from the initial conditions (6.6). Resolving the Hamiltonian system and Riccati equation, we get

$$p^{\pm}(t) \equiv p^\circ, \quad q^{\pm}(t) \equiv q^\circ, \quad b^{\pm}(t) \equiv b^\circ \tag{6.95}$$

for any $t \geq 0$.

We define the polynomials $P_0^{\pm}$ according to (6.74):

$$P_0^{\pm}\left(\zeta, t; c_m^{\pm}\right) = \sum_{m=0}^{M_0} c_m^{\pm} \chi_m^{\pm}(t) \mathcal{H}_m\left[\varrho^{\pm}(t)\zeta\right]. \tag{6.96}$$

Here,

$$\begin{aligned} \varrho^{\pm} &= \varrho^\circ = \left(H_{qq}/H_{pp}\right)^{1/4}, \\ \chi_m^{\pm} &= \left(H_{pp}/H_{qq}\right)^{m/4} \exp\left[-\mathrm{i}\left(m + \frac{1}{2}\right)\sqrt{H_{pp}H_{qq}}\, t\right], \end{aligned} \tag{6.97}$$

where the derivatives H_{pp}, H_{qq} are calculated for $p = p^\circ, q = 0$. The substitution of (6.96), (6.97) into (6.76), (6.77) yields

$$\begin{aligned} w = z(s,0) \sum_{m=0}^{M_0} \exp\left[\mathrm{i}\varepsilon^{-1}\left(\omega^\circ t + p^\circ \varphi + \frac{1}{2} b^\circ \varphi^2\right)\right] \\ \times \left(c_m^{+} + c_m^{-}\right)\left(H_{pp}/H_{qq}\right)^{m/4} \mathcal{H}_m\left[\varepsilon^{-1/2}\left(H_{qq}/H_{pp}\right)^{1/4} \varphi\right], \end{aligned} \tag{6.98}$$

where the constants $c_m^{\pm}$ are found by (6.81). The function (6.98) is the superposition of $M_0 + 1$ numbers of eigenmodes (6.90), where M_0 is the degree of polynomials $w_0^\circ(\zeta)$, $v_0^\circ(\zeta)$ appearing in the initial conditions (6.6), (6.10), (6.11).

So, if $a_0 = p^\circ$, $b_0 = b^\circ$, then the initial nth WP with the center on the weakest generatix does not split into the n^+th and the n^-th packets. In this case, solution (6.76) takes form (6.98) and defines the stationary WP with the center on the unmoving (weakest) generatrix $\varphi = 0$.

6.2.2 Properties of the Hamiltonian system solutions

The analysis of the Hamilton system allows one to explore some dynamic characteristics of non-stationary WPs. We rewrite system (6.47) with the initial conditions (6.48) in the following form:

$$\dot{q}^{\pm} = \pm\frac{2\left[\left(p^{\pm}\right)^{8} - g\left(q^{\pm}\right)\right]}{\left(p^{\pm}\right)^{5} H_0}, \quad q^{\pm}(0) = 0, \tag{6.99}$$

$$\dot{p}^{\pm} = \mp\frac{g'\left(q^{\pm}\right)}{2H_0\left(p^{\pm}\right)^{4}}, \quad p^{\pm}(0) = a_0. \tag{6.100}$$

We recall that relation (6.49)

$$H\left[p^{\pm}(t), q^{\pm}(t)\right] = H_0 \tag{6.101}$$

is the integral for the Hamiltonian system (6.99), (6.100).

Let $a_0 > 0$. It is obvious that

$$p^{\pm}(t) > 0 \tag{6.102}$$

at least for some initial period of time. From all possible solutions of Eq. (6.101), the functions

$$p^{\pm}(t) = p_j\left[q^{\pm}(t)\right], \quad j = 1, 2, \tag{6.103}$$

where

$$p_1(\varphi) = \sqrt[4]{\frac{H_0^2 + \sqrt{H_0^4 - 4g(\varphi)}}{2}}, \quad p_2(\varphi) = \sqrt[4]{\frac{H_0^2 - \sqrt{H_0^4 - 4g(\varphi)}}{2}}, \tag{6.104}$$

satisfy inequality (6.102).

The substitution of (6.103) into (6.99) gives the relations for the group velocities $v_j^{\pm}(t)$ and the functions $\omega^{\pm}(t)$:

$$v_g^{\pm} = \dot{q}^{\pm}(t) = \pm v_j\left[q^{\pm}(t)\right], \quad \omega^{\pm}(t) = \pm\omega_j\left[q^{\pm}(t)\right], \quad j = 1, 2 \tag{6.105}$$

with

$$\begin{aligned} v_1(\varphi) &= \frac{2^{5/4}\sqrt{H_0^4 - 4g(\varphi)}}{H_0\sqrt[4]{H_0^2 + \sqrt{H_0^4 - 4g(\varphi)}}}, \\ v_2(\varphi) &= -\frac{2^{5/4}\sqrt{H_0^4 - 4g(\varphi)}}{H_0\sqrt[4]{H_0^2 - \sqrt{H_0^4 - 4g(\varphi)}}}, \\ \omega_1(\varphi) &= \frac{2\sqrt{H_0^4 - 4g(\varphi)} - H_0^2}{H_0}, \\ \omega_2(\varphi) &= -\frac{2\sqrt{H_0^4 - 4g(\varphi)} + H_0^2}{H_0}. \end{aligned} \tag{6.106}$$

Obviously, the following inequality

$$H_0^4 \geq 4g\left[q^{\pm}(t)\right] \tag{6.107}$$

should hold for the $n^{\pm}$th WPs at any $t \geq 0$.

Integrating the first equation in (6.105), we arrive at the relation

$$t = \pm \int_0^{q^\pm} \frac{dq}{v_j(q)}, \quad j = 1, 2, \tag{6.108}$$

implicitly defining the functions $q^\pm(t)$.

We differentiate Eq. (6.99) with respect to t and take into account (6.100). As a result, we obtain the formula for the center acceleration of the $n^\pm$th WPs:

$$\dot{v}_g^\pm = \ddot{q}^\pm(t) = -\frac{\left[7\left(p^\pm\right)^8 + g\left(q^\pm\right)\right] g'\left(q^\pm\right)}{\left(p^\pm\right)^{10} H_0^2}, \tag{6.109}$$

which is valid for two possible variants (6.103) of the wave parameter $p^\pm(t)$.

The obtained relations for the functions $p^\pm(t), q^\pm(t)$ and their derivatives make it possible to investigate the dynamics of the $n^\pm$th packets depending on properties of the function $g(\varphi) = \lambda(\varphi)k^2(\varphi)$. Of interest are two different cases where WPs run in the direction of decreasing or increasing the function $g(\varphi)$.

Without loss of generality, we shall study the WPs for which

$$v_g^\pm = \dot{q}^\pm(t) \geq 0 \tag{6.110}$$

at least for some interval of time, $0 \leq t < t'$.

Case 1. Let

$$g'(\varphi) < 0 \quad \text{for} \quad 0 < \varphi < \varphi_2. \tag{6.111}$$

We note that $H_0^4 \geq 4g_0$, where $g_0 = g(0)$. Hence, by virtue of (6.110), inequality (6.107) holds for any $t \geq 0$.

Let $a_0 \geq g_0^{1/8}$. Then, as follows from Eq. (6.99), inequality (6.110) holds at $t = 0$ for the n^+th WP. From two relations (6.105) for the n^+th packet, one can choose only one $v_g^+ = v_1\left[q^+(t)\right]$ that satisfies inequality (6.110) for any $t \geq 0$ (at least as long as the inequality $q^+(t) < \varphi_2$ holds). Here, $p^+(t) = p_1\left[q^+(t)\right], \omega^+(t) = \omega_1\left[q^+(t)\right]$. Analyzing Eqs. (6.100), (6.105) and (6.109), one arrives at the inequalities

$$\dot{p}^+ > 0, \quad \dot{v}_g^+ > 0, \quad \dot{\omega}^+ > 0, \tag{6.112}$$

which are fulfilled for $t \geq 0$.

Similar analysis in the case when $0 < a_0 < g_0^{1/8}$ gives $p^-(t) = p_2\left[q^-(t)\right]$, $v_g^- = -v_2\left[q^-(t)\right]$, $\omega^-(t) = -\omega_2\left[q^-(t)\right]$ for the n^-th packet. Here, one obtains the inequalities

$$\dot{p}^- < 0, \quad v_g^- = -v_2\left[q^-(t)\right] > 0, \quad \dot{v}_g^- > 0, \quad \dot{\omega}^- > 0, \tag{6.113}$$

for $t \geq 0$.

Inequalities (6.112), (6.113) indicate that the WP motion in the direction of decreasing the function $g(\varphi)$ is always accompanied by increasing both its group velocity $v_g^{\mp}(t)$ and the function $\omega^{\pm}(t)$ (we note that $|\omega^{\pm}(t)|$ is the frequency of the travelling WP), the wave parameters $p^+(t)$, $p^-(t)$ being increasing and decreasing functions for $a_0 \geq g_0^{1/8}$ and $0 < a_0 < g_0^{1/8}$, respectively.

Case 2. Now, we assume that

$$g'(\varphi) > 0 \quad \text{for} \quad 0 < \varphi < \varphi_2. \tag{6.114}$$

Here, inequality (6.107) requires consideration of three possible options:

$$a_0^8 \neq g_0, \quad H_0^4 > 4 \sup_{0<\varphi<\varphi_2} g(\varphi), \tag{6.115}$$

$$a_0^8 \neq g_0, \quad H_0^4 \leq 4 \sup_{0<\varphi<\varphi_2} g(\varphi), \tag{6.116}$$

$$a_0^8 = g_0. \tag{6.117}$$

Consider each of these cases.

Case 2a. Let inequality (6.115) be satisfied. Then condition (6.107) holds for any $t \geq 0$. The analysis of the Hamiltonian system (6.99), (6.100) and of relations (6.103)-(6.106), (6.109) leads to the inequalities

$$\begin{gathered}\dot{p}^+ < 0, \quad v_g^+ = v_1\left[q^+(t)\right] > 0, \quad \dot{v}_g{}^+ < 0, \quad \dot{\omega}^+ < 0 \quad \text{if} \quad a_0 > g_0^{1/8}, \\ \dot{p}^- > 0,\, v_g^- = -v_2\left[q^-(t)\right] > 0,\, \dot{v}_g{}^- < 0,\, \dot{\omega}^- < 0 \quad \text{if} \quad 0 < a_0 < g_0^{1/8}\end{gathered} \tag{6.118}$$

which hold for any $t \geq 0$. Here,

$$p^+ = p_1\left[q^+(t)\right], \quad \omega^+ = \omega_1\left[q^+(t)\right] \quad \text{for} \quad a_0 > g_0^{1/8} \tag{6.119}$$

$$p^- = p_2\left[q^-(t)\right], \quad \omega^- = -\omega_2\left[q^-(t)\right], \quad \text{for} \quad 0 < a_0 < g_0^{1/8}. \tag{6.120}$$

Inequalities (6.118) show that in case (6.115), when the "energy"of the initial WP is enough large, the both WPs (denoted by n^+ and n^-) run in the direction of increasing the function $g(\varphi)$ with decreasing the group velocities. At that, in the case of $a_0 > g_0^{1/8}$ the wave parameter p^+ decreases, and for $0 < a_0 < g_0^{1/8}$ the parameter p^- grows.

Case 2b. Let now inequality (6.116) hold. We consider in more detail the case when $a_0 > g_0^{1/8}$. Here, $\dot{q}^+ > 0$ what means that the n^+th WP moves in the direction of increasing the function $g(\varphi)$. From Eq. (6.100), it follows that

$$\dot{p}^+ < 0 \quad \text{for} \quad t \geq 0. \tag{6.121}$$

On the other hand, there can exist such $0 < t_r^+ < +\infty$ and $q_r^+ = q^+(t_r^+) > 0$ so that

$$H_0^4 = 4g(q_r^+) \tag{6.122}$$

and none of functions (6.103) for the n^+th packet satisfies inequality (6.121) for any $t \geq 0$.

Let us consider the function

$$p^+(t) = \begin{cases} p_1\left[q^+(t)\right] & \text{for} \quad 0 \leq t < t_r^+, \\ \left(H_0^4/2\right)^{1/4} & \text{for} \quad t = t_r^+, \\ p_2\left[q^+(t)\right] & \text{for} \quad t_r^+ < t \leq t_0^+, \end{cases} \tag{6.123}$$

where

$$t_r^+ = \int_0^{q_r^+} \frac{dq}{v_1(q)}, \quad t_0^+ = t_r^+ + \int_{q_r^+}^{0} \frac{dq}{v_2(q)}. \tag{6.124}$$

So to calculate the functions $q^+(t) \in [0, q_r^+]$, one needs to use the relations

$$t = \int_0^{q^+} \frac{dq}{v_1(q)} \quad \text{and} \quad t = t_r^+ + \int_{q_r^+}^{q^+} \frac{dq}{v_2(q)} \tag{6.125}$$

for the intervals $0 \leq t \leq t_r^+$ and $t_r^+ < t \leq t_0^+$, respectively.

Remark 6.4. Due to Eqs. (6.106), (6.122), the integrand functions in formulas (6.108), (6.124), (6.125) have a singularity at the point $q = q_r^+$. Let $g(\varphi)$ be an infinitely differentiable function. Then

$$4g(\varphi) - H_0^4 \sim q - q_r^+ \quad \text{as} \quad q \to q_r^+ \tag{6.126}$$

and, as a consequence, we have the estimates

$$\left\{\frac{1}{v_1(q)}, \quad \frac{1}{v_2(q)}\right\} = O\left(|q - q_r^+|^{-1/2}\right) \quad \text{as} \quad q \to q_r^+, \tag{6.127}$$

which guarantee the convergence of the improper integral (6.124), (6.125).

The substitution of (6.123) into Eqs. (6.39) and (6.99) results in the expressions for the group velocities and frequencies of the n^+th packet:

$$v_g^+(t) = \begin{cases} v_1\left[q^+(t)\right] > 0 & \text{for} \quad 0 \leq t < t_r^+, \\ 0 & \text{for} \quad t = t_r^+, \\ v_2\left[q^+(t)\right] < 0 & \text{for} \quad t_r^+ < t \leq t_0^+, \end{cases} \tag{6.128}$$

$$\omega^+(t) = \begin{cases} \omega_1\left[q^+(t)\right] & \text{for} \quad 0 \leq t < t_r^+, \\ -H_0 & \text{for} \quad t = t_r^+, \\ \omega_2\left[q^+(t)\right] & \text{for} \quad t_r^+ < t \leq t_0^+. \end{cases} \tag{6.129}$$

When taking into account signs for the group velocities in relations (6.125) in accordance to (6.128), one can notice that the wave parameter (6.123) satisfies inequality (6.121) at all time interval $0 \le t \le t_0^+$. The functions $p^+(t), q^+(t)$ constructed in such a way, where $q^+(t)$ is defined as the function inverse to (6.125), are continuous and smooth at the segment $0 \le t \le t_0^+$ and satisfy the Hamiltonian system.

From the analysis of functions (6.123)-(6.129) we have the following conclusions about the dynamic behavior of the n^+th WP:

- for $0 \le t < t_r^+$, the motion of the WP in the direction of growing the function $g(\varphi)$ is accompanied with decreasing the group velocity v_g^+, wave parameter p^+ and frequency ω^+ as well;
- at time $t = t_r^+$, one observes the reflection of the WP from the generatrix $\varphi = q_r^+ = q^+(t_r^+)$ which is determined from Eq. (6.122);
- in the interval $t_r^+ < t \le t_0^+$, the n^+th WP runs in the opposite direction, the absolute values of functions $v_g^+ < 0$, $\omega^+ < 0$ increasing and $p^+ > 0$ decreasing;
- at time $t = t_0^+$, the WP center returns in its original position $\varphi = 0$, the subsequent behavior of the packet at $t > t_0^+$ being determined by the properties of the function $g(\varphi)$ at $\varphi < 0$.

Similarly, considering the case $0 < a_0 < g_0^{1/8}$, one can construct the functions

$$p^-(t) = \begin{cases} p_2\left[q^-(t)\right] & \text{for} \quad 0 \le t < t_r^-, \\ \left(H_0^4/2\right)^{1/4} & \text{for} \quad t = t_r^-, \\ p_1\left[q^-(t)\right] & \text{for} \quad t_r^- < t \le t_0^-, \end{cases} \tag{6.130}$$

$$v_g^-(t) = \begin{cases} -v_2\left[q^-(t)\right] > 0 & \text{for} \quad 0 \le t < t_r^-, \\ 0 & \text{for} \quad t = t_r^-, \\ -v_1\left[q^-(t)\right] < 0 & \text{for} \quad t_r^- < t \le t_0^-, \end{cases} \tag{6.131}$$

$$\omega^-(t) = \begin{cases} -\omega_2\left[q^-(t)\right] & \text{for} \quad 0 \le t < t_r^-, \\ H_0 & \text{for} \quad t = t_r^-, \\ -\omega_1\left[q^-(t)\right] & \text{for} \quad t_r^- < t \le t_0^-, \end{cases} \tag{6.132}$$

indicating the reflection of the n^-th WP from the generatrix $\varphi = q^-(t_r^-)$. Here, $t_r^- = -\int\limits_0^{q_r^-} \dfrac{dq}{v_2(q)}$ is the moment of reflection of the n^--th packet from the generatrix $\varphi = q_r^-$, which is found from Eq. (6.122), $t_0^- = t_r^- - \int\limits_{q_r^-}^{0} \frac{dq}{v_1(q)}$ is the moment of returning the WP center to its original position $\varphi = 0$, and the function $q^- \in [0,\, q_r^-]$ is found from the equations

$$t = -\int\limits_{0}^{q^-} \frac{dq}{v_2(q)} \quad \text{and} \quad t = t_r^- - \int\limits_{q_r^-}^{q^-} \frac{dq}{v_1(q)} \tag{6.133}$$

at the intervals $0 \le t \le t_r^-$ and $t_r^- < t \le t_0^-$, respectively. Under the introduced above assumptions with regard to the function $g(\varphi)$ (see Remark 6.4), the last integrals are convergent at the point $q = q_r^-$.

Let the shell edges be simply supported. Then $\lambda = \pi^4 n^4 k^2(\varphi)/l^4(\varphi)$, where $l = s_2(\varphi) - s_1(\varphi)$, and the reflection condition (6.116) for the $n^{\pm}$th WP with n semi-waves along the generatrix can be written as

$$n_1 \le n \le n_2, \tag{6.134}$$

where

$$n_1 = 1 + \mathrm{Int}\left(a_0^2 \sqrt{\frac{K l_0^4}{\pi^2 k_0^2}} - \frac{l_0^2}{\pi^2 k_0} \sqrt{\frac{K^2 l_0^4}{k_0^2} - 1} \right),$$

$$n_2 = 1 + \mathrm{Int}\left(a_0^2 \sqrt{\frac{K l_0^4}{\pi^2 k_0^2}} + \frac{l_0^2}{\pi^2 k_0} \sqrt{\frac{K^2 l_0^4}{k_0^2} - 1} \right),$$

$$K = \sup_{0<\varphi<\varphi_2} k^2(\varphi)/l^4(\varphi), \quad k_0 = k(0), \quad l_0 = l(0),$$

and $\mathrm{Int}(x)$ is the integer part of a number x. This shows that the reflection of the $n^{\pm}$th packets occurs only if their number n falls into segment (6.134), whose length

$$n_2 - n_1 = 1 + \mathrm{Int}\left[\sqrt{2a_0^4 \left(\frac{K l_0^4}{\pi^2 k_0^2} - 1 \right)} \right] \tag{6.135}$$

depends on the parameters a_0, k_0, l_0, K. For example, under constant parameters k_0, l_0, K, the boundaries of segment (6.134), as well as its length, are non-decreasing functions of a_0.

Remark 6.5. Let the function $g(\varphi)$ defined on the set $-\varphi_1 < \varphi < \varphi_1$ be even and satisfies the conditions:

$$g'(0) = 0, \quad g'(\varphi) < 0 \quad \text{for} \quad -\varphi_1 < \varphi < 0$$

and

$$g'(\varphi) > 0 \quad \text{for} \quad 0 < \varphi < \varphi_1. \tag{6.136}$$

Let the reflection conditions (6.116) be also satisfied. Then the functions $p^{\pm}(t)$, $q^{\pm}(t)$ are periodic with period $2t_0^{\pm}$, where

$$t_0^{\pm} = t_r^{\pm} + \int_{q_r^{\pm}}^{0} \frac{dq}{v_2(q)}, \quad t_r^{\pm} = \int_{0}^{q_r \pm} \frac{dq}{v_1(q)},$$

and a value $q_r^{\pm}$ is a solution of the equation $H_0^4 = 4g(q)$.

Indeed, let for example, $a_0^8 > g_0$. Consider the function $p^+(t)$ which is given by formulas (6.123), (6.125) in the interval $0 \le t \le t_0^+$. Taking into account the evenness of the function $g(\varphi)$, we shall construct a function $p^+(t)$ in the interval $t_0^+ < t \le 2t_0^+$ as follows:

$$p^+(t) = \begin{cases} p_2\left[q^+(t)\right] & \text{for} \quad t_0^+ \le t < 2t_0^+ - t_r^+, \\ \left(H_0^4/2\right)^{1/4} & \text{for} \quad t = 2t_0^+ - t_r^+, \\ p_1\left[q^+(t)\right] & \text{for} \quad 2t_0^+ - t_r^+ < t \le 2t_0^+. \end{cases}$$

Here, $[-q_r^+, 0] \ni q^+(t)$ is found from the equations

$$t = t_0^+ + \int_0^{q^+} \frac{dq}{v_2(q)} \quad \text{and} \quad t = 2t_0^+ - t_r^+ + \int_{-q_r^+}^{q^+} \frac{dq}{v_1(q)}$$

on the sets $t_0^+ < t \le 2t_0^+ - t_r^+$ and $2t_0^+ - t_r^+ < t \le 2t_0^+$, respectively. Note that $p^+(0) = p^+(2t_0^+) = a_0$, $q^+(0) = q^+(2t_0^+) = 0$. We extend the functions $p^+(t)$, $q^+(t)$ to be periodical on the set $t \ge 0$ with the period $2t_0^+$. It can be readily verified that the functions obtained are differentiable and satisfy the Hamiltonian system. The periodicity of the functions $p^+(t)$, $q^+(t)$ clearly implies the periodicity of the functions $\omega^{\pm}(t)$, $v_g^{\pm}(t)$.

Case 2c. And finally, let condition (6.117) be fulfilled. Here, $H_0^4 = 4g(0)$. Hence, taking into account increasing the function $g(\varphi)$ on the set $0 < \varphi < \varphi_2$, one concludes that inequality (6.107) does not hold for $t > 0$. This means that for $a_0^8 = g_0$, the movement of the $n^{\pm}$-th packets in the direction of increasing the function $g(\varphi)$ is impossible. In this case, either both packets travel in the negative direction of varying the coordinate φ (their dynamics are determined by the properties of the function $g(\varphi)$ on the set $\varphi < 0$) or both packets are localized near the weakest generatrix $\varphi = 0$. In particular, if conditions (6.89) are fulfilled, the regime of stationary vibrations localized in the vicinity of the line $\varphi = 0$ and being the superposition (6.116) of eigenmodes (6.90) is possible.

Remark 6.6. The Riccati equation, as well as formulas (6.65) for the coefficients $A_k(t)$ of the polynomial (6.63) do not allow establishing the properties of the functions $b^{\pm}(t)$, $A_k^{\pm}(t)$. Some dependences of the packet width and wave amplitude on the shell geometry will be revealed below when considering examples.

Remark 6.7. The obtained relations for the functions $p^{\pm}(t)$, $v_g^{\pm}(t)$, $\omega^{\pm}(t)$ permit to determine conditions for holding estimates (6.20), (6.84) at least for the specified functions. The least favorable case is (6.111) for $0 < a_0 < g_0^{1/8}$. Let $\inf\limits_{0<\varphi<\varphi_2} g(\varphi) \ll 1$. Then the function $p^- = p_1[q^-(t)] > 0$ can take very small values, but this contradicts (6.84). In particular, the case when the shell curvature $k(\varphi)$ vanishes at some generatrix $0 < \varphi < \varphi_2$ falls out of consideration. In this case, the constructed solution (6.76), (6.77) should be considered only on a certain time interval where condition (6.84) holds. In all other cases, the estimates (6.20), (6.84) for the functions $p^{\pm}(t)$, $v_g^{\pm}(t)$, $\omega^{\pm}(t)$ are valid if $a_0, g_0, g'(\varphi) \sim 1$. Another solution in the form of a running WPs, which is valid for a very small $p^{\pm}(t)$, will be constructed in Chapter 7.

6.3 The influence of shell geometry on the propagation of wave packets

Let us turn to examples allowing us to study in more detail the effect of the shell geometric parameters on the dynamics of WPs.

Example 6.1. At first, we consider the simplest case when all shell parameters are constant: $k = 1$, $s_1 = 0$, $s_2 = l$. In this case, all functions appearing in solution (6.76), (6.77) can be found explicitly. Let the edges be simply supported. Then solutions of the Hamiltonian system and the Riccati equation read

$$p^{\pm} = a_0, \quad q^{\pm}(t) = \pm H_p t, \quad \omega^{\pm} = \pm H_p a_0 \mp H_0, \quad b^{\pm}(t) = \frac{b_0}{1 \pm b_0 H_{pp} t} \tag{6.137}$$

with

$$H_0 = \sqrt{a_0^4 + \frac{\lambda}{a_0^4}}, \quad H_p = \frac{2(a_0^8 - \lambda)}{a_0^5 H_0}, \quad H_{pp} = 2\frac{a_0^{16} + 12\lambda a_0^8 + 3\lambda^2}{a_0^{10} H_0^3},$$

where $\lambda = \pi^4 n^4 l^{-4}$ and the associated eigenfunction of the boundary-value problem (6.8), (6.9) is $z(s) = \sin(\pi n s/l)$.

The coefficients $h_j(t)$ of the amplitude equation (6.59) have a simple form:

$$h_0^{\pm} = \pm\frac{1}{2}H_{pp}, \quad h_1^{\pm}(t) = \pm \mathrm{i} b^{\pm}(t) H_{pp}, \quad h_2^{\pm} = \mathrm{i}, \quad h_3^{\pm}(t) = \pm\frac{\mathrm{i}}{2} b^{\pm}(t) H_{pp}.$$

Inserting the last relations into recurrent formulas (6.65), we find the coefficients $A_k^{\pm}(t)$ of the Mth degree polynomials (6.63),

$$A_M^{\pm}(t) = d_M^{\pm}\Psi_M^{\pm}(t), \quad A_{M-1}^{\pm}(t) = d_{M-1}^{\pm}\Psi_{M-1}^{\pm}(t),$$
$$A_{M-r}^{\pm}(t) = \Psi_{M-r}^{\pm}(t)d_{M-r}^{\pm} + \Psi_{M-r}^{\pm}(t)$$
$$\times \sum_{m=1}^{N} \frac{\mathrm{i}^m(M-r+1)(M-r+2)\ldots(M-r+2m)d_{M-r+2m}^{\pm}\left[C^{\pm}(t)-1\right]^m}{2^m m!\left[b_0C^{\pm}(t)\right]^m} \tag{6.138}$$

with

$$C^{\pm}(t) = 1 \pm b_0H_{pp}t, \quad N = \mathrm{Int}(r/2), \quad r = 2, 3, \ldots, M,$$
$$\Psi_{\varsigma}^{\pm}(t) = \frac{1}{\sqrt{\left[C^{\pm}(t)\right]^{2\varsigma+1}}}, \quad \varsigma = 0, 1, 2, \ldots, M,$$

where $M = M_0$ is the degree of the polynomials $w_0^{\circ}(\zeta), v_0^{\circ}(\zeta)$ in (6.11), and $\mathrm{Int}(x)$ denotes the integer part of a number x.

In this case (for $a_0^8 \neq \lambda$), the initial WP splits into two identical packets running in the opposite directions with the group velocities

$$v_g^{\pm} = \pm\frac{2\left(a_0^8 - \lambda\right)}{a_0^5H_0}. \tag{6.139}$$

The wave number $p^{\pm}$ and the frequency $|\omega^{\pm}|$ are constant. Over time, the packet width, which is proportional to $\varepsilon^{1/2}/\Im b^{\pm}(t)$, increases and the amplitude of waves decreases (here $A_{\varsigma}^{\pm}(t) \to 0$ as $t \to \infty$ for any fixed ς). This means that the running WPs "spread" over the shell surface.

If $a_0^8 = \lambda$, then the initial WP does not split, but spreads over the shell, the shell oscillating with the maximal amplitude at the generatrix $\varphi = 0$.

In Fig. 6.2, the plots of functions $v_g^+(a_0)$, $\omega^+(a_0)$ for $n = 1$ and various l are shown. The numbers from 1 to 4 mark the curves for shells of length $l = 1.5; 1; 2; 4$, respectively. The negative value of the group velocity v_g^+ means that the 1^+st packet runs in the negative direction of varying the φ -coordinate. As can be seen, with increasing the wave number a_0, the influence of the shell length (as well as of the number of semi-waves along the generatrix) on the group velocity and the frequency of oscillations decreases. Note also that decreasing a_0 results in increasing the error of solution (6.76), (6.77) because, as the plots show, the necessary estimates (6.20) for the functions v_g^+, ω^+ do not hold.

The simplest example considered allows comparing our solution (after some simplifications) with the classical Boussinesq beam solution. Consider the cross-section $s = l/(2n)$. In this section, the constructed solution identically satisfies (in the leading approximation) the equation

$$\varepsilon^8\frac{\partial 4}{\partial\varphi^4}\left[\frac{1}{k(\varphi)}\frac{\partial^4U_n}{\partial\varphi^4}\right] + \varepsilon^6\frac{\partial^4}{\partial\varphi^4}\left[\frac{1}{k(\varphi)}\frac{\partial^2U_n}{\partial t^2}\right] + \lambda_nk(\varphi)U_n = 0, \tag{6.140}$$

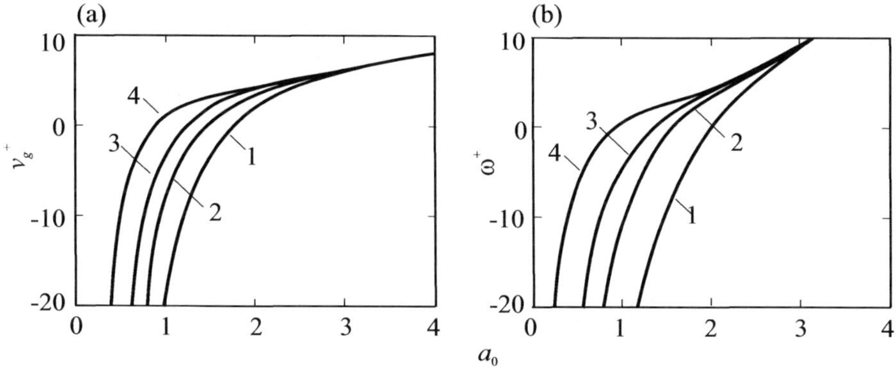

FIGURE 6.2: Group velocity v_g^+ (a) and frequency ω^+ (b) versus wave parameter a_0 for different values of length: curve 1 – $l = 1$, 2 – $l = 1.5$, 3 – $l = 2$, 4 – $l = 4$.

where $n = 1, 2, \ldots$ We assume $k \equiv 1$, $\lambda_n = 0$ and integrate Eq. (6.140) four times with respect to the variable φ. Requiring the conditions

$$U_n \to 0, \ \partial U_n/\partial\varphi \to 0 \quad \text{as} \quad |\varphi| \to +\infty$$

be fulfilled, we arrive at the classical equation (the subscript n is below omitted)

$$\frac{\partial^2 U}{\partial t^2} + \varepsilon^2 \frac{\partial^4 U}{\partial \varphi^4} = 0 \tag{6.141}$$

describing free vibrations of the Bernoulli-Euler beam. Its solution with arbitrary conditions

$$U|_{t=0} = \Theta_1(\varphi), \quad \dot{U}|_{t=0} = \varepsilon\Theta_2''(\varphi) \tag{6.142}$$

is known as the Boussinesq solution [212]

$$\begin{aligned} U(\varphi, t) = &\frac{1}{2(2\pi\varepsilon t)^{1/2}} \int\limits_{-\infty}^{+\infty} \Theta_1(\varphi - \eta) \left(\cos\frac{\eta^2}{4\varepsilon t} + \sin\frac{\eta^2}{4\varepsilon t} \right) d\eta \\ &+ \frac{1}{2(2\pi\varepsilon t)^{1/2}} \int\limits_{-\infty}^{+\infty} \Theta_2(\varphi - \eta) \left(\sin\frac{\eta^2}{4\varepsilon t} - \cos\frac{\eta^2}{4\varepsilon t} \right) d\eta. \end{aligned} \tag{6.143}$$

For simplicity, we set $w_0^\circ = 1, v_0^\circ = 0$ and compare the asymptotic solution (6.76), (6.77) with the Boussinesq one (6.143), where

$$\Theta_1 = c_{0,0}^\circ \exp\left[\mathrm{i}\varepsilon^{-1} \left(a_0\varphi + \frac{1}{2} b_0 \varphi^2 \right) \right], \quad \Theta_2 = 0. \tag{6.144}$$

We note that passage to the limit, $\lambda \to 0$, means the transition to an infinitely long cylindrical shell or its replacement by an infinite plane strip.

Figure 6.3 shows the transverse deflection of the shell for $h/R = 10^{-3}, R = 100$ sm, $\nu = 0.3, a_0 = 2, b_0 = 2\mathrm{i}, n = 1$ at the time moment $t = 0.1$. The

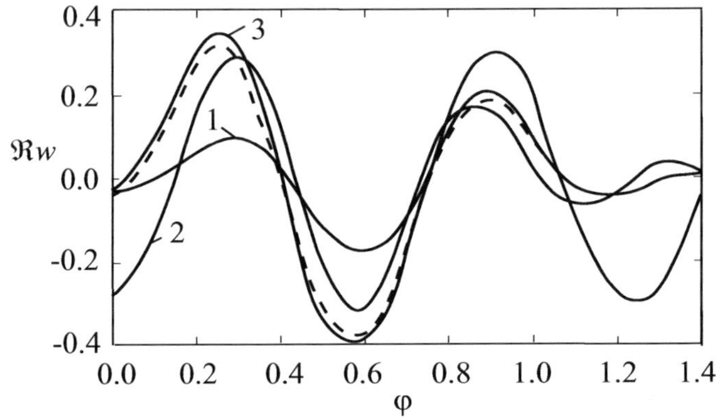

FIGURE 6.3: Transverse deflection in the section $s = l/(2n)$ for $l = 0.5; 1.2; 3.3$ (curves 1, 2, 3, respectively). Dotted curve corresponds to the Boussinesq solution.

numbers 1, 2, 3 correspond to the real part of the asymptotic solution, $\Re w$, for shells of length $l = 0.5; 1.2; 3.3$, respectively, and the dashed line does to the Boussinesq solution for the beam. It can be seen that, as the shell length increases, the difference between asymptotic solution in the form of a running WP and the classical one (6.143) decreases. We also point out that for $\lambda = 0$ formula (6.139) for the group velocity coincides with a similar formula, which can be easily derived from the Boussinesq solution (6.143).

Example 6.2. Let us now consider a non-circular cylindrical shell which is simply supported along the straight ends ($s_1 = 0$, $s_2 = l$) and has an ellipse in the cross-section with the semi-axes e_1, e_2 ($e_1 \neq e_2$). It is convenient to pass to the polar coordinate system (Figure 6.4), where the ellipse is given by the formula

$$r(\theta) = \sqrt{\frac{e_1^2}{1 - \delta^2 \sin^2 \theta}}, \quad -\pi < \theta \leq \pi, \quad \delta = \sqrt{1 - e_2^2/e_2^1},$$

and the shell curvature is

$$k = \frac{r^2 + 2r'^2 - rr''}{\left(r^2 + r'^2\right)^{3/2}}.$$

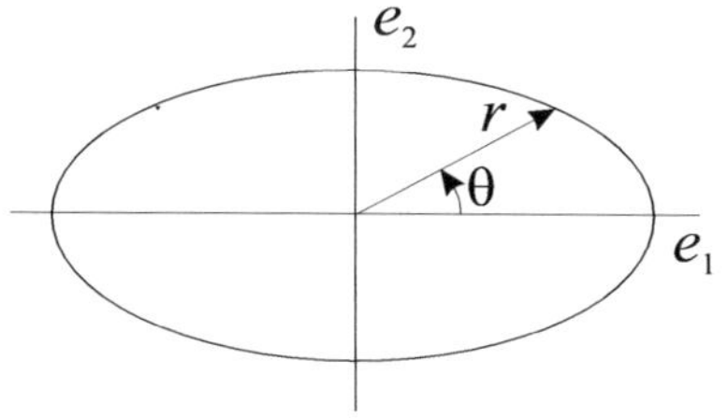

FIGURE 6.4: Ellipse with semi-axes e_1, e_2 in the shell cross-section.

In order to pass in expansions (6.76), (6.77) to the new coordinate system, one needs to make the substitutions

$$\varphi = \int_0^\theta \kappa(\theta)d\theta, \quad q^\pm = \int_0^{\theta^\pm} \kappa(\theta)d\theta \tag{6.145}$$

with $\kappa(\theta) = \sqrt{r^2(\theta) + r'^2(\theta)}$ and to take into account the following relations:

$$\frac{dq^\pm}{dt} = \kappa\frac{d\theta^\pm}{dt}, \quad \frac{\partial H^\pm}{\partial q^\pm} = \frac{1}{\kappa}\frac{\partial H^\pm}{\partial \theta^\pm}. \tag{6.146}$$

In Eqs. (6.145), the functions $\theta = \theta^\pm(t)$ define the center of the $n^\pm$th WP in the polar coordinate system.

In this example, the function $g(\varphi) = \pi^4 n^4 k^2(\varphi)/l^4$ is even on the set $-\varphi_1 < \varphi < \varphi_1$, where $\varphi_1 = \kappa(\pi)$, and satisfies all the conditions formulated in Remark 6.5. The center of the initial WP is on the weakest generatrix $\varphi = 0$ $(\theta = 0)$, on which the function $g(\varphi)$ has a minimum.

Figure 6.5 shows the results of numerical integration of the Hamiltonian system for the shell with parameters $l = 1.5$, $e_1 = 2$, $e_2 = 1$ at $n = 1$ (one semi-wave along the generatrix) and for different values of the initial wave number a_0. The plots of the functions $v_g^\pm(t), \omega^\pm(t)$ are shown in Fig. 6.6. The numbers 1, 2, 3 indicate solutions for $a_0 = 0.8$, 1.5, 2.2, respectively. Curves 1 and 2 correspond to the cases $0 < a_0 < g_0^{1/8}$ and $a_0 > g_0^{1/8}$, respectively, under the reflection conditions (6.116), and curve 3 relates to the case $a_0 > g_0^{1/8}$ under condition (6.115).

The plots depicted in Figs. 6.5, 6.6 completely correspond to the above qualitative analysis of the constructed solution. In particular, if $a_0 = 0.8$, then at the time moment $t_r^- \approx 0.30$ it is observed the effect of reflecting the 1^-st packet from the generatrix $\theta = \theta_r^- \approx 0.65$; at $t = t_0^- \approx 0.73$, the center of the 1^-st packet returns to its original position. With the increase of a parameter a_0, the moment of reflection, $t_r^\pm$, occurs later, the generatrix $\theta = \theta_r^\pm$ shifts towards the increase in the shell curvature (for $a_0 = 1.5$, one has $t_r^+ \approx 0.52$, $\theta_r + \approx 0.85$), and for sufficiently large values of a_0 (in this case, at

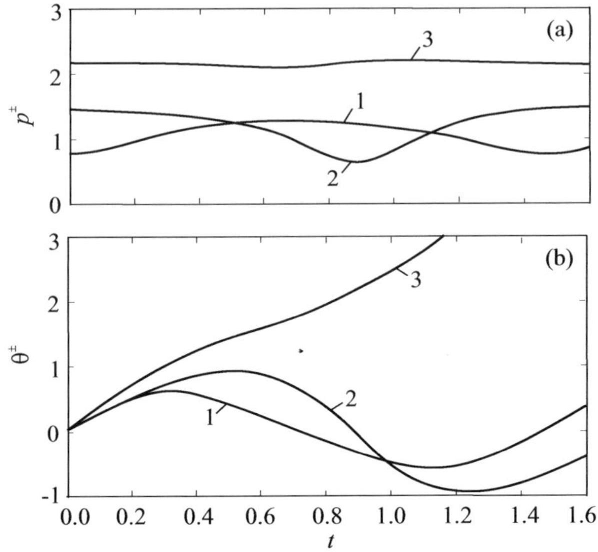

FIGURE 6.5: Solutions of the Hamiltonian system for different $a_0 = 0.8; 1.5; 2.2$ (curves 1, 2, 3, respectively).

$a_0 = 2.2$) under conditions (3.32), the effect of reflection of the $1^{\pm}$st WP is absent.

In the cases considered, the boundaries n_1 and n_2 of segment (6.134) determining the reflection condition of the $n^{\pm}$th packets are as follows: $n_1 = 1$, $n_2 = 2$ with $a_0 = 0.8$ and $n_1 = 1$, $n_2 = 8$ for $a_0 = 1.5$. Thus, in the first case, only the $1^{\pm}$st and $2^{\pm}$nd WPs satisfy the conditions of reflection, while the remaining packets with $n \geq 3$ travelling in the circumferential direction without obstacle.

The calculations performed at large time intervals confirms the above conclusion about the periodicity of the functions $p^{\pm}(t)$, $q^{\pm}(t)$, $v_g^{\pm}(t)$, $\omega^{\pm}(t)$ for the cases $a_0 = 0.8$ and $a_0 = 1.5$. This means that the wave packets perform oscillating motions in the vicinity of the generatrix $\theta = 0$, being repeatedly reflected from the lines $\theta = \theta_r^{\pm}$.

Figures 6.7, 6.8 demonstrate the plots of the functions $\Im b^{\pm}(t)$, $w_{\max}^{\pm}(t) = \max_{-\pi<\theta\leq\pi} w^{\pm}(t)$ for the shell with parameters $h/R = 10^{-3}$, $\nu = 0.3$ for $b_0 = i$ and $w_0^{\circ} = 1$ and $v_0^{\circ} = 0$. Here, $1/\mathrm{Im}\, b^{\pm}(t)$ specifies the width of WPs, and $w_{\max}^{\pm}(t)$ is the maximal amplitude of waves in these packets. As can be seen, in the cases of $a_0 = 0.8$ and $a_0 = 1.5$, it is observed strong focusing of the WPs, which is accompanied by increasing the wave amplitudes. For $a_0 = 2.2$, when the reflection effect is absent, focusing is absent. One can see the obvious dependence of the degree of both focusing and growing amplitudes of WPs on the initial wave number a_0.

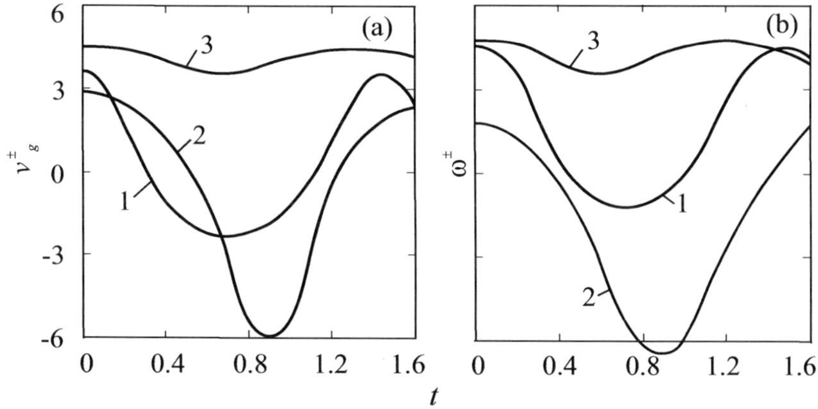

FIGURE 6.6: Functions $v_g^\pm(t), \omega^\pm(t)$ for $a_0 = 0.8, 1.5, 2.2$ (curves 1, 2, 3, respectively).

Example 6.3. In this example, we consider a circular simply supported cylindrical shell with an oblique edge (see Fig. 6.9). Here

$$k = 1, \quad s_1 = 0, \quad s_2(\varphi) = l_0 + (\cos\varphi - 1)\tan\beta,$$
$$\lambda(\varphi) = \frac{\pi^4 n^4}{s_2^4(\varphi)}, \quad z(s, \varphi) = \sin\frac{\pi n s}{s_2(\varphi)},$$

where β is the angle between the planes of upper and lower edges.

Let the center of initial WP be at the generatrix $\varphi = 0$ of length l_0. If $0 < \beta < \arctan(l_0/2)$, then this generatrix is the longest and, therefore, the weakest. This line is the center of the stationary WPs being the localized eigenforms corresponding to low-frequency vibrations of the shell (see Example 4.2 in Section 4.3). For $\beta < 0$, the generatrix $\varphi = 0$ is the shortest, and the opposite line $\varphi = \pi$ is the weakest. If $\beta = 0$, then the weakest line is absent.

The outcomes of the numerical solution of the Hamiltonian system is given in Fig. 6.10. The calculations were performed for $l_0 = 2$, $a_0 = 1.5$, $n = 1$ and different values of the cut angle $\beta = -30°, 0°, 6°, 30°, 44°$. The plots of functions $\omega^\pm(t), v_g^\pm(t)$ for the same values of β is shown in Fig. 6.11. For the assumed values of parameters a_0, l_0, the inequality $a_0 > g_0^{1/8}$ holds.

It may be seen that for $\beta = -30°, 0°, 6°$ the functions $q^\pm(t)$ grow unrestrictedly with increasing time t. This means that all WPs propagate in the circumferential direction without the reflection effects. If $\beta = 30°, 44°$, then the condition of reflection (6.116) holds for the $1^\pm$st WPs. The figures show that as the cut angle β increases, the reflection of the $1^\pm$st packets occurs earlier, and the reflection lines $\varphi = q_r^\pm$ approach the weakest generatrix $\varphi = 0$. We note also that for $\beta = 30°$, one gets $n_1 = 1$, $n_2 = 5$, i.e., only WPs with numbers $n = 1, ..., 5$ satisfy the conditions of reflection, while the remaining packets behave according to the scheme described above when $\beta = 6°$.

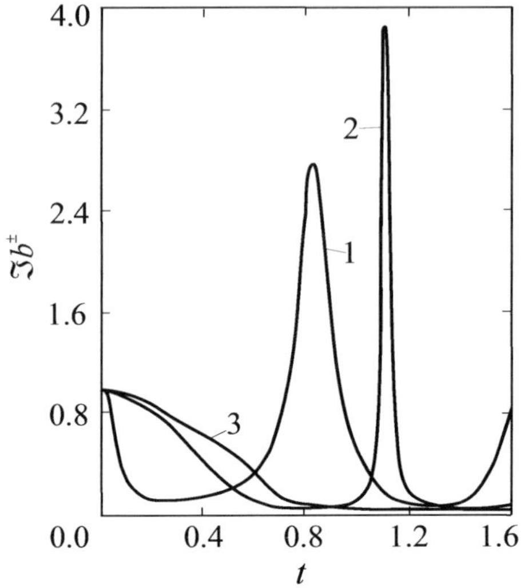

FIGURE 6.7: Function $\Im b^{\pm}(t)$ for $a_0 = 0.8, 1.5, 2.2$ (curves 1, 2, 3, respectively).

If $\beta = 44°$, then $n_1 = 1$, $n_2 = 860$ which means that almost all WPs satisfying the estimate $n \sim 1$ (see above conditions (6.7) for amplitudes in the initial conditions (6.5)) are reflected from some generatrix $\varphi = q_r^{\pm}$. In the cases where the reflection effects take place, the functions $p^{\pm}(t), q^{\pm}(t), v_g^{\pm}(t), \omega^{\pm}(t)$ are periodic (see Remark 6.5).

In Figure 6.12, the plots of the functions $\mathrm{Im}\, b^{\pm}(t), w_{\max}^{\pm}(t)$ for different angles β and fixed $b_0 = \mathrm{i}$, $h/R = 10^{-3}, \nu = 0.3$, $w_0^\circ = 1$, v_0° are presented. The values of parameters a_0, l_0 are the same. For $\beta = 6°$, when the effects of reflection are absent, the qualitative pattern of localized wave processes is characterized by a rapid "spreading" of packets over the shell surface. An increase in the cut angle leads to a qualitative change in the shell dynamics. In the case of reflection of wave packets, one can observe their focusing which is accompanied by noticeable increasing amplitudes. In this case, the larger the cut angle, the stronger the focusing and the greater the amplitude of waves at the moment of focusing.

Figure 6.13 displays a wave pattern on the shell surface with the slant angle $\beta = 44°$ in the cross section $s = s_2(\varphi)/2$ in different points of time: for $t = 0$ (the initial instant time), $t \approx 0.5$ (the point of time corresponding to the reflection of the $1^{\pm}$st WP from the generatrix $\varphi \approx 0.53$), $t \approx 1.55$ (the instant time corresonding to the second focusing), $t \approx 1.75$ (the instant time of returning the WP center to the initial position $\varphi = 0$ after repeated reflections from the generatrix $\varphi \approx -0.53$). The numbers 1 and 2 denote the solutions with the initial amplitudes $w_0^\circ = 1$, $v_0^\circ = 0$ and $w_0^\circ = 1$, $v_0^\circ = 2$,

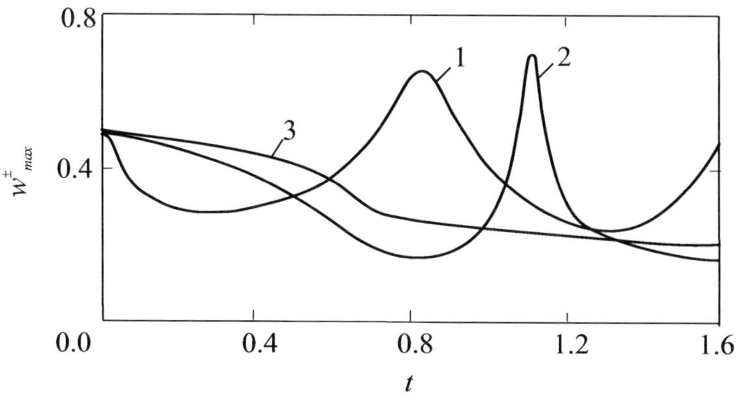

FIGURE 6.8: The maximal amplitude $w^{\pm}_{max}$ for $a_0 = 0.8$, 1.5, 2.2 (curves 1, 2, 3, respectively).

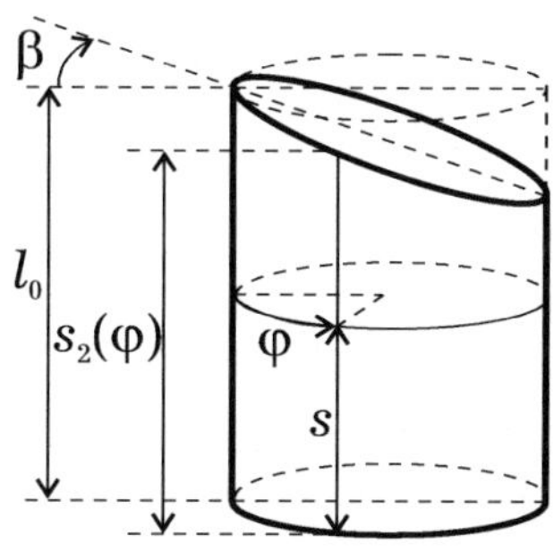

FIGURE 6.9: Circular cylindrical shell with oblique edge.

respectively. As expected, the presence of the initial positive velocities leads to an increase in the wave amplitudes.

The examples considered in this section allow concluding that the dynamic characteristics of the travelling WPs are strongly dependent of the shell geometry. In particular, the presence of a variable curvature or oblique edges can lead to a strong localization of non-stationary bending vibrations in areas adjacent to the weakest generatrix, as well as to the effects of the reflection being accompanied by focusing and increasing amplitudes. The technological imperfections in the shell geometry (the initial deviations of the shell surface from the cylindrical one) may also cause localization of non-stationary vibrations in thin shells.

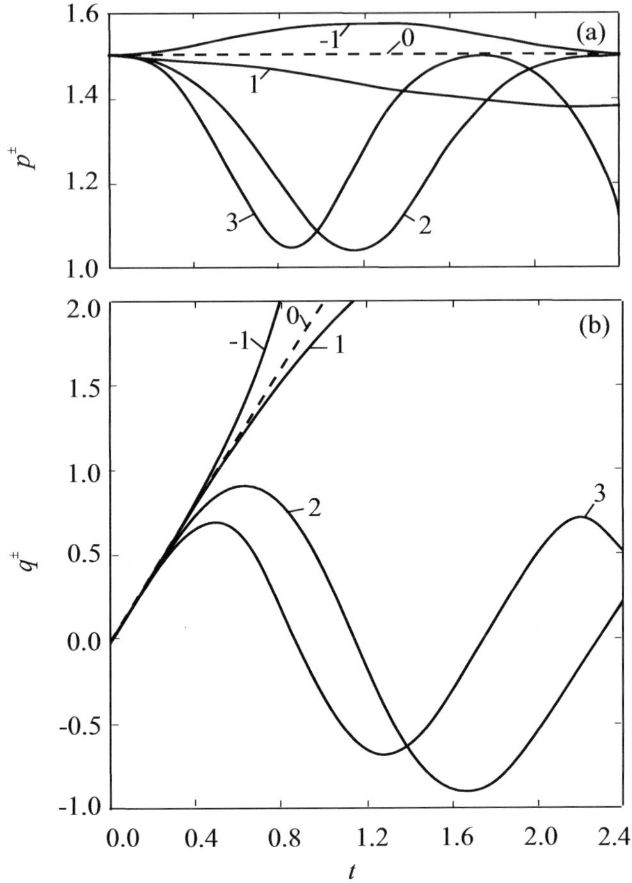

FIGURE 6.10: Solutions of the Hamiltonian system at the cut angles $\beta = -30^\circ, 0^\circ, 6^\circ, 30^\circ, 44^\circ$ (curves $-1, 0, 1, 2, 3$, respectively).

6.4 Constructing solutions in the form of running wave packets based on Maslov's method

Here, we consider the special case [142, 144], when the shell is bounded by two plane edges perpendicular to the generatrix:

$$0 \leq s \leq l, \quad \varphi_1 \leq \varphi \leq \varphi_2.$$

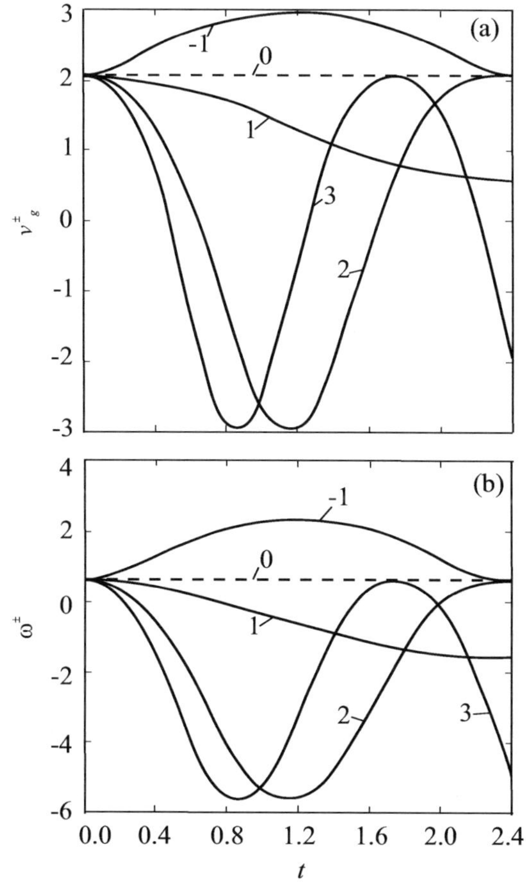

FIGURE 6.11: Group velocity $v_g^{\pm}(t)$ and frequency $\omega^{\pm}(t)$ at the cut angles $\beta = -30°,\ 0°,\ 6°,\ 30°,\ 44°$ (curves $-1, 0, 1, 2, 3$, respectively).

Let $S_0(\varphi) = S_{01}(\varphi) + \mathrm{i}S_{02}(\varphi)$ be the phase function, where $S_{01} = \Re S_0(\varphi)$, and the function $S_{02} = \Im S(\varphi)$ appearing in (6.5) satisfies the conditions

$$S_{02}(0) = 0, \quad \frac{dS_{02}(0)}{d\varphi} = 0, \quad \frac{d^2 S_{02}(0)}{d\varphi^2} > 0. \tag{6.147}$$

In this case, the solution of the initial-boundary problem (6.3)-(6.5) with expansions (6.10) taking into account can be represented in the form of series

$$\begin{aligned} W &= \sum_{n=1}^{\infty} U_n(\varphi, t, \varepsilon) z_n(s), \\ \Phi &= \sum_{n=1}^{\infty} F_n(\varphi, t, \varepsilon) z_n(s) \end{aligned} \tag{6.148}$$

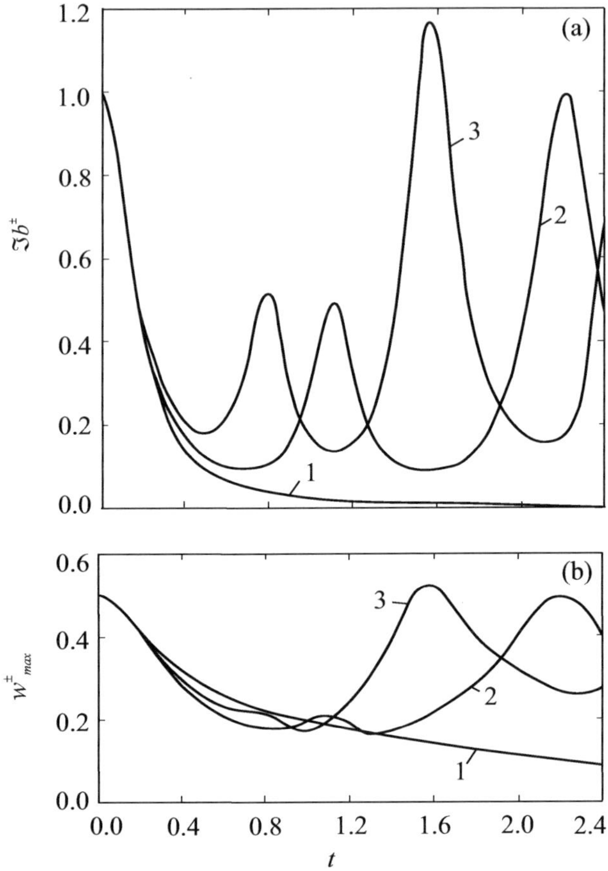

FIGURE 6.12: Plots of the functions $\Im b^{\pm}(t), w^{\pm}_{\max}(t)$ for the cut angles $\beta = -30°, 0°, 6°, 30°, 44°$ (curves $-1, 0, 1, 2, 3$, respectively).

in the beam functions $z_n(s)$. We substitute (6.148) into Eqs. (6.3). Then, taking into account the orthogonality of the functions $z_n(s)$ on the segment $0 \le s \le l$, we arrive at the sequence of equations

$$\varepsilon^8 \frac{\partial^4}{\partial \varphi^4}\left[\frac{1}{k(\varphi)}\frac{\partial^4 U_n}{\partial \varphi^4}\right] + \varepsilon^6 \frac{\partial^4}{\partial \varphi^4}\left[\frac{1}{k(\varphi)}\frac{\partial^2 U_n}{\partial t^2}\right] + \lambda_n k(\varphi) U_n + \mathbf{L}_\varepsilon U_n = 0, \quad (6.149)$$

with respect to the functions $U_n (n = 1, 2, \ldots)$ In Eq. (6.149), L_ε is the differential operator which in the asymptotic construction used below yields terms of order $O\left(\varepsilon^2\right)$ with respect to the leading ones. When constructing the leading approximation of the asymptotic solution, these terms can be omitted.

The initial conditions for U_n read

$$U_n|_{t=0} = w_n^{\circ}(\varphi, \varepsilon) \exp\left[i\varepsilon^{-1} S_0(\varphi)\right], \quad (6.150)$$

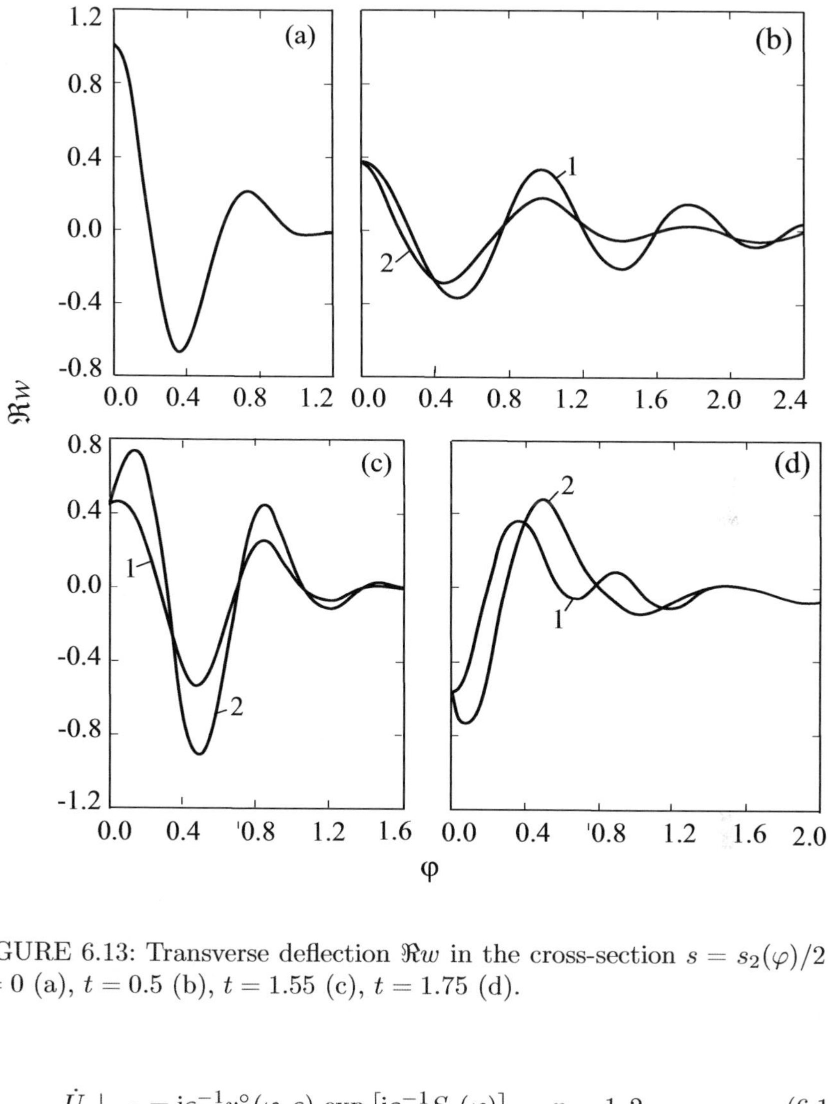

FIGURE 6.13: Transverse deflection $\Re w$ in the cross-section $s = s_2(\varphi)/2$ for $t = 0$ (a), $t = 0.5$ (b), $t = 1.55$ (c), $t = 1.75$ (d).

$$\dot{U}_n|_{t=0} = \mathrm{i}\varepsilon^{-1} v_n^\circ(\varphi, \varepsilon) \exp\left[\mathrm{i}\varepsilon^{-1} S_0(\varphi)\right], \quad n = 1, 2, \ldots \tag{6.151}$$

In what follows, the subscript n in all functions will be omitted.

Consider problem (6.149)-(6.150) for a fixed n and under particular assumptions for the amplitudes w°, v° from the initial conditions (6.150), (6.151). Let

$$\begin{aligned} w^\circ &= \varepsilon^{-1} c_{0,2}^\circ \varphi^2 + \varepsilon^{-1/2} c_{0,1}^\circ \varphi + c_{0,0}^\circ, \\ v^\circ &= \varepsilon^{-1} d_{0,2}^\circ \varphi^2 + \varepsilon^{-1/2} d_{0,1}^\circ \varphi + d_{0,0}^\circ, \end{aligned} \tag{6.152}$$

where $c^\circ_{0,m}$, $d^\circ_{0,m}$ ($m = 0, 1, 2, \ldots$) are complex numbers such that $|c^\circ_{0,m}|$, $|d^\circ_{0,m}| \sim 1$. A much wider class of functions $w^\circ(\varphi, \varepsilon)$, $v^\circ(\varphi, \varepsilon)$ will be introduced at the end of this section.

Under the above assumptions, the Cauchy problem (6.149)-(6.151) can be solved using the method developed by V.P. Maslov [131] for approximate integration of the quantum mechanics equations.

6.4.1 Canonical system of equations and auxiliary initial conditions

In view of the initial condition (6.150), the solution of equation (6.149) will be sought in the form

$$U(\varphi, t, \varepsilon) = u(\varphi, t, \varepsilon) \exp\left[\mathrm{i}\varepsilon^{-1} S(\varphi, t)\right],$$
$$u = \sum_{j=-2}^{\infty} \varepsilon^{j/2} u_j(\varphi, t), \tag{6.153}$$

where $u(\varphi, t), S(\varphi, t)$ are complex-valued and sufficiently differentiable functions with respect to φ and t, and $\Im S(\varphi, t) > 0$ in some two-sided punctured neighborhood of the generatrix $\varphi = q(t)$ on which $\Im S = 0$. This inequality guarantees a decrease in the amplitude of waves with a distance from the asymptotic line $\varphi = q(t)$.

Similar formal solutions, oscillating in a small neighborhood of the space-time ray and exponentially decaying outside this small neighborhood, were constructed (by other methods) in acoustics [16, 107]. The solutions for wave equations constructed in this form have been called *"quasi-photons"* [19].

We assume that

$$\frac{\partial S}{\partial \varphi} \neq 0 \tag{6.154}$$

in the region $\mathfrak{D} = \{\varphi_1 \le \varphi \le \varphi_2,\ 0 \le t \le t' < +\infty\}$. Let us substitute (6.153) into Eq. (6.149) and equate the coefficients with the same powers of ε. As a result, we get the canonical system of equations for the unknowns S, u_j:

$$\varepsilon^{-1} \exp(\mathrm{i}\varepsilon^{-1} S) u_{-2} \left[S_t \mp H\left(S_\varphi,\, \varphi\right)\right] = 0, \tag{6.155}$$

$$\varepsilon^{0} \exp(\mathrm{i}\varepsilon^{-1} S) \left(\dot{u}_{-2} \pm \hat{\mathbf{\Pi}} u_{-2}\right) = 0,$$
$$\varepsilon^{1/2} \exp(\mathrm{i}\varepsilon^{-1} S) \left(\dot{u}_{-1} \pm \hat{\mathbf{\Pi}} u_{-1}\right) = 0, \tag{6.156}$$
$$\varepsilon \exp(\mathrm{i}\varepsilon^{-1} S) \left(\dot{u}_{0} \pm \hat{\mathbf{\Pi}} u_{0} \mp \hat{\mathbf{R}} u_{-2}\right) = 0, \ \ldots,$$

where

$$H(p, \varphi) = \sqrt{p^4 + \frac{\lambda k^2(\varphi)}{p^4}}, \tag{6.157}$$

$$
\begin{aligned}
\hat{\mathbf{\Pi}} &= H_p\left(S_\varphi,\, \varphi\right) \frac{\partial}{\partial \varphi} + \frac{1}{2} H_{pp}\left(S_\varphi,\, \varphi\right) S_{\varphi\varphi} + G(\varphi, t), \\
G(\varphi, t) &= -\tfrac{4H_\varphi(S_\varphi,\varphi)}{S_\varphi} - \frac{H_p\left(S_\varphi,\, \varphi\right) H_\varphi\left(S_\varphi,\, \varphi\right)}{2H\left(S_\varphi,\, \varphi\right)} \\
&\quad - \tfrac{2k'(\varphi)\left[S_\varphi^4 - H^2(S_\varphi,\varphi)\right]}{k(\varphi) S_\varphi H(S_\varphi,\varphi)}, \\
\hat{\mathbf{R}} &= \frac{\mathrm{i}}{2} H_{pp}\left(S_\varphi,\, \varphi\right) \frac{\partial^2}{\partial \varphi^2}.
\end{aligned} \tag{6.158}
$$

Here, the subscripts t, φ, p mean differentiation with respect to the corresponding variables, and signs $\mp$, $\pm$ indicate the availability of the positive and negative branches of solution being under construction. In what follows, U^+, S^+, u^+ denote functions corresponding to the positive branch (in Eqs. (6.155), (6.156), the upper signs correspond to them), and U^-, S^-, u^- are functions for the negative branch.

Following the tradition, equation (6.155) will be called the Hamilton-Jacobi equation, and the sequence of equations (6.156) is the sequence of transfer equations. Equations for determining the functions u_j with $j \geq 1$ are not given here. To analyze the error of the constructed asymptotic solution of problem (6.149)-(6.151), the left-hand sides of the Hamilton-Jacobi equation and the transport equations are written out together with their coefficients $\varepsilon^{-1} \exp(\mathrm{i}\varepsilon^{-1} S) u_{-2}$ and $\varepsilon^j \exp(\mathrm{i}\varepsilon^{-1} S)$, respectively, with which they appear in Eq. (6.149) after introducing ansatz (6.153). When writing the transport equations (6.156), we disregarded summands which do not affect an accuracy of the functions u_j defined below with $j = -2, -1, 0$. Omitted terms, in particular, generated by the operator $\mathbf{L}_\varepsilon$ in (6.149), should be considered only when finding u_j for $j \geq 1$.

The method of integration of Eqs. (6.155), (6.156) used below does not directly satisfy the initial condition for velocities (6.151). Bearing in mind the presence of two branches of solutions of Eq. (6.149), we introduce the auxiliary initial conditions

$$
U^\pm|_{t=0} = w_0^\pm \exp\left[\mathrm{i}\varepsilon^{-1} S_0(\varphi)\right], \tag{6.159}
$$

where

$$
\begin{aligned}
w_0^\pm &= \varepsilon^{-1} c_{0,2}^\pm \varphi^2 + \varepsilon^{-1/2} c_{0,1}^\pm \varphi + c_{0,0}^\pm, \\
c_{0,j}^+ &+ c_{0,j}^- = c_{0,j}^\circ, \quad j = 0, 1, 2.
\end{aligned} \tag{6.160}
$$

The substitution of (6.153) into (6.159), (6.160) results in the sequence of conditions for the functions $S^\pm, u_j^\pm$:

$$
S^\pm|_{t=0} = S_0(\varphi), \tag{6.161}
$$

$$
u_j^\pm|_{t=0} = c_{0,-j}^\pm \varphi^{-j}, \quad j = -2, -1, 0. \tag{6.162}
$$

Next, we consider the Cauchy problems (6.155), (6.161) and (6.156), (6.162). The superposition of positive and negative branches of solutions (6.159) will allow satisfying to the both initial conditions, (6.150) and (6.151). However, before proceeding to the construction of approximate solutions of these problems, we introduce into consideration the concept of asymptotically equivalent functions.

6.4.2 Class of asymptotically equivalent functions

Let $f(x) \geq 0$, $g(x)$ be smooth real-valued functions defined on the set of real numbers $\mathbb{R}$.

Definition 6.1 We write $g = O_f\left(\varepsilon^m\right)$ if

$$\frac{\partial^\alpha g}{\partial x^\alpha}\,\mathrm{e}^{-f/\varepsilon} = O\left(\varepsilon^{m-\alpha/2}\right) \tag{6.163}$$

for $m - \alpha/2 \geq 0$ at any compact set of real numbers.

For example, $x^m = O_{x^2}\left(\varepsilon^{m/2}\right)$, where m is a non-negative integer number.

We point out some basic properties of functions from the class $O_f\left(\varepsilon^m\right)$:

$$\begin{gathered} O\left(\varepsilon^m\right) \subset O_f\left(\varepsilon^m\right), \\ O_f\left(\varepsilon^{m_1}\right) O_f\left(\varepsilon^{m_2}\right) = O_f\left(\varepsilon^{m_1+m_2}\right), \\ \sqrt{\varepsilon}\partial/\partial x\, :\, O_f\left(\varepsilon^m\right) \to O_f\left(\varepsilon^m\right). \end{gathered} \tag{6.164}$$

The last of these properties indicates that the operator $\sqrt{\varepsilon}\partial/\partial x$ does not change the class of a function.

6.4.3 Approximate solution of the canonical system (Maslov's method)

Equation (6.155) is non-linear and does not admit an explicit solution. To solve problems (6.155), (6.161) and (6.156), (6.162), we use the method proposed by Maslov [131]. The idea of the method is as follows. Let $S(\varphi, t) = S_1(\varphi, t) + \mathrm{i}\, S_2(\varphi, t)$, where $\mathrm{i} = \sqrt{-1}$, $S_1 = \Re\, S$, $S_2 = \Im\, S$. Then in the region where $S_2(\varphi, t) > 0$, the function $\exp\left[\mathrm{i}\varepsilon^{-1} S(\varphi, t)\right]$ is of the order $O(\varepsilon^\infty)$. Therefore, to construct an approximate solution of Eq. (6.149) in the form of (6.153), the canonical system of Eqs. (6.155), (6.156) is to be solved only in a small neighborhood of the region

$$\tilde{\mathfrak{D}} = \{(\varphi, t) \in \mathbb{R}^2 : \operatorname{Im} S(\varphi, t) = 0\}, \tag{6.165}$$

at that everywhere beyond $\tilde{\mathfrak{D}}$, the inequality $S_2 > 0$ has to be satisfied.

In what follows, we will omit the subscripts $\pm$ bearing in mind the positive branch of a solution. The unknown u_j will be sought in the class of asymptotically equivalent functions $O_f\left(\varepsilon^m\right)$. Let $u_j \in O_{S_2}\left(\varepsilon^{|j|/2}\right)$, where $j = -2, -1, 0$. Then taking into account the properties of functions form $O_f\left(\varepsilon^m\right)$, one can conclude that to construct an approximate solution of Eq. (6.149) with an accuracy up to $O\left(\varepsilon^{3/2}\right)$, the Hamilton-Jacobi equation (6.155) and the sequence of Eqs. (6.156) with respect to $u_j (j = -2, -1, 0)$ may be solved with an accuracy up to $O_{S_2}\left(\varepsilon^{3/2}\right)$ and $O_{S_2}\left(\varepsilon^{(|j|+1)/2}\right)$, respectively.

We shall use the theorems proved in book [131] and allowing us to seek approximate solutions of Eqs. (6.155), (6.156) in the class of asymptotically equivalent functions.

Let $p(t), q(t)$ be a solution of the Hamiltonian system

$$\begin{gathered}\dot{p} = -H_q(p,q), \quad \dot{q} = H_p(p,q), \\ p(0) = \frac{dS_0(0)}{d\varphi}, \quad q(0) = 0,\end{gathered} \tag{6.166}$$

and $B(t), C(t)$ be a solution of the system in variations,

$$\begin{gathered}\dot{B} = -H_{pq}B - H_{qq}C, \quad \dot{C} = H_{pp}B + H_{pq}C, \\ B(0) = \frac{d^2S_0(0)}{d\varphi^2}, \quad C(0) = 1\end{gathered} \tag{6.167}$$

with coefficients calculated on the solution of problem (6.166).

Theorem 6.1 Let

$$\frac{dS_{01}}{d\varphi} \neq 0 \quad \text{for} \quad \varphi = 0. \tag{6.168}$$

Then there exists $t' \in (0, +\infty)$ such that for any $t \in [0, t']$ the inequality $C(t) \neq 0$ is satisfied and there exists an approximate with accuracy up to $O_{S_2}\left(\varepsilon^{3/2}\right)$ solution of the Hamilton-Jacobi equation (6.155),

$$\begin{gathered}S(\varphi, t) = S_0(0) + \int_0^t \{p(\tau)\dot{q}(\tau) - H[p(\tau), q(\tau)]\}d\tau \\ +p(t)[\varphi - q(t)] + \frac{1}{2}\frac{B(t)}{C(t)}[\varphi - q(t)]^2,\end{gathered} \tag{6.169}$$

which satisfies the inequality $\Im S^+(\varphi, t) > 0$ in the region $\tilde{\mathfrak{D}}$ and the initial condition (6.161) with an accuracy up to $O_{S_{02}}\left(\varepsilon^{3/2}\right)$.

Remark 6.8. Theorem 6.1 follows from Theorem 2.1 proved in [131] (see page 46) to more general case where $H(p, x, t) \in C^\infty(\mathbb{R}^{2n+1})$, $p, x \in \mathbb{R}^n$. In our case, $n = 1, x \equiv \varphi$, and the Hamiltonian function (6.157) is considered at the set

$$\mathbb{K} = \{0 < p < +\infty, \ \varphi_1 \leq \varphi \leq \varphi_2\}.$$

From Theorem 6.1, it follows that requirement (6.154) is always satisfied.

Before proceeding to solving the sequence of equations (6.156) with the initial conditions (6.160), we write down the inhomogeneous transport equation in the general case:

$$\frac{\partial y}{\partial t} + \hat{\mathbf{\Pi}}y = F(\varphi, t), \tag{6.170}$$

where the operator $\hat{\mathbf{\Pi}}$ is introduced in accordance with (6.158), and $F(\varphi,t)$ is a smooth function. Consider the initial condition

$$y|_{t=0} = y_0(\varphi) \tag{6.171}$$

for Eq. (6.170), where $y_0(\varphi)$ is also a smooth function.

Let a function $S(\varphi,t)$ be an approximate solution (6.169) of problem (6.155), (6.161) for $t \in [0,t']$. Let also the following estimates

$$F = O_{S_2(\varphi,t)}\left(\varepsilon^{N/2}\right), \quad y_0 = O_{S_{02}(\varphi)}\left(\varepsilon^{N/2}\right), \tag{6.172}$$

hold, where N is some natural number.

Definition 6.2 The smooth function $y(\varphi,t)$ satisfying Eq. (6.170) and the boundary condition (6.171) with an accuracy up to functions $O_{S_2(\varphi,t)}\left(\varepsilon^{(N+1)/2}\right)$ and $O_{S_{02}(\varphi)}\left(\varepsilon^{(N+1)/2}\right)$, respectively, will be called the approximate solution of the Cauchy problem (6.170), (6.171) in the interval $t \in [0,t']$.

We introduce the functions

$$\begin{gathered}\tilde{\Psi}(t) = \tfrac{1}{2}H_{pq}[p(t),\, q(t)] - G[q(t),\, t],\\ \gamma(\varphi,t) = \frac{\varphi - q(t)}{C(t)}.\end{gathered} \tag{6.173}$$

In what follows, $\sqrt{C(t)}$ will denote the continues branch of a function such that $\sqrt{C(0)} = 1$.

Under the conditions assumed above, the following result holds.

Theorem 6.2. An approximate, with an accuracy up to $O_{S_2(\varphi,t)}\left(\varepsilon^{(N+1)/2}\right)$, a solution of problem (6.170)-(6.172) at the interval $t \in [0,t']$ exists and has the form:

$$y(\varphi,t) = \frac{1}{\sqrt{C(t)}}\exp\int_0^t \tilde{\Psi}(t_1)dt_1 \times \left[\frac{1}{N!}\frac{\partial^N y_0}{\partial\varphi^N}(0)\,\gamma^N(\varphi,t)\right.$$
$$\left. + \int_0^t \sqrt{C(t_1)}\exp\left(-\int_0^{t_1}\tilde{\Psi}(t_2)dt_2\right) \times \frac{1}{N!}\frac{\partial^N F}{\partial\varphi^N}\, q(t_1)(C(t_1)\gamma(\varphi,t_1))^N dt_1\right]. \tag{6.174}$$

Remark 6.9. Theorem 6.2 is a particular case of Theorem 3.1 proved in [131] (see p. 93) for the case, where

$$H(p,x,t) \in C^\infty(\mathbb{R}^{2n+1}),\ p,x \in \mathbb{R}^n.$$

Applying Theorem 6.2 step by step to the first two equations from (6.156) with the initial conditions (6.162), we find that

$$\begin{gathered}u_{-2} = c^+_{0,2}\frac{[\varphi - q(t)]^2}{C^{5/2}(t)} \in O_{S_2}(\varepsilon),\\ u_{-1} = c^+_{0,1}\frac{\varphi - q(t)}{C^{3/2}(t)} \in O_{S_2}\left(\varepsilon^{1/2}\right).\end{gathered} \tag{6.175}$$

Note that

$$H_{pp}\left(S_\varphi,\varphi\right)=\frac{12H^2\left(S_\varphi,\varphi\right)S_\varphi^4+6H^4\left(S_\varphi,\varphi\right)-16S_\varphi^8}{H^3\left(S_\varphi,\varphi\right)S_\varphi^2}\in O_{S_2}(1). \qquad (6.176)$$

Then, taking into account the property of the differentiation operator of functions from the class $O_f(\varepsilon^m)$,

$$\partial/\partial\varphi:\quad O_{S_2}\left(\varepsilon^m\right)\to O_{S_2}\left(\varepsilon^{m-1/2}\right),\quad m\geq 1/2,$$

one gets $\partial^2 u_{-2}/\partial\varphi^2\in O_{S_2}(1)$. Hence,

$$\hat{\mathbf{R}}u_{-2}=\frac{\mathrm{i}}{2}H_{pp}\left(S_\varphi,\varphi\right)\frac{\partial^2 u_{-2}}{\partial\varphi^2}\in O_{S_2}(1). \qquad (6.177)$$

Thus, all the conditions of Theorem 6.2 are satisfied. Applying it once more to the third equation of system (6.156), we find that

$$u_0=\frac{c_{0,0}^+}{C^{1/2}(t)}+\frac{\mathrm{i}c_{0,2}^+H_{pp}[p(t),q(t)]\,t}{C^{3/2}(t)}. \qquad (6.178)$$

The constructed functions $u_j(\varphi,t)$ satisfy Eqs. (6.156) and the initial conditions (6.162) with an accuracy up to $O_{S_2}\left[\varepsilon^{(|j|+1)/2}\right]$ and $O_{S_{02}}\left[\varepsilon^{(|j|+1)/2}\right]$, respectively.

Remark 6.10. Let the amplitudes w°, v° from the initial conditions (6.150), (6.151) be defined by the expansions

$$w^\circ=\sum_{j=-M}^{0}\varepsilon^{j/2}u_j^\circ(\varphi),\quad v^\circ=\sum_{j=-M}^{0}\varepsilon^{j/2}r_j^\circ(\varphi), \qquad (6.179)$$

where $u_j^\circ(\varphi), r_j^\circ(\varphi)\in O_{S_{02}}\left(\varepsilon^{|j|/2}\right)$. In particular, one can consider $u_j^\circ = c_{0,-j}^\circ\varphi^{-j}$, $r_j^\circ=d_{0,-j}^\circ\varphi^{-j}$. Then the function $U^+(\varphi,t,\varepsilon)$ should to be sought in the form (6.153), where

$$u(\varphi,t,\varepsilon)=\sum_{j=-M}^{\infty}\varepsilon^{j/2}u_j(\varphi,t). \qquad (6.180)$$

Moreover $u_{-M},u_{-M+1},\ldots,u_0$ can be found from the system of $M+1$ recurrent transfer equations:

$$\begin{aligned}&\dot{u}_j+\hat{\mathbf{\Pi}}u_j-\hat{\mathbf{R}}u_{j-2}=0,\\ &u_j|_{t=0}=u_j^\circ(\varphi),\quad j=-M,-M+1,\ldots,0,\end{aligned} \qquad (6.181)$$

where $\hat{\mathbf{R}}_{-M}=\hat{\mathbf{R}}_{-M+1}=0$, $\hat{\mathbf{R}}_j=\hat{\mathbf{R}}$ for $j\geq -M+2$.

Equations (6.181) can be solved in two different ways. The first method consists in the consequence application of Theorem 6.2 and therefore means calculating series of cumbersome integrals as well as proving not evident estimates $\hat{\mathbf{R}}u_{j-2} \in O_{S_2}\left(\varepsilon^{|j|/2}\right)$ for $j \geq -M+4$. The second way is to change the system of equations (6.181) by the one equation with respect to function (6.180), which is called the generalized transfer equation. The method of solution of similar equations based on using the birth and annihilation operators is given in monograph [131].

6.4.4 The leading approximation

Consider the superposition of the positive and negative branches of the constructed solutions,

$$\begin{aligned} U = &\left[\sum_{j=-2}^{0} \varepsilon^{j/2} u_j^+(\varphi, t) + O(\varepsilon^{1/2})\right] \exp\left[\mathrm{i}\varepsilon^{-1} S^+(\varphi, t)\right] \\ &+ \left[\sum_{j=-2}^{0} \varepsilon^{j/2} u_j^-(\varphi, t) + O(\varepsilon^{1/2})\right] \exp\left[\mathrm{i}\varepsilon^{-1} S^-(\varphi, t)\right], \end{aligned} \tag{6.182}$$

where signs $^+$ and $^-$ indicate that the functions u_j^+, S^+ and u_j^-, S^- are calculated by the aforementioned formulas for the Hamiltonians $H(p,q)$ and $-H(p,q)$, respectively.

Inserting (6.182) into the initial conditions (6.150), (6.151) and equating coefficients by the same powers of ε, we get

$$c_{0,j}^{\pm} = \frac{1}{2}\left(c_{0,j}^{\circ} \mp \frac{d_{0,j}^{\circ}}{H_0}\right), \quad H_0 = H\left[S_0'(0),\, 0\right], \quad j = 0, 1, 2. \tag{6.183}$$

Function (6.182) is the leading approximation in the formal asymptotic expansion for the solution of problem (6.149)-(6.151). The functions u_j for $j \geq 0$ (see Eqs. (6.153), (6.180)) remains undefined in this approximation. In this case, they may be considered as approximate smooth functions of variables φ, t. The leading approximation (6.182) satisfies Eq. (6.149) with an accuracy up to $O\left(\varepsilon^{3/2}\right)$.

So, the asymptotic relations for prediction of the normal displacements of points at the middle surface of the shell with the initial conditions

$$\begin{aligned} W|_{t=0} &= w^{\circ}(\varphi, \varepsilon) z(s) \exp\left[\mathrm{i}\varepsilon^{-1} S_0(\varphi)\right], \\ \dot{W}|_{t=0} &= \mathrm{i}\varepsilon^{-1} v^{\circ}(\varphi, \varepsilon) z(s) \exp\left[\mathrm{i}\varepsilon^{-1} S_0(\varphi)\right], \end{aligned} \tag{6.184}$$

where w°, v° are given by (6.152), are as follows,

$$W = \frac{1}{2}\,z(s)\sum\nolimits^{\pm}\left\{\varepsilon^{-1}\left(c^\circ_{0,2} \mp \frac{d^\circ_{0,2}}{H_0}\right)\frac{[\varphi - q^{\pm}]^2}{[C^{\pm}(t)]^{5/2}}\right.$$
$$+\,\varepsilon^{-1/2}\left(c^\circ_{0,1} \mp \frac{d^\circ_{0,1}}{H_0}\right)\frac{\varphi - q^{\pm}}{[C^{\pm}(t)]^{3/2}} + \left(c^\circ_{0,0} \mp \frac{d^\circ_{0,0}}{H_0}\right)\frac{1}{[C^{\pm}(t)]^{1/2}}$$
$$\left.\pm \mathrm{i}\left(c^\circ_{0,2} \mp \frac{d^\circ_{0,2}}{H_0}\right)\frac{H_{pp}\,[p^{\pm}(t),\, q^{\pm}(t)]\,t}{[C^{\pm}(t)]^{3/2}}\right\}\exp\{\mathrm{i}\varepsilon^{-1}S^{\pm}(\varphi,t)\},$$
$$S^{\pm} = S_0(0) + \int_0^t\{p^{\pm}(\tau)\dot{q}^{\pm}(\tau) \mp H\,[p^{\pm}(\tau), q^{\pm}(\tau)]\}d\tau$$
$$+\,p^{\pm}(t)\,[\varphi - q^{\pm}(t)] + \frac{1}{2}\,\frac{B^{\pm}(t)}{C^{\pm}(t)}\,[\varphi - q^{\pm}(t)]^2 . \tag{6.185}$$

Here, the symbol $\sum^{\pm}$ means the summation of the expression following it with the superscripts (+) and (−), respectively. If $q^{\pm} \neq 0$, then function (6.185) defines the pair of WPs, the n^+th and n^-th, with the centers on the generatrices $\varphi = q^+(t)$ and $\varphi = q^-(t)$. These packets correspond to the eigenfunction $z_n(s)$ and the associated eigenvalue λ_n with a number n of the boundary-value problem (6.8), (6.9).

Finally, to construct a solution of the governing Eqs. (6.3) with the initial conditions (6.5), one needs to supply all the functions in (6.185) with the subscribe n and perform summing up in accordance with expansion (6.148). An asymptotic solution constructed in such a way will describe wave process being the superposition of a set of localized families of bending waves, i.e., the $n^{\pm}$th wave packets ($n = 1, 2, \ldots$), travelling in the circumferential direction.

The difference of the two methods, as developed in Section 6.1 and used in Section 6.4, is the following. Solution (6.185) is the expansion by the beam functions in the case where the shell edges are perpendicular to the generatrix and is not valid for cases when the generatrix length is variable. And the method proposed in Section 6.1 allows performing the asymptotic decomposition of the initial 2D boundary-value problem (6.3)-(6.6) into the sequence of 1D boundary-value problems at moving generatrices (generally, of a variable length) being the centers of travelling WPs; this approach may be applied to shells with arbitrary edges.

6.4.5 Comparison of different solutions

It is of interest to compare the solution constructed in Section 6.4 with the one from Section 6.2 for the case where $s_1 = 0$, $s_2 = l$ are constants. Here, $z(s)$ and λ are independent of φ, and the Hamiltonian (6.40) has the form (6.157).

The coincidence of Eqs. (6.169) and (6.77) for the phase function $S^{\pm}$ becomes obvious if we pass in (6.77) to the initial coordinate accordantly with (6.75) and take into account relations (6.39), (6.86).

Let the functions $w_0^\circ(\zeta)$, $v_0^\circ(\zeta)$ in (6.11) be the quadratic trinomials:

$$w_0^\circ = c_{0,0}^\circ + c_{0,1}^\circ\zeta + c_{0,2}^\circ\zeta^2, \quad v_0^\circ = d_{0,0}^\circ + d_{0,1}^\circ\zeta + d_{0,2}^\circ\zeta^2. \tag{6.186}$$

Note that

$$bH_{pp} = \dot{C}/C - H_{pq}. \tag{6.187}$$

The last equation allows without difficulty calculating coefficients (6.65) in polynomials (6.65). In our case ($s_1 = 0$, $s_2 = l$), the functions $\Psi_m^{\pm}(t)$ appearing in (6.65) are as follows:

$$\Psi_m^{\pm}(t) = \left[C^{\pm}(t)\right]^{m+1/2}, \quad m = 0, 1, 2. \tag{6.188}$$

Taking into account the last equation and using the recurrent relations (6.65), we arrive at the required polynomial

$$\begin{aligned} P_0^{\pm}\left(\xi^{\pm}, t; d_m^{\pm}\right) &= \frac{d_2^{\pm}}{\left[C^{\pm}(t)\right]^{5/2}}\xi^{\pm 2} + \frac{d_1^{\pm}}{\left[C^{\pm}(t)\right]^{3/2}}\xi^{\pm} \\ &+ \frac{d_0^{\pm}}{\left[C^{\pm}(t)\right]^{1/2}} \pm \frac{\mathrm{i}d_2^{\pm}H_{pp}t}{\left[C^{\pm}(t)\right]^{3/2}}. \end{aligned} \tag{6.189}$$

The constants $d_m^{\pm}$ are calculated by (6.80). The straightforward substitution of constructed polynomials (6.189) into (6.76), (6.77) results in formula (6.185), derived by using the Maslov method.

Chapter 7

Effect of External Forces on Wave Packets in Zero Curvature Shells

In this chapter, the asymptotic method developed in Chapter 6 is applied to investigate localized families of flexural waves running in the zero curvature shells with taking into account an external load. In a general case, external forces are non-stationary and result in appearance of slowly varying membrane stress resultants. The wave forms of motion of a shell in the neighborhood of momentless dynamic strain-stress state (SSS) are studied. It is assumed that the dynamism index of SSS caused by external forces is less than that of a solution under construction.

The effect of membrane non-uniform and non-stationary forces as well as of the response of an inhomogeneous elastic isotropic foundation (or elastic filler) on the travelling WPs in cylindrical shells is analyzed. The calculation relations for non-circular conical shells with variable thickness and/or elastic properties are given.

The case when the cylindrical shell subjected to compressive or axially tensile forces is considered separately. The axial forces may depend on the circumferential coordinate and time. In contrast to the previously found solutions, the construction of running WPs having a large variability both in the circumferential and axial directions is proposed.

7.1 Classification of dynamical stress-strain state

Let x, y be curvilinear coordinates on the reference surface of a shell, where x is a distance to the cone vertex (or a distance measured along the shell axis in the case of a cylinder), and y is a coordinate on the directrix.

We assume that the shell is under external slowly varying forces causing non-stationary SSS. This state depends on the nature of loading, the ways of fixing the shell edges and can be classified as the membrane SSS, the purely general state, or a superposition of the main membrain SSS and the integrals of the edge effect [67]. We shall study here a motion of the shell in the neighborhood of the main SSS which is membrane and characterized by the in-plane stress resultants $T_i^\circ(x, y, t^*)$, $i = 1, 2, 3$. The last are found from equations of the membrane theory of thin shells [173]. We assume also that the stress resultants T_i° do not reach their critical values resulting in the buckling of the initial membrane SSS.

Let F be one of unknown functions characterizing the shell dynamics. We introduce the estimates for derivatives with respect to the coordinates x, y and time t^*:

$$\frac{\partial F}{\partial x} \sim R^{-1} h_*^{-\varsigma_1} F \quad \frac{\partial F}{\partial y} \sim R^{-1} h_*^{-\varsigma_2} F \quad \text{as} \quad h_* \to 0, \tag{7.1}$$

$$\frac{\partial F}{\partial t^*} \sim \omega_c \, h_*^{-\varsigma_t} F \quad \text{for} \quad h_* \to 0, \tag{7.2}$$

where $\omega_c = \sqrt{E/(\rho R^2)}$ is the characteristic frequency, ς_1, ς_2 are the indices of variation in the $x-$ and $y-$ directions, respectively, and ς_t is the index of dynamism of the shell SSS [63, 64]. Alongside with estimates (7.1), (7.2), we introduce those for the forces T_i°:

$$\begin{aligned}&\frac{\partial T_i^\circ}{\partial x}, \frac{\partial T_i^\circ}{\partial y} \sim T_i^\circ \quad \text{as} \quad h_* \to 0,\\&\frac{\partial T_i^\circ}{\partial t^*} \sim \omega_c h_*^{-\varsigma_L} T_i^\circ \quad \text{as} \quad h_* \to 0,\end{aligned} \tag{7.3}$$

where ς_L is the index of dynamism of external forces.

In this chapter, we consider the two classes of problems differing in the indices of variation and dynamism of the shell SSS:

$$\varsigma_1 = 0, \quad \varsigma_2 = 1/4, \quad \varsigma_t = -1/2, \quad \varsigma_L \le -3/4; \tag{7.4}$$

$$\varsigma_1 = 1/2, \quad \varsigma_2 = 1/2, \quad \varsigma_t = 0, \quad \varsigma_L \le -1/2. \tag{7.5}$$

The solutions in the form of travelling WPs with indices (7.4) without taking into account external forces were constructed in Chapter 6. Below, we study similar solutions accounting for the membrane forces which satisfy inequality (7.4). Solutions with higher indices $\varsigma_1, \varsigma_2, \varsigma_t$ satisfying (7.5) will be constructed in Section 7.4. We note that in the both cases, the inequality $\varsigma_L < \varsigma_t$ is a necessary condition for the solutions to be asymptotically correct. This inequality means that the rate of variation of membrane (in-plane) stress resultants is less than the group velocity and instantaneous frequency of vibrations in travelling WPs.

7.2 Effect of external forces on non-stationary localized vibrations of a medium-length cylindrical shell

We consider a medium-length non-circular cylindrical shell of a constant thickness with constant physical characteristics resting on an inhomogeneous elastic isotropic foundation (or containing an inhomogeneous elastic filler). Let $x = Rs, y = R\varphi$. The shell is not necessarily closed in the circumferential direction and occupies the region

$$s_1(\varphi) \le s \le s_2(\varphi), \quad 0 \le \varphi \le \varphi_1.$$

To predict the impact of an elastic foundation on the shell, we use the simplest Winkler model, in accordance to which the response of the foundation is defined as

$$q_n = -c_f w. \tag{7.6}$$

In the general case, the coefficient of subgrade reaction c_f depends on the both coordinates on the shell surface. We consider here the specific case, where $c_f(\varphi)$ as well as the hoop stress resultant $T_2^\circ(\varphi, t)$ are functions of the circumferential coordinate φ.

To describe the shell behavior in the vicinity of non-stationary membrane SSS, one can use the coupled equations (2.51). We rewrite them as

$$\begin{gathered}\varepsilon^4 \Delta\Delta W + \varepsilon^2 \Delta_{\mathrm{T}} W + cW + \Delta_k \Phi + \varepsilon^2 \frac{\partial^2 W}{\partial t^2} = 0, \\ \varepsilon^4 \Delta\, \Delta\Phi - \Delta_k W = 0,\end{gathered} \tag{7.7}$$

where $\varepsilon^{-4} RW$ and $R^2 hE\Phi$ are the normal deflection and stress functions, respectively, $t = t^*/t_c$ is dimensionless time, $t_c = \varepsilon^{-3}\omega_c^{-1}$ is the characteristic time, ε is the small parameter introduced in accordance with (4.13), and $c = R^3 E^{-1} h^{-2} c_f$ is the dimensionless spring "constant"of the Winkler foundation. The differential operators Δ, Δ_{T}, Δ_k are introduced as

$$\begin{gathered}\Delta z = \frac{\partial^2 z}{\partial \varphi^2} + \frac{\partial^2 z}{\partial s^2}, \quad \Delta_k z = k(\varphi)\frac{\partial^2 z}{\partial s^2}, \\ \Delta_{\mathrm{T}} z = \frac{\partial}{\partial \varphi}\left(T_2 \frac{\partial z}{\partial \varphi}\right) + \frac{\partial}{\partial s}\left(T_3 \frac{\partial z}{\partial \varphi}\right) + \frac{\partial}{\partial \varphi}\left(T_3 \frac{\partial z}{\partial s}\right) + \frac{\partial}{\partial s}\left(T_1 \frac{\partial z}{\partial s}\right),\end{gathered} \tag{7.8}$$

where $T_i^\circ = \varepsilon^6 R^{-1} EhT_i$.

The functions $k(\varphi), s_j(\varphi), T_2(\varphi, t), T_1(s, \varphi, t), T_3(s, \varphi, t)$ are assumed to be sufficiently differentiable with respect to their arguments and together with their derivatives are values of order $O(1)$ as $\varepsilon \to 0$.

At the edges $s = s_j(\varphi)$, we consider the boundary conditions (6.4) from a simply supported group or a clamped group. If one edge is clamped, conditions from a free edge group (6.4) can be chosen on the second edge. Conditions (6.5) are assumed as initial ones for the function W.

To construct an asymptotic solution of the initial-boundary-value problem (7.7), (6.4), (6.5), one can use all the principal equations and formulas derived in Section 6.1 after some modifications. Now, the operator $\mathbf{L}_0$ reads

$$\mathbf{L}_0 = \frac{k^2[q(t)]}{p^4(t)}\frac{\partial^4}{\partial s^4} + p^4(t) - T_2[q(t), t]p^2 - \left[\omega(t) - \dot{q}(t)\, p(t)\right]^2, +c[q(t)] \tag{7.9}$$

and the operators $\mathbf{L}_m$ for $m \geq 1$ have the same structure as (6.23), where

$$\mathbf{\Gamma} = HH_{pq} - \frac{2pHH_p}{\eta}\int\limits_{s_1[q(t)]}^{s_2[q(t)]} zT_3 \frac{\partial z}{\partial s} ds, \quad \eta = \int\limits_{s_1[q(t)]}^{s_2[q(t)]} z^2 ds. \tag{7.10}$$

Let $\lambda_n(\varphi)$, $z_n(s, \varphi)$ be the eigenvalues and the associated eigenfunctions of the boundary-value problem for Eq. (6.8) with one of variants of the boundary conditions (6.9) at the edges $s = s_i(\varphi)$. In particular, in the case of a simple support of both edges, one has

$$\lambda_n = \frac{\pi^4 n^4}{l^4(\varphi)}, \qquad z_n = \sin\{\lambda_n^{1/4}(\varphi)\,[s - s_1(\varphi)]\}, \tag{7.11}$$

where $l(\varphi) = s_2(\varphi) - s_1(\varphi)$. Then the Hamilton function is as follows:

$$H(p, q, t) = \sqrt{p^4 + \frac{\lambda(q)k^2(q)}{p^4} - T_2(q, t)p^2 + c(q)}. \tag{7.12}$$

It is obvious that now solution (6.76), (6.77) should be considered on the interval $0 \leq t < t_b$ where the inequality

$$p^4 + \frac{\lambda(q)k^2(q)}{p^4} - T_2(q, t)p^2 + c(q) > 0 \tag{7.13}$$

holds for both branches of the functions $p^{\pm}(t), q^{\pm}(t)$.

If for any $t \geq 0$ on each generatrix the hoop stress resultant is not compressive, i.e. $T_2 \leq 0$, then inequality (7.13) is satisfied. Let now the inequality $T_2(\varphi, t) > 0$ hold on a part of the shell surface beginning from some moment of time. It is clear that the function $T_2 > 0$ satisfying inequality (7.13) should not exceed the critical value resulting in the shell buckling. We consider here the particular case,

$$T_2(\varphi, t) = \Lambda(t)\tau_2(\varphi), \tag{7.14}$$

where $\Lambda(t) \geq 0$ and the function $\tau_2(\varphi)$ is positive at least on the part Ω_+ of the shell surface. Here, Λ may be considered as a load parameter which does not necessarily depend on time t. For a certain critical value of Λ, the shell may loss stability in the neighborhood of some generatrix from the region Ω_+ [222]. Let

$$F_1(p, q) = \frac{1}{\tau_2(q)}\left[p^2 + \frac{\lambda_1(q)k^2(q)}{p^6} + \frac{c(q)}{p^2}\right], \tag{7.15}$$

where $\lambda_1(q)$ is the least from all eigenvalues $\lambda_n(q)$ of the boundary-value problem (6.8), (6.9) for a fixed value of q. We have the estimate

$$\begin{gathered} p^4 + \frac{\lambda(q)k^2(q)}{p^4} - T_2(q, t)p^2 + c(q) = p^2\tau_2(q)\left[\frac{p^2}{\tau_2(q)} + \frac{\lambda(q)k^2(q)}{p^6\tau_2(q)} - \Lambda\right] \geq \\ p^2\tau_2(q)\left[F_1(p, q) - \Lambda\right] \geq p^2\tau_2(q)\left(\Lambda_b - \Lambda\right). \end{gathered} \tag{7.16}$$

Here,

$$\begin{gathered} \Lambda_b = \min_{q\in\Omega_+,\, p>0} F_1(p, q) = F_1(p_b, q_b), \\ p_b = \sqrt[4]{\frac{c(q_b) + \sqrt{c^2(q_b) + 12\lambda_1(q_b)k^2(q_b)}}{2}}, \end{gathered} \tag{7.17}$$

and the parameter q_b can be found from the equation

$$-\tau_2'\left(p^2 + \frac{\lambda_1 k^2}{p^6} + \frac{c}{p^2}\right) + \tau_2\left[\frac{(\lambda_1 k^2)'}{p^6} + \frac{c'}{p^2}\right] = 0, \tag{7.18}$$

where the prime means differentiation with respect to q.

Estimate (7.16) shows that the inequality

$$\Lambda < \Lambda_b \tag{7.19}$$

is the sufficient condition for inequality (7.13) to be satisfied. In particular, if $\Lambda(t)$ is a function of time such that $\Lambda(0) = 0$, $\dot{\Lambda}(t) > 0$ for $t > 0$, then solution (6.18) is valid only in the interval $0 \leq t < t_b$, where t_b is the root of the equation $\Lambda(t) = \Lambda_b$. If s_i, k, τ_2, c are not dependent of φ (here, $k = 1, \tau = 1, s_1 = 0, s_2 = l, c, \lambda_1$ are constants, and minimization in (7.17) should be carried out only by the variable p), then $\Lambda = \Lambda_b$ is the critical value of the load parameter Λ [77,222], which results in the buckling of the membrane SSS of a cylindrical shell with constant parameters under the normal external pressure $Q_n = \Lambda_b EhR^{-1}\varepsilon^6\tau_2$. If s_i, k, τ_2, c are functions of the circumferential coordinate φ, then formula (7.17) gives the leading approximation $\Lambda_b^{(0)}$ for the critical parameter $\Lambda_b = \Lambda_b^{(0)} + \varepsilon\Lambda_b^{(1)} + \ldots$ corresponding to the buckling of a shell with variable parameters under the non-uniform external pressure $Q_n = \Lambda_b EhR^{-1}\varepsilon^6\tau_2(\varphi)$ in the neighborhood of the weakest generatrix $\varphi = q_b$ (see Chap. 7 in [222]).

7.2.1 Stationary wave packets

In Section 6.2, we considered the specific case when under some assumptions with respect to the functions $s_i(\varphi), k(\varphi)$ and the parameters a_0, b_0 from the initial conditions (6.6), the constructed solution in the form of travelling WPs degenerated into the stationary solution which is the superposition of eigenmodes localized in the neighborhood of the weakest generatrix at the shell surface. Here, we shall study the analogous case when: 1) a circular cylindrical shell of constant length is pre-stressed by the non-uniform hoop stress resultant $T_2(\varphi)$; 2) a circular cylindrical shell with constant geometrical parameters rests at an inhomogeneous elastic foundation with the variable coefficient $c(\varphi)$.

7.2.1.1 Stationary WPs in the case of non-uniform hoop stress resultant

Let $c = 0, s_1, s_2, k$ be constants, and $T_2 = T_2(\varphi)$ is the function of φ. Here,

$$H(p, q) = \sqrt{p^4 + \frac{\lambda k^2}{p^4} - p^2 T_2(q)} \tag{7.20}$$

and the Hamilton system reads

$$\begin{aligned} \dot{q} &= \frac{2p^8 - 2\lambda k^2 - T_2(q)p^6}{p^5 H}, \\ \dot{p} &= \frac{p^2 T_2'(q)}{2H}, \end{aligned} \tag{7.21}$$

with the boundary conditions

$$q(0) = 0, \quad p(0) = a_0. \tag{7.22}$$

Let the hoop stress resultant $T_2(\varphi)$ satisfy the conditions

$$T_2'(0) = 0, \quad T_2''(0) < 0, \tag{7.23}$$

that is, there is a local maximum on the generatrix $\varphi = 0$. Let also the parameter $a_0 = p^\circ$ be found as a solution of the equation

$$2a_0^8 - 2\lambda k^2 - T_2(0)a_0^6 = 0, \tag{7.24}$$

and

$$b_0 = b^\circ = \mathrm{i}\sqrt{H_{qq}/H_{pp}}. \tag{7.25}$$

It can be readily shown that the parameters $p^\pm = p^\circ, q^\pm = 0$ and $b^\pm = b^\circ$ identically satisfy the Hamilton system (7.21) and the Riccati equation (6.57), respectively.

We construct the polynomial $P_0^\pm$ in the form of superposition (6.96) of the Hermite polynomials. The substitution of (6.96) into Eqs. (6.76), (6.77) gives

$$\begin{aligned} w = z(s,0) \sum_{m=0}^{M_0} \exp\left[\mathrm{i}\varepsilon^{-1}\left(\omega^\circ t + p^\circ\varphi + \frac{1}{2}b^\circ\varphi^2\right)\right] \\ \times \left(c_m^+ + c_m^-\right)\left(H_{pp}/H_{qq}\right)^{m/4}\mathcal{H}_m\left[\varepsilon^{-1/2}\left(H_{qq}/H_{pp}\right)^{1/4}\varphi\right]. \end{aligned} \tag{7.26}$$

This function is the superposition of the M_0+1 number of eigenmodes localized in the neighborhood of the generatrix $\varphi = 0$ with the frequencies

$$\omega^\circ = \omega_0^\circ + \varepsilon\omega_1^{(m)}, \tag{7.27}$$

where

$$\omega_0^\circ = H\left(p^\circ, 0\right), \quad \omega_1^{(m)} = \left(m + \frac{1}{2}\right)\sqrt{H_{pp}H_{qq}}\Big|_{\substack{p=p^\circ \\ q=0}}, \qquad m = 0, 1, 2, \ldots \tag{7.28}$$

7.2.1.2 Stationary WPs in the case of non-uniform elastic foundation

Let now T_2, s_1, s_2, k be constants, and $c = c(\varphi)$ be the given function satisfying the conditions

$$c'(0) = 0, \quad c''(0) > 0. \tag{7.29}$$

Here, the Hamiltonian system takes the form,

$$\begin{aligned} \dot{q} &= \frac{2p^8 - 2\lambda k^2 - T_2 p^6}{p^5 H}, \quad q(0) = 0, \\ \dot{p} &= -\frac{c'(q)}{2H}, \quad p(0) = a_0, \end{aligned} \tag{7.30}$$

with the Hamiltonian

$$H(p,q) = \sqrt{p^4 + \frac{\lambda k^2}{p^4} - p^2 T_2 + c(q)}. \tag{7.31}$$

Let a_0, b_0 again be found from Eqs. (7.24), (7.25). Then $p^{\pm} = p^{\circ}, q^{\pm} = 0$, $b^{\pm} = b^{\circ}$ and the previously constructed solution (6.76), (6.77) degenerates into the superposition of eigenmodes (7.26) localized in the neighborhood of the generatrix, on which the coefficient $c(\varphi)$ of the subgrade reaction is minimal.

7.2.2 Effect of stationary non-uniform hoop stress resultants on dynamics of WPs

Here, we consider the case, where $k = 1, s_1 = 0, s_2 = l, c = 0$, and the hoop membrane force $T_2(\varphi)$ is the function of an angle φ.

From the analysis of equations (7.21), similar to that performed in Section 6.3, one can conclude that the behavior of WPs is affected by the distribution of stress resultant $T_2(\varphi)$ in the shell middle surface. Omitting the details of this analysis, one notes that there are two essentially different cases. If a WP moves in the direction of increasing the compressive force $T_2(\varphi)$, then its movement is accompanied by an increase in speed. But if it moves in the direction of decreasing $T_2(\varphi)$, then under the inequality

$$\inf_{\varphi} T_2(\varphi) < 2p_r^2 - 2\lambda p_r^{-6}, \tag{7.32}$$

where

$$p_r = \sqrt[4]{0,5\left(\sqrt{H_0^4 + 12\lambda} - H_0^2\right)}, \tag{7.33}$$

there exists a time point $t = t_r$, at which the WP reflects from the generatrix $\varphi = q_r$. As expected, the flexural vibrations tend to propagate towards the weakest generatrix, on which the compressive hoop stress resultant T_2 is maximum.

To study in details the effect of the static forces T_2 on the dynamics of non-stationary WPs, we consider the following

Example 7.1. Let $k = 1, s_1 = 0, s_2 = l, c = 0, \lambda = (\pi n/l)^4$, $T_2 = \Lambda\tau_2(\varphi)$, where $\tau_2 = 1 + \alpha_1 \cos\varphi$. Such hoop forces appear in a cantilever cylindrical shell under the non-uniformly distributed in the circumferential direction normal pressure $Q_n = \varepsilon^6 R^{-1} E h T_2(\varphi)$. We assume that the center of the initial WP is situated on the generatrix $\varphi = 0$, which is the weakest at $\alpha_1 > 0$. In the vicinity of this line, when $\Lambda = \Lambda_b = 4\pi 3^{-3/4}/[l(1+\alpha_1)]$, the shell buckles [222] with the formation of dents whose wavenumber is proportional to the parameter $p_b = 3^{1/8}(\pi/l)^{1/2}$.

Figure 7.1 demonstrates the outcomes of a numerical solution of the Hamiltonian system for $l = 1, n = 1, a_0 = 2, b_0 = \mathrm{i}, \Lambda = 2$ and different values of the

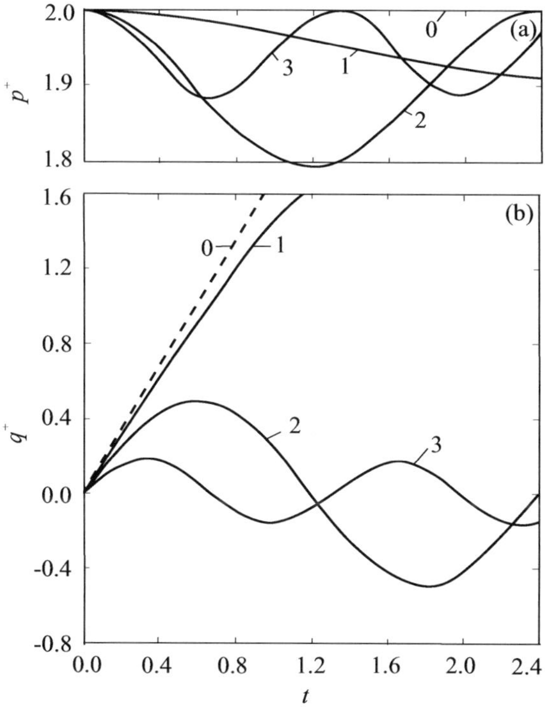

FIGURE 7.1: Solutions of the Hamiltonian system: functions $p^+(t)$ (a) and $q^+(t)$ (b) for $\alpha_1 = 0$; 0.05; 0.5; 1 (curves 0, 1, 2, 3 respectively).

parameter $\alpha_1 = 0$; 0.05; 0.5; 1 (curves 0, 1, 2, 3, respectively). Plots of the functions $\omega^\pm(t), v_g^\pm(t)$ are shown in Figure 7.2. Here, $\Lambda < \Lambda_b$ for all values α_1 under consideration. For $\alpha = 0.05$, wave packets run over the shell surface unhindered (without reflection effects). If $\alpha_1 = 0.5$ or $\alpha_1 = 1$, then one holds condition (7.26) resulting in the reflection of the WP from some generatrix, the functions $p^\pm(t), q^\pm(t), \omega^\pm(t), v_g^\pm(t)$ being periodic. The dependence of the WP width as well as of the maximum amplitude of waves on time is shown in Fig. 7.3. In those cases where there are no reflections ($\alpha_1 = 0$, $\alpha_1 = 0.05$), the WP rapidly spreads over the shell surface. If the conditions of reflection are satisfied (at $\alpha_1 = 0.5$, $\alpha_1 = 1$), then on the period $2t_0^\pm$ of the solution of the Hamilton system it is observed focusing the WP accompanied by an increase of the wave amplitudes.

7.2.3 Effect of inhomogeneous elastic foundation on dynamics of WPs

Let $k = 1, s_1 = 0, s_2 = l, T_2$ be constants, and $c(\varphi)$ be a function of φ. In this case, the analysis of system (7.30) reveals that if $2a_0^8 - 2\lambda - T_2 a_0^6 > 0$,

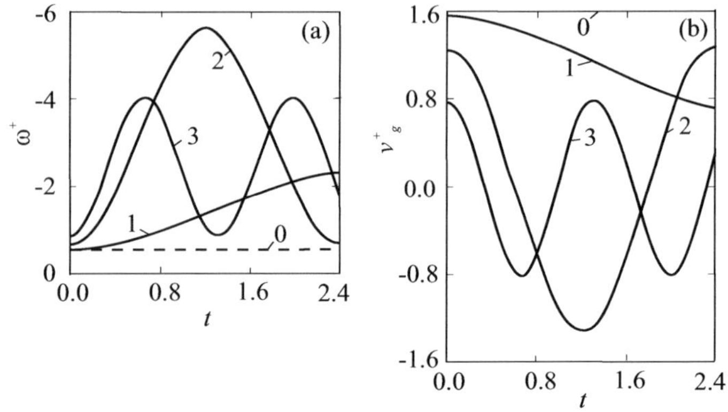

FIGURE 7.2: Plots of functions $\omega^+(t)$ (a) and $v_g^+(t)$ for $\alpha_1 = 0; 0.05; 0.5; 1$ (curves 0, 1, 2, 3, respectively).

then the WP movement in the direction of decreasing the function $c(\varphi)$ is accompanied by an increase in the wave number p and group velocity $v_g = \dot{q}$. On the contrary, if the WP runs in the direction of increasing the "spring constant" $c(\varphi)$, then its movement is accompanied by the decrease in parameters p and $v_g = \dot{q}$ so that at some time point t_r there occurs a reflection of the center WP from some generatrix $\varphi = q_r$. This generatrix is found from the equation

$$p_r^4 + \frac{\lambda}{p_r^4} - T_2 p_r^2 - c(q_r) - H_0^2 = 0, \tag{7.34}$$

where $H_0 = H(a_0, 0, 0)$, and the parameter $p_r \in (0, a_0)$ is a root of the equation

$$2p_r^8 - T_2 p_r^6 - 2\lambda = 0. \tag{7.35}$$

Thus, just as in the case of non-uniform distribution of the hoop stress resultant, the inhomogeneity of an elastic foundation can lead to localization of the flexural modes of vibrations in those areas of the shell that are adjacent to the weakest generatrix.

7.2.4 Effect of dynamic hoop stress resultants on dynamics of WPs

Here we consider the more general case when the force $T_2(\varphi, t)$ depends on time. A similar case occurs when the shell is loaded by external or internal unsteady pressure $Q_n = \varepsilon^6 R^{-1} E h T_2(\varphi, t)$. The problem in this formulation was first studied in [155].

Example 7.2. First, we consider a cylindrical shell with constant parameters ($k = 1, s_1 = 0, s_2 = l$) subject to the action of the dynamic pressure $Q_n = \varepsilon^6 R^{-1} E h \Lambda(t)$.

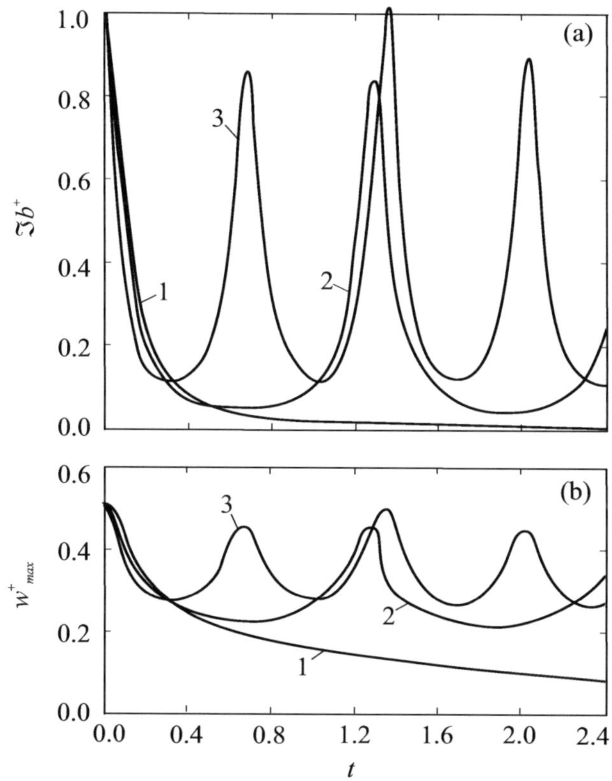

FIGURE 7.3: Plots of functions $\mathrm{Im}b^+(t)$ (a) and $w^+_{max}(t)$ (b) for $\alpha_1 =$ 0; 0.05; 0.5; 1 (curves 0, 1, 2, 3, respectively).

Solving the Hamiltonian system and the Riccati equation, we get

$$p^{\pm} = a_0, \quad q^{\pm}(t) = \int_0^t H_p(t)dt, \quad b^{\pm} = \frac{b_0}{C^{\pm}(t)},$$
$$C^{\pm}(t) = 1 + b_0 \int_0^t H_{pp}(t)dt \tag{7.36}$$

with

$$H_0(t) = \sqrt{a_0^4 + \lambda/a_0^4 - \Lambda(t)a_0^2}, \quad H_p(t) = \frac{2a_0^8 - 2\lambda - \Lambda(t)a_0^6}{a_0^5 H_0(t)},$$
$$H_{pp}(t) = \frac{2(a_0^{16} + 12\lambda a_0^8 + 3\lambda^2)}{a_0^{10}H_0^3(t)} - \frac{3a_0^6(a_0^8 + 5\lambda)\Lambda(t)}{a_0^{10}H_0^3(t)}. \tag{7.37}$$

The coefficients h_j in the amplitude equation (6.62) are equal to

$$h_0^{\pm}(t) = \frac{1}{2}H_{pp}(t), \quad h_1(t) = \pm \mathrm{i}b^{\pm}(t)H_{pp}(t), \quad h_2(t) = \mathrm{i},$$
$$h_3^{\pm}(t) = i\left[\pm\frac{1}{2}b^{\pm}(t)H_{pp}(t) + \frac{\partial H}{\partial \Lambda}\frac{\dot{\Lambda}(t)}{2H_0(t)}\right]. \tag{7.38}$$

We substitute (7.37), (7.38) into the recurrence relations (6.65). When deriving the equation for $A^{\pm}_{M-r}(t)$, we use the method of induction, in so doing the integral $\int\limits_0^t H_{pp}C^{-(\varsigma+2)}(C-1)^{\varsigma}dt$, where $\varsigma = \mathrm{Int}(r/2-1)$ is the integer part of the number $r/2-1$, is calculated using the identity $\dot{C} \equiv b_0 H_{pp}$ (we note that the function $C(t)$ is known only with an accuracy up to $\Lambda(t)$). As a result, we arrive again at equations (6.138). However, here

$$\Psi^{\pm}_m(t) = \sqrt{\frac{H_0(0)}{[C^{\pm}(t)]^{2m+1}\,H(t)}}. \tag{7.39}$$

Since $\Lambda < \Lambda_b$, from (7.37) it follows the inequality $H_{pp} > 0$. Hence, $\mathrm{Im}\,\dot{C}^{\pm} > 0$ for any $t \geq 0$. This means that for any law of pressure variation that satisfies the previously introduced restrictions for membrane forces, the width of WPs grows indefinitely. In other words, a wave packet excited in some way is always "spreading" over the shell surface. The dependence of the wave amplitude on the function $\Lambda(t)$ is more complicated. For instance, if $\Lambda < 0$, $\dot{\Lambda} < 0$ (growing internal pressure), then $\dot{\Psi}^{\pm}_m < 0$, that is the amplitude of waves decreases; and if $\Lambda > 0$, $\dot{\Lambda} > 0$ (growing external pressure), then it is observed growing amplitudes, in particular at $\Lambda \to \Lambda_b$.

In the following example, we study the effect of a slowly growing uniform pressure on the natural modes, localized in the vicinity of the weakest generatrix.

Example 7.3. Consider a circular cylindrical shell with oblique edge (see Figure 6.9 and the notations of Example 6.3). The shell edges are simply supported. Let the shell be loaded with the slowly varying pressure $Q_n = \varepsilon^6 R^{-1}Ehc_t^* t^*$, where $c_t^* = c_t/t_c$, $c_t \sim 1$, $t_c = \varepsilon^3 R^{-1}(E/\rho)^{1/2}$ is the characteristic time. Then the membrane hoop stress resultant $T_2 = \Lambda(t) = c_t t$. As the initial conditions, we consider one of the natural modes of low-frequency vibrations of the shell free of stresses ($T_2 = 0$) near the weakest generatrix $\varphi = 0$. To do this, we assume that $a_0 = p^{\circ} = \sqrt{\pi n/l_0}$, $b_0 = b^{\circ} = \mathrm{i}\sqrt{H_{qq}/H_{pp}}$ (see Eqs. (7.24), (7.25)) in formulas (6.6), (6.16), where l_0 is the length of the longest generatrix, and the derivatives H_{pp}, H_{qq} are calculated for $p = p^{\circ}$, $\varphi = 0$.

If $c_t = 0$, then the solution of the problem under consideration is determined by formula (7.26), which specifies a stationary WP on the shell surface in the form of superposition of $M_0 + 1$ eigenmodes of low-frequency vibrations near the line $\varphi = 0$.

Now suppose $c_t \neq 0$. As the calculation formulas, we take (6.76), (6.77). Figures 7.4-7.6 represent plots of the functions $p^{\pm}(t)$, $q^{\pm}(t)$, $\omega^{\pm}(t)$, $v_g^{\pm}(t)$, $\Im b^{\pm}(t)$, $w_{max}^{\pm}(t)$. Calculations were performed for $l_0 = 2$, $\beta = 30°$, $n = 1$, $w_0^{\circ} = 1$, $v_0^{\circ} = 0$ and different values of $c_t = -3; -2; 0; 1.5; 2.5$. In the case when the pressure is external ($c_t > 0$), the calculations were carried out on the finite time intervals $0 < t < t_b$, where the condition $\Lambda < \Lambda_b$ is satisfied. Here $t_b \approx 1.833$ and $t_b \approx 1.020$ for $c_t = 1.5$ and $c_t = 2.5$, respectively. Curves in Figs. 7.4–7.6 show that the pressure growing in absolute

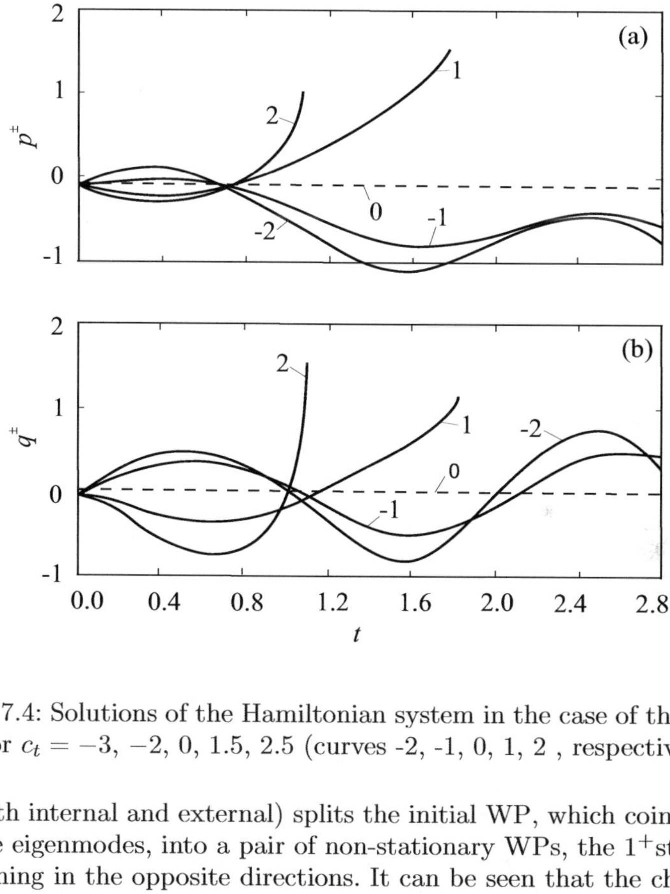

FIGURE 7.4: Solutions of the Hamiltonian system in the case of the dynamic loading for $c_t = -3, -2, 0, 1.5, 2.5$ (curves -2, -1, 0, 1, 2 , respectively).

value (both internal and external) splits the initial WP, which coincides with one of the eigenmodes, into a pair of non-stationary WPs, the 1^+st and 1^-st ones, running in the opposite directions. It can be seen that the character of wave processes at internal and external pressures is different. Under a slowly growing internal pressure ($c_t < 0$), multiple reflections of the WPs from some generatrices are observed, accompanied by focusing and a slight increase in the amplitude of waves. If the pressure is external and grows slowly enough, then reflections of the WPs are also possible, but the further behavior of the shell is characterized by rapid increasing the functions $|\omega^{\pm}(t)|$, $|v_g^{\pm}(t)|$, $|w_{max}^{\pm}(t)|$ as $t \to t_b$.

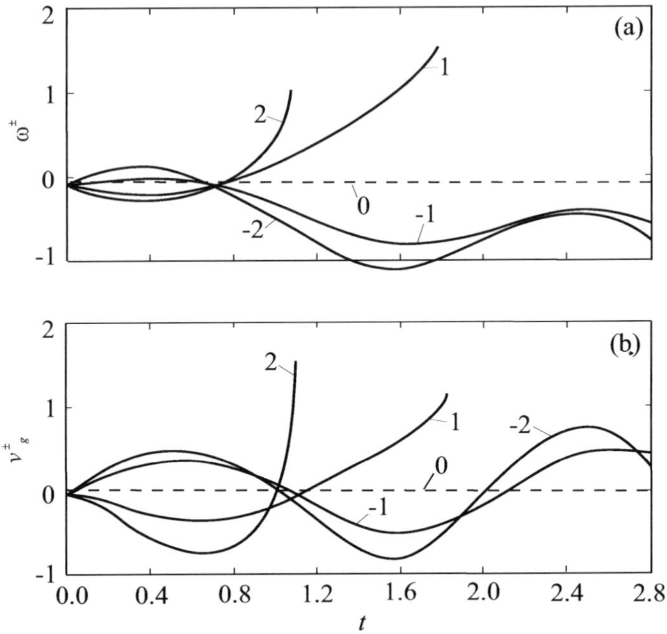

FIGURE 7.5: Plots of the functions $\omega^{\pm}(t)$ (a) and $v_g^{\mp}(t)$ (b) for $c_t = -3, -2, 0, 1.5, 2.5$ (curves -2, -1, 0, 1, 2 , respectively).

Note that for t closed to t_b, that is, at values of external pressure close to the critical one, the aforementioned functions contradicts the assumption about a small index of dynamism α_t of the constructed solution and, consequently, the error of solution (6.76), (6.77) increases.

7.3 Localized families of bending waves in medium-length conical shells

Consider a thin truncated conical shell of not necessarily circular cross-section resting on isotropic inhomogeneous elastic Winkler foundation. Let $x = Rs, y = R\varphi$ be the curvilinear coordinates on the reference (middle) surface, where x is a distance to the cone apex, y is a coordinate on the cone directrix (see Section 4.6), and R is the characteristic dimension of the shell. In the general case, the shell thickness h, Young's modulus E, Poisson's ratio ν, the material density ρ and the spring constant c_f are functions of the axial and circumferential coordinates s and φ, respectively.

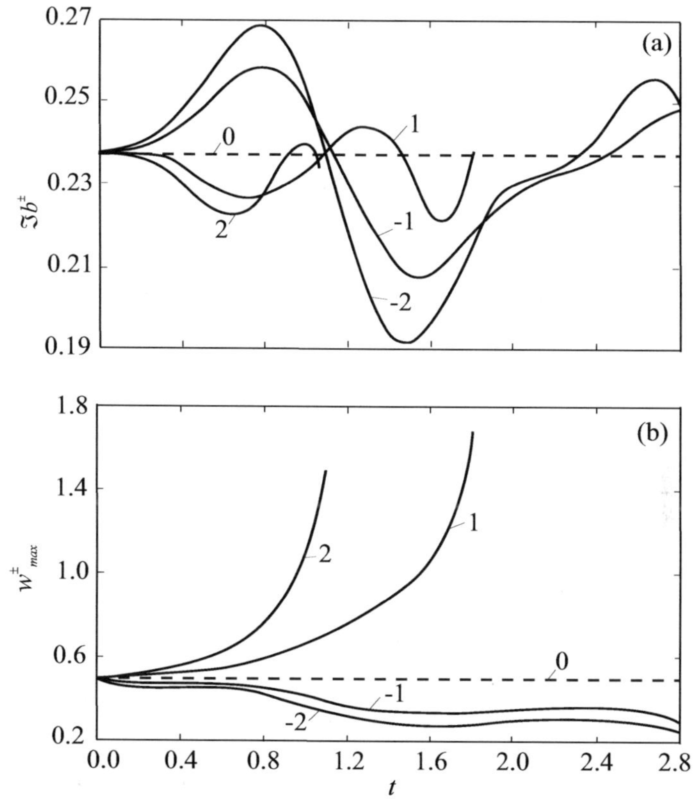

FIGURE 7.6: Plots of the functions $\Im b^{\pm}(t)$ (a) and $w^{\pm}_{max}(t)$ (b) for $c_t = -3, -2, 0, 1.5, 2.5$ (curves -2, -1, 0, 1, 2 , respectively).

We assume that the shell is under the action of external slowly varying forces resulting in the non-stationary membrane stress state. Let this state be specified by the in-plane stress resultants $T_i^{\circ}(x, y, t^*)$, $i = 1, 2, 3$. Furthermore, it is assumed that the forces T_i° do not reach their critical values which lead to the buckling of the initial membrane SSS.

As in Section 6.1, we shall study wave processes which are characterized by the indices of variation and dynamism satisfying conditions (7.4).

We consider the differential equations (2.51) as the governing ones. In the dimensionless form they read

$$\begin{gathered}\varepsilon^4 \Delta\left(d\Delta W\right) + \varepsilon^2 \Delta_{\mathrm{T}} W + cW + \Delta_k \Phi + \varepsilon^2 \gamma \frac{\partial^2 W}{\partial t^2} = 0, \\ \varepsilon^4 \Delta\left(g^{-1}\Phi\right) - \Delta_k W = 0\end{gathered} \tag{7.40}$$

with

$$d=\frac{Eh^3\left(1-\nu_0^2\right)}{E_0h_0^3\left(1-\nu^2\right)},\quad g=\frac{Eh}{E_0h_0},\quad \gamma=\frac{\rho h}{\rho_0h_0},\quad c=\frac{R^3}{E_0h_0^2}c_f,\tag{7.41}$$

where h_0, E_0, ρ_0 are the characteristic values of the shell thickness, Young's modulus and the material density, respectively. The differential operators in Eqs. (7.40) are introduced as

$$\Delta z=\frac{1}{s^2}\frac{\partial^2 z}{\partial\varphi^2}+\frac{1}{s}\frac{\partial z}{\partial s}+\frac{\partial^2 z}{\partial s^2},\quad \Delta_k z=\frac{k(\varphi)}{s}\frac{\partial^2 z}{\partial s^2},$$

$$\Delta_{\mathrm{T}}z=\frac{1}{s}\left[\frac{1}{s}\frac{\partial}{\partial\varphi}\left(T_2\frac{\partial z}{\partial\varphi}\right)+\frac{\partial}{\partial s}\left(T_3\frac{\partial z}{\partial\varphi}\right)+\frac{\partial}{\partial\varphi}\left(T_3\frac{\partial z}{\partial s}\right)+\frac{\partial}{\partial s}\left(sT_1\frac{\partial z}{\partial s}\right)\right],\tag{7.42}$$

where $T_i^\circ=\varepsilon^6R^{-1}E_0h_0T_i$. The dimensionless time t, the radial deflection W and the stress function Φ are introduced as in Section 6.1.

All the functions appearing in Eqs. (7.40)-(7.42) are assumed to be sufficiently differentiable with respect to their arguments and together with the derivatives are quantities of order $O(1)$ as $\varepsilon\to 0$.

On the edges $s=s_j(\varphi)$, we consider the boundary conditions (6.4) from the simple or clamped support groups. Just as in the case of a cylindrical shell, we accept the initial conditions (6.5) for the function W.

7.3.1 Eigenmodes of bending vibrations of a beam with variable characteristics

Consider the boundary-value problem consisting of the ordinary differential equation

$$\frac{d^2}{ds^2}\left[g_2(s)\frac{d^2z}{ds^2}\right]+\frac{d}{ds}\left[g_1(s)\frac{dz}{ds}\right]+g_0(s)z-\lambda r(s)z=0\tag{7.43}$$

and one of the variants of boundary conditions (6.9), where $g_0(s), g_1(s), g_2(s)$ are real sufficiently differentiable functions, and $r(s)$ is a positive and continuous function on the segment $[s_1, s_2]$.

The boundary-value problem (7.43), (6.9) is self-adjoint [169]. In view of the above assumptions about the functions entering (7.43), it has a countable set of real eigenvalues $\lambda_1, \lambda_2, \ldots$ [169]. Let $z_1(s)$, $z_2(s), \ldots$ be the corresponding complete orthonormal system of real eigenfunctions with the orthonormality condition

$$\int_{s_1}^{s_2} r(s)z_i(s)z_j(s)ds=\delta_{ij},$$

where δ_{ij} is the Kronecker delta.

Note that the ordinary differential equation (7.43) can be interpreted as describing free transverse vibrations of the Euler-Bernoulli beam with variable characteristics. Thus, the sequence of functions $z_1(z), z_2(x), \dots$ is the set of eigenforms of free bending vibrations of this beam.

Let the functions W_0, V_0 from the initial conditions (6.5), (6.6), satisfy one of the considered variants of boundary conditions (6.4). Then on the segment $[s_1(\varphi), s_2(\varphi)]$, they can be expanded [95, 169] into absolutely and uniformly convergent series (6.10) in the eigenforms $z_n(s)$. However, now the functions w_n°, v_n° can be found by the formulas

$$\begin{aligned} w_n^\circ &= \int_{s_1(\varphi)}^{s_2(\varphi)} r(s) W_0(s, \varphi, \varepsilon) z_n(s, \varphi) ds, \\ v_n^\circ &= \int_{s_1(\varphi)}^{s_2(\varphi)} r(s) V_0(s, \varphi, \varepsilon) z_n(s, \varphi) ds \end{aligned} \tag{7.44}$$

and admit expansions in accordance with (6.11).

Taking into account the dependence of $s_1(\varphi), s_2(\varphi)$ on the coordinate φ, we have assumed in relations (6.40) the parametric dependence of the eigenfunctions z_n on φ. Obviously, $\lambda_n(\varphi)$ are also functions of φ.

7.3.2 Algorithm for constructing solutions and principal relations

Following the algorithm given in Chapter 6, we seek a solution of the initial boundary-value problem (7.40), (6.4), (6.5) in the form of superposition (6.12) of travelling WPs (6.18). The substitution of expansion (6.18) into the original equations (7.40) generates the sequence of differential equations (6.22), where the differential operator $\mathbf{L}_0$ reads

$$\begin{aligned} \mathbf{L}_0 &= \mathbf{L}_0^* - \gamma s \left[\omega(t) - \dot{q}(t)\, p(t)\right]^2, \\ \mathbf{L}_0^* &= \frac{\partial^2}{\partial s^2}\left(\frac{s^3 g k^2}{p^4}\frac{\partial^2}{\partial s^2}\right) + \left(\frac{dp^4}{s^3} - \frac{p^2 T_2}{s}\right) + c, \end{aligned} \tag{7.45}$$

and the operators $\mathbf{L}_m$ for $m \geq 1$ are defined in accordance with (6.23). However, now the operator $\mathbf{N}$ appearing in (6.23) is given as follows,

$$\mathbf{N} = s\gamma p\ddot{q} - s\gamma' \omega \dot{q} - p\frac{\partial T_3}{\partial s} - 2pT_3\frac{\partial}{\partial s}. \tag{7.46}$$

Here, as above, the prime means differentiation with respect to the variable φ, and all functions of this argument are calculated for $\varphi = q(t)$.

The substitution of (6.18) into the boundary conditions (6.17) results in the sequence of boundary conditions, which for the simply supported edges has the form of (6.32)-(6.34).

All unknown functions included into ansatz (6.18) can be found by using the algorithm developed in Section 6.1.

First, we consider the boundary value problem arising in the leading approximation:

$$\begin{aligned}\mathbf{L}_0 w_0 \equiv \mathbf{L}_0^* w_0 - \gamma s\left[\omega(t) - \dot{q}(t)\,p(t)\right]^2 w_0 = 0,\\ w_0 = 0, \quad \frac{\partial^2 w_0}{\partial s^2} = 0 \quad \text{for} \quad s = s_i[q(t)].\end{aligned} \tag{7.47}$$

This problem, in contrast to the similar one, (6.35), (6.36), for a cylindrical shell, does not admit solutions in the explicit form. However, as before, its solution can be expressed in terms of the eigenmodes $z_n(s,\varphi)$ for the Euler-Bernoulli beam.

We specify the form of Eq. (7.43), assuming

$$g_2 = \frac{s^3 g k^2}{p^4}, \quad g_1 = 0, \quad g_0 = \frac{dp^4}{s^3} - \frac{p^2 T_2}{s} + c, \quad r(s) = \gamma s. \tag{7.48}$$

Under the above assumptions regarding the parameters included in the problem, the coefficients of Eq. (7.43) satisfy the previously formulated properties. Then the solution to problem (7.47) can be represented as

$$w_0 = P_0(\xi, t) z_n\left[s, p(t), q(t), t\right], \tag{7.49}$$

where $P_0(\xi,t)$ is a polynomial in ξ. Substituting (7.49) into Eq. (7.47) yields the relation for the instantaneous frequency of vibrations,

$$\omega = \dot{q}(t)p(t) - H^{\pm}\left[p(t), q(t), t\right], \tag{7.50}$$

where

$$H^{\pm}(p,q,t) = \pm\sqrt{\lambda_n\left(p,q,t\right)} \tag{7.51}$$

is the Hamilton function of the problem under consideration.

It should be noted that now, by virtue of relations (7.48), the eigenfunction z_n and the associated eigenvalue λ_n of the boundary value problem (7.43), (6.9) parametrically depend on the wave number $p = p_n(t)$ and the center $q = q_n(t)$ of the nth travelling WP. We also allow an explicit dependence of z_n and λ_n on time t, which is possible in the case of slow dynamic loading of the shell (when T_2 is a function of time).

Considering boundary value problems arising in the leading and second-order approximations leads to the Hamiltonian system (6.47), the Riccati equation (6.57) and the amplitude equation (6.62) with respect to the functions $p(q), q(t), b(t), P_0(\xi,t)$, which are solved step by step. We note that these equations, up to coefficients, are invariant with respect to the shape of a shell and the way of its loading.

We indicate the available differences for a conical shell pre-stressed by the membrane stress resultants $T_i(s,\varphi,t)$. As before, the coefficients $h_i(t)$ of the amplitude equation (6.62) are calculated by formulas (6.56), but here

$$\mathbf{\Gamma} = HH_{pq} + \frac{1}{\eta}\left[HH_p \int\limits_{s_1[q(t)]}^{s_2[q(t)]} \left(\gamma' s z^2 - 2pzT_3\frac{\partial z}{\partial s} - pz^2\frac{\partial T_3}{\partial s}\right) ds\right],$$

$$\eta = \int\limits_{s_1[q(t)]}^{s_2[q(t)]} \gamma s z^2 ds. \tag{7.52}$$

7.4 Wave packets in a cylindrical shell pre-stressed by axial forces

Here, we present the solution of a problem [15] on WPs in a non-circular cylindrical shell compressed by non-uniform and, in a general case, non-stationary axial forces.

The curvilinear orthogonal coordinate system on the reference surface is introduced as above (see, e.g., in Section 6.1): x is the axial coordinate, and y is the coordinate on the directrix. Let R be the characteristic dimension of the shell mid-surface, $R_2 = R/k(\varphi)$ be the radius of curvature, the shell edges be not necessary plane curves, and the thickness $h(\varphi)$, Young's modulus $E(\varphi)$, Poisson's ratio $\nu(\varphi)$ and the material density $\rho(\varphi)$ are functions of φ.

Suppose that the shell is subject to a nonhomogeneous axial force $T_1^\circ(y, t^*)$ slowly varying in time, and the compressive force does not reach its critical value at which the shell loses stability [222].

Let the stress resultant $T_1^*(y, t^*)$ satisfy conditions (7.2). We shall study non-stationary oscillations of the shell, for which the dynamic SSS is characterized by the indices of variation $\varsigma_1 = \varsigma_2 = 1/2$ and the dynamism index $\varsigma_t = 0$ satisfying conditions (7.5). Then the following equations can be used as the governing ones:

$$\mu^2\Delta\left(d\Delta W\right) + T_1(\varphi, t)\frac{\partial^2 W}{\partial s^2} - k(\varphi)\frac{\partial^2 \Phi}{\partial s^2} + \gamma\frac{\partial^2 W}{\partial t^2} = 0,$$

$$\mu^2\Delta\left(g^{-1}\Delta\Phi\right) + k(\varphi)\frac{\partial^2 W}{\partial s^2} = 0. \tag{7.53}$$

Here, W, Φ, T_1 are the dimensionless normal deflection, the stress function and the axial force, respectively, t is the dimensionless time, which are coupled with their physical counterparts W^*, Φ^*, T^*, t^* by means of the following relations

$$W = \frac{\mu^2 W^*}{R}, \quad \Phi = \frac{\Phi^*}{\mu^2 h_0 E_0 R^2}, \quad T_1 = -\frac{T_1^*}{\mu^2 E_0 h_0}, \quad t = \mu t^*/t_c, \tag{7.54}$$

$$\mu^4 = \frac{h_0^2}{12\,(1-\nu_0^2)\,R^2}, \qquad t_c = R\sqrt{\rho_0/E_0}, \tag{7.55}$$

where μ is a small parameter, t_c is the characteristic time, h_0, E_0, ν_0, ρ_0 are the characteristic values of the functions $h(\varphi), E(\varphi), \nu(\varphi), \rho(\varphi)$, and d, g, γ are the functions introduced by Eqs. (5.133) and depending only on the circumferential coordinate φ.

Let the functions W, Φ satisfy the Navier conditions:

$$W = \frac{\partial^2 W}{\partial s^2} = \Phi = \frac{\partial^2 \Phi}{\partial s^2} = 0 \quad \text{for} \quad s = s_1(\varphi), s_1(\varphi). \tag{7.56}$$

We introduce the initial conditions

$$\begin{aligned} W|_{t=0} &= W_0(s,\varphi,\mu)\,\exp\left[\mathrm{i}\mu^{-1}S_0(\varphi)\right], \\ \dot{W}|_{t=0} &= \mathrm{i}\mu^{-1}V_0(s,\varphi,\mu)\,\exp\left[\mathrm{i}\mu^{-1}S_0(\varphi)\right], \end{aligned} \tag{7.57}$$

where the function $S_0(\varphi)$ is defined in accordance with (6.6), and W_0, V_0 are complex-valued functions such that

$$\frac{\partial^m W_0}{\partial s^m}, \frac{\partial^m V_0}{\partial s^m} \sim \mu^{-m} \qquad \mu \to 0, \quad m = 0, 1, \ldots \tag{7.58}$$

and satisfying the boundary conditions (7.56).

We note that heretofore the initial conditions (6.5) defined the initial WP which had the large variability only in the φ-coordinate direction. In our case, relations (7.57) specify the initial WP in which the wavelength in both the circumferential and axial directions is a value of order $h_*^{-1/2}$, where $h_* = h_0/R$.

7.4.1 Algorithm for constructing a solution in the form of travelling WPs

Taking into account a high variability of the SSS in the axial direction, we zoom in by assuming

$$x = \mu^{-1}s. \tag{7.59}$$

Consider the sequence of functions

$$\begin{aligned} z_n(x,\varphi) &= \sin\{\delta_n(\varphi)[x - x_1(\varphi)]\}, \\ \delta_n(\varphi) = \lambda_n^4(\varphi) &= \frac{\pi n}{x_2(\varphi) - x_1(\varphi)}, \qquad n = 1, 2, \ldots \end{aligned} \tag{7.60}$$

Here and below $x_i = \mu^{-1}s_i(\varphi)$, $i = 1, 2, \ldots$

For any $\varphi \in [\varphi_1, \varphi_2]$, the functions

$$W_0(\mu x, \varphi; \mu) = \tilde{W}_0(x, \varphi; \mu) \quad V_0(\mu x, \varphi; \mu) = \tilde{V}_0(x, \varphi; \mu)$$

can be expanded in the system of functions $z_n(x,\varphi)$ into uniformly convergent on the segment $[x_1(\varphi), x_2(\varphi)]$ series [116]:

$$\begin{aligned}
\tilde{W}_0 &= \sum_{n=1}^{\infty} w_n^\circ(\varphi,\mu) z_n(x,\varphi), \quad w_n^\circ = \int_{x_1(\varphi)}^{x_2(\varphi)} \tilde{W}_0(x,\varphi;\mu) z_n(x,\varphi)dx, \\
\tilde{V}_0 &= \sum_{n=1}^{\infty} v_n^\circ(\varphi,\mu) z_n(x,\varphi), \quad v_n^\circ = \int_{x_1(\varphi)}^{x_2(\varphi)} \tilde{V}_0(x,\varphi;\mu) z_n(x,\varphi)dx.
\end{aligned} \tag{7.61}$$

Let w_n°, v_n° be the polynomials of the argument $\mu^{-1/2}\varphi$ with coefficients continuously depending on $\mu^{1/2}$. Then w_n°, v_n° can be represented as series (6.11), where $\zeta = \mu^{-1/2}\varphi$, and instead of the parameter ε, one needs to write μ. In what follows in this section, when referring to the formulas of Chapter 6, we mean everywhere the replacement of the small parameter ε with μ.

The functions W, Φ will be sought as the superposition of WPs (6.12):

$$W = \sum_{n=1}^{\infty} w_n(s,\varphi,t,\mu), \quad \Phi = \sum_{n=1}^{\infty} f_n(s,\varphi,t,\mu), \tag{7.62}$$

We obey each of the nth wave packets to the initial conditions (7.57) and pass to the moving coordinate system associated with the center of the nth packet accordingly to the formula:

$$\varphi = q(t) + \mu^{1/2}\xi. \tag{7.63}$$

Then the system of equations (7.53) for the nth WP can be rewritten as (hereinafter, the subscript n is omitted everywhere):

$$\begin{aligned}
& d\frac{\partial^4 w}{\partial x^4} + 2\mu d\frac{\partial^4 w}{\partial x^2 \partial \xi^2} + \mu\frac{\partial^2 d}{\partial \xi^2}\frac{\partial^2 w}{\partial x^2} + 2\mu\frac{\partial d}{\partial \xi}\frac{\partial^3 w}{\partial x^2 \partial \xi} + \mu^2\frac{\partial^2 d}{\partial \xi^2}\frac{\partial^2 w}{\partial \xi^2} \\
& \quad + 2\mu^2\frac{\partial d}{\partial \xi}\frac{\partial^3 w}{\partial \xi^3} + \mu^2 d\frac{\partial^4 w}{\partial \xi^4} + T_1\frac{\partial^2 w}{\partial x^2} - k\frac{\partial^2 f}{\partial x^2} + \mu^2\gamma\frac{\partial^2 w}{\partial t^2} \\
& \quad - 2\mu^{3/2}\gamma\dot{q}\frac{\partial^2 w}{\partial t \partial \xi} + \mu\gamma\dot{q}^2\frac{\partial^2 w}{\partial \xi^2} - \mu^{3/2}\gamma\ddot{q}\frac{\partial w}{\partial \xi} = 0, \\
& g^{-1}\frac{\partial^4 f}{\partial x^4} + 2\mu g^{-1}\frac{\partial^4 f}{\partial x^2 \partial \xi^2} + \mu\frac{\partial^2 g^{-1}}{\partial \xi^2}\frac{\partial^2 f}{\partial x^2} + 2\mu\frac{\partial g^{-1}}{\partial \xi}\frac{\partial^3 f}{\partial x^2 \partial \xi} \\
& \quad + \mu^2\frac{\partial^2 g^{-1}}{\partial \xi^2}\frac{\partial^2 f}{\partial \xi^2} + 2\mu^2\frac{\partial g^{-1}}{\partial \xi}\frac{\partial^3 f}{\partial \xi^3} + \mu^2 g^{-1}\frac{\partial^4 f}{\partial \xi^4} + k\frac{\partial^2 w}{\partial x^2} = 0.
\end{aligned} \tag{7.64}$$

On the edges $x = x_i(\varphi)$, the boundary conditions for the functions w, f have the same form (7.56) with the replacement of the argument s with x.

The solution of the formulated problem is sought in the form analogous to (6.18):

$$w = \sum_{j=0}^{\infty} \mu^{j/2} w_j(s,\xi,t) \exp\left[i\mu^{-1}S(\xi,t,\mu)\right],$$

$$f = \sum_{j=0}^{\infty} \mu^{j/2} f_j(s,\xi,t) \exp\left[i\mu^{-1}S(\xi,t,\mu)\right], \tag{7.65}$$

$$S = \int_0^t \omega(\tau)d\tau + \mu^{1/2}p(t)\xi + \tfrac{1}{2}\mu b(t)\xi^2,$$

where

$$\Im\, b(t) > 0 \quad \text{for any} \quad 0 \le t \le t' < +\infty, \tag{7.66}$$

$$\omega,\ p,\ b,\ \dot\omega,\ \dot p,\ \dot b,\ \dot q,\ w_j,\ f_j,\ \frac{\partial w_j}{\partial x},\ \frac{\partial f_j}{\partial x} = O(1) \quad \text{as} \quad \mu \to 0. \tag{7.67}$$

Substituting (7.65) into Eqs. (7.64) and equating coefficients with the same powers of $\mu^{1/2}$, we arrive at the sequence of equations:

$$\sum_{j=0}^{m} \mathbf{L}_j \mathbf{X}_{m-j} = 0, \quad m = 0, 1, 2, \ldots \tag{7.68}$$

Here

$$\mathbf{L}_0 = \begin{pmatrix} l_{11} & l_{12} \\ l_{21} & l_{22} \end{pmatrix}, \quad \mathbf{X}_j = (w_j, f_j)^T,$$

$$l_{11} = d\left(\frac{\partial^2}{\partial x^2} - p^2\right)^2 + T_1\frac{\partial^2}{\partial x^2} - \gamma(\omega - \dot q p)^2, \tag{7.69}$$

$$l_{12} = -l_{21} = -k\frac{\partial^2}{\partial x^2}, \quad l_{22} = g^{-1}\left(\frac{\partial^2}{\partial x^2} - p^2\right)^2$$

and the operators $\mathbf{L}_j$ for $j \ge 1$ have the same structures as in (6.23), but

$$\mathbf{N} = \frac{1}{2}\mathbf{L}_{pq} + \begin{pmatrix} -i\ddot q p & 0 \\ 0 & 0 \end{pmatrix}. \tag{7.70}$$

In this case, in (6.23), the operator $\mathbf{L}$ with the subscripts p, q, ω is the matrix operator defined by differentiating the operator $\mathbf{L}_0$ with respect to corresponding parameters. The superscript T means transpose. The functions k, d, g, γ, T_1 and their derivatives are taken at $\varphi = q(t)$.

The substitution of series (7.65) into conditions (7.56) and expansion of the functions x_i into the Taylor series in powers of $\mu^{1/2}\xi$ leads to the boundary conditions for the vector function $\mathbf{X}_j$ with $x = x_i[q(t)]$:

$$\mathbf{X}_0 = 0, \quad \frac{\partial^2 \mathbf{X}_0}{\partial x^2} = 0, \tag{7.71}$$

$$\mathbf{X}_1 + \xi\, x_i^{'} \frac{\partial \mathbf{X}_0}{\partial x} = 0, \qquad \frac{\partial^2 \mathbf{X}_1}{\partial x^2} + \xi\, x_i^{'} \frac{\partial^3 \mathbf{X}_0}{\partial x^3} = 0, \tag{7.72}$$

$$\mathbf{X}_2 + \xi\, x_i^{'} \frac{\partial \mathbf{X}_1}{\partial x} + \frac{1}{2}\xi^2 \left(x_i^{''} \frac{\partial \mathbf{X}_0}{\partial x} + x_i^{'2} \frac{\partial^2 \mathbf{X}_0}{\partial x^2} \right) = 0,$$

$$\frac{\partial^2 \mathbf{X}_2}{\partial x^2} + \xi\, x_i^{'} \frac{\partial^3 \mathbf{X}_1}{\partial x^3} + \frac{1}{2}\xi^2 \left(x_i^{''} \frac{\partial^3 \mathbf{X}_0}{\partial x^3} + x_i^{'2} \frac{\partial^4 \mathbf{X}_0}{\partial x^4} \right) - \frac{4i\, x_i^{'}}{p} \frac{\partial^3 \mathbf{X}_0}{\partial x^3} = 0, \ldots \tag{7.73}$$

Thus, the initial-boundary-value problem (7.53), (7.56), (7.57) being originally two-dimensional with respect to coordinates, has been reduced to the sequence of 1D boundary-value problems for the coupled differential equation with coefficients being evaluated at the moving generatrix $\varphi = q(t)$.

7.4.2 Resolving equations and relations

In the leading approximation ($m = 0$), one has the homogeneous boundary value problem consisting of the vector differential equation

$$\mathbf{L}_0 \mathbf{X}_0 = 0 \tag{7.74}$$

and the boundary conditions (7.71). Satisfying this conditions, we seek the components of the vector $\mathbf{X}_0 = (w_0, f_0)^T$ in the form

$$w_0 = P_0(\xi,\, t)\, z[x, q(t)], \quad f_0 = Q_0(\xi,\, t)\, z[x, q(t)], \tag{7.75}$$

where P_0, Q_0 are unknown polynomials in ξ. Substituting the vector $\mathbf{X}_0$ with components (7.75) into Eqs. (7.74), one arrives at

$$Q_0 = \Lambda P_0, \quad \Lambda = \frac{kg\delta^2}{(\delta^2 + p^2)^2}, \qquad \omega = \dot{q} p \mp H(p, q, t), \tag{7.76}$$

where

$$H = \gamma^{-1/2}(q) \sqrt{d(q)\,[\delta^2(q) + p^2]^2 + \frac{g(q) k^2(q) \delta^4(q)}{[\delta^2(q) + p^2]^2} - \delta^2(q) T_1(q, t)} \tag{7.77}$$

is the Hamiltonian function.

In first-order approximation, we get the non-homogeneous system of differential equations

$$\mathbf{L}_0 \mathbf{X}_1 = -\mathbf{L}_1 \mathbf{X}_0 \tag{7.78}$$

with the inhomogeneous boundary conditions (7.72) . From the compatibility condition for this problem, one obtains (see Section 6.1) the Hamiltonian system (6.47) with respect to the functions $p(t), q(t)$. Then the solution of system (7.78) can be represented as

$$\mathbf{X}_1 = P_1(\xi, t)\, z[x, q(t)]\mathbf{Y}[p(t), q(t)] + \mathbf{X}_1^{(p)}(x, \xi, t), \tag{7.79}$$

where P_1 is an undefined polynomial in ξ, $\mathbf{Y} = (1,\, \Lambda)^T$, and

$$\mathbf{X}_1^{(p)} = \xi b \mathbf{X}_p + \xi \mathbf{X}_q - \mathrm{i}\frac{\partial \mathbf{X}_p}{\partial \xi} \tag{7.80}$$

is a particular solution of problem (7.78), (7.72). As before, the subscripts p and q mean differentiation with respect to correspondent variables.

In the second-order approximation, one arrives again at the inhomogeneous boundary-value problem consisting of the equation

$$\mathbf{L}_0\mathbf{X}_2 = -\mathbf{L}_1\mathbf{X}_1 - \mathbf{L}_2\mathbf{X}_0 \tag{7.81}$$

and the boundary conditions (7.73). Considering the compatibility condition of this problem, just as it was done in Section 6.1, we arrive at the Riccati equation (6.57) for the function $b(t)$ and the amplitude equation (6.59) for determining the polynomial P_0. In Eq. (6.59), the coefficients h_0, h_1, h_2 are defined by formulas (6.56), while

$$\begin{aligned} h_3(t) = \frac{\mathrm{i}}{2H}\Bigg[& bHH_{pp} - \dot{\omega} - 2H_q H_p + \ddot{q}p + \frac{l'}{l} H\, H_p \\ & + \frac{2p}{\gamma}\left(d' + \frac{g'}{g^2}\Lambda^2\right)(p^2 + \delta^2) + \frac{2}{l\gamma}\int\limits_{x_1(q)}^{x_2(q)} \left(\frac{\partial l_{11}}{\partial p} - \Lambda^2 \frac{\partial l_{22}}{\partial p}\right) z_q z dx \Bigg] \end{aligned} \tag{7.82}$$

with $l(q) = x_2(q) - x_1(q)$.

7.4.3 Solution analysis

The resulting Hamiltonian system, the Riccati equation and the amplitude equation are invariant with respect to both the variability of the constructed asymptotic solution along the shell axis and the way of loading. These equations are the principal ones and allow constructing the main, with accuracy up to $O\left(\mu^{1/2}\right)$, approximate solution of the initial-boundary-value problem (7.53), (7.56), (7.57):

$$w = w^+ + w^-, \quad f = f^+ + f^-, \tag{7.83}$$

$$\begin{aligned} w^{\pm} &= \left[w_0^{\pm} + O\left(\mu^{1/2}\right)\right] \exp\left(\mathrm{i}\mu^{-1} S^{\pm}\right), \\ f^{\pm} &= \left[f_0^{\pm} + O\left(\mu^{1/2}\right)\right] \exp\left(\mathrm{i}\mu^{-1} S^{\pm}\right), \\ w_0^{\pm} &= P_0^{\pm}\left(\xi^{\pm}, t; d_m^{\pm}\right) z\left[s, q^{\pm}(t)\right], \\ f_0^{\pm} &= P_0^{\pm}\left(\xi^{\pm}, t; d_m^{\pm}\right) \frac{k\left[q^{\pm}(t)\right]}{\left[p^{\pm}(t)\right]^4} \frac{\partial^2 z}{\partial s^2}\left[s, q^{\pm}(t)\right], \\ S^{\pm} &= \int_0^t \omega^{\pm}(\tau) d\tau + \mu^{1/2} p^{\pm}(t)\xi^{\pm} + \frac{1}{2}\mu b^{\pm}(t){\xi^{\pm}}^2. \end{aligned} \tag{7.84}$$

It should be noted that Hamiltonian (7.77) has no singularities at $p(t) = 0$. Thus, the constructed solution (7.83), (7.84), which takes into account large variability along the shell axis (here $\varsigma_1 = 1/2$), is a generalization of the analogous solution (6.76), (6.77) with the index of variation $\varsigma_1 = 0$ and can be used to study travelling WPs for small (or zero) values of the wave number $p(t)$ (see Remark 6.7).

The method proposed here for construction of solutions in the form of travelling WPs with large variability in the axial direction is capable of finding eigenmodes with large number of waves in both directions and concentrated in the vicinity of the weakest generatrix. We recall that similar forms of free localized vibrations with large number of waves in the axial direction were found above (as a particular case) when studying quasi-stationary problems on parametric vibrations of the shell (see Section 5.1).

We consider some properties of non-stationary solutions in the form of travelling WP for a cylindrical shell with constant geometric and physical parameters subjected to the action of stationary inhomogeneous axial forces $T_1(\varphi)$. Here $k = d = g = \gamma = 1$. The analysis of the Hamiltonian system shows that if

$$\begin{gathered} a_0^2 + \delta^2 - \delta < 0, \quad T_1'(\varphi) < 0 \quad \text{for} \quad 0 < \varphi < \varphi_2, \\ T_1(\varphi) = T_1(-\varphi) \quad \text{on the set} \quad \varphi \in (-\varphi_2, \varphi_2), \\ \inf_{(0,\varphi_2)} T_1(\varphi) < A, \quad A = 2 - H^2(0, a_0, 0)\delta^{-2}, \end{gathered} \tag{7.85}$$

then there exists such t_r that

$$\begin{gathered} v_g(0)v_g(t) > 0 \quad \text{for} \quad 0 < t < t_r, \\ v_g(t_r) = 0, \\ v_g(0)v_g(t) < 0 \quad \text{for} \quad t_r < t < 3t_r, \\ v_g(3t_r) = 0, \\ v_g(0)v_g(t) > 0 \quad \text{for} \quad 3t_r < t < 4t_r, \\ v_g(4t_r) = v_g(0), \end{gathered} \tag{7.86}$$

where $v_g = \dot{q}(t)$. In this case, t_r is found from the equation

$$T_1\left[q(t_r)\right] = A \tag{7.87}$$

and the following identity holds:

$$q(3t_r) \equiv -q(t_r).$$

Relations (7.86) indicate the presence of the reflection effect of the WP running in the direction of decreasing the compressive axial force $T_1(\ f)$ from some generatrix $f = q(t_r)$. It is also easy to show (the proof is similar to that given in Remark 6.5) that in the case of periodicity of $T_1(\ f)$, the functions $p(t), q(t),\ o(t)$ are also periodic with period $4t_r$.

Thus, the non-uniform distribution of the axial compressive forces T_1 along the edges can lead to the effect of localization of excited WPs in the regions adjacent to the weakest generatrix on which the force T_1 is maximum.

7.4.4 Examples

To study the effect of axial forces on the dynamic characteristics of running WPs, let us turn to examples.

Example 7.4. Consider a simply supported circular cylindrical shell with constant geometric and physical parameters under the action of stationary, non-uniformly distributed along the edge axial forces

$$T_1 = 1 + \cos 2\varphi. \tag{7.88}$$

Let

$$\begin{gathered} x_1 = 0, \quad x_2 = 4, \quad k = 1, \quad a_0 = 2/3, \quad b_0 = \mathrm{i}, \\ \tilde{W}_0(x, \varphi; \mu) = w_n^\circ z_n, \quad \tilde{V}_0(x, \varphi; \mu) = v_n^\circ z_n. \end{gathered} \tag{7.89}$$

In this case, the initial WP is located on the weakest generatrix $\varphi = 0$, where the compressive force (7.88) is maximum. Calculations were carried out at $h = 0.02$, $R = 50$, $\nu = 0.3$, $w_n^\circ = 1$, $v_n^\circ = 1$, $n = 1$. Note that for the considered values of parameters (7.89), conditions (7.85) are fulfilled. In Fig 7.7, the plots of the functions $p^+(t), q^+(t), \omega^+(t)$ (curves 1–3, respectively) indicate that the behavior of the 1^+st WP is fully consistent with the results above analysis: 1) there is a multiple reflection of the WP, 2) the functions $p^+(t), q^+(t), \omega(t)$ periodic. The plots of the functions $\Im\, b^+(t)$, $\Re\, w^+_{max}(t)$ (curves 4, 5, respectively) show that each reflection is accompanied by strong focusing the WP (see increasing the function $\Im\, b(t)$) and growing the amplitude of waves in this packet.

Note that since the functions $p^+(t)$, $q^+(t)$ are periodic, the coefficients of the Riccati equation are also periodic functions. An approximate calculation of the monodromy matrix for the linear system of equations (6.85) equivalent to the Riccati equation revealed that the function $b^+(t)$ is not periodic.

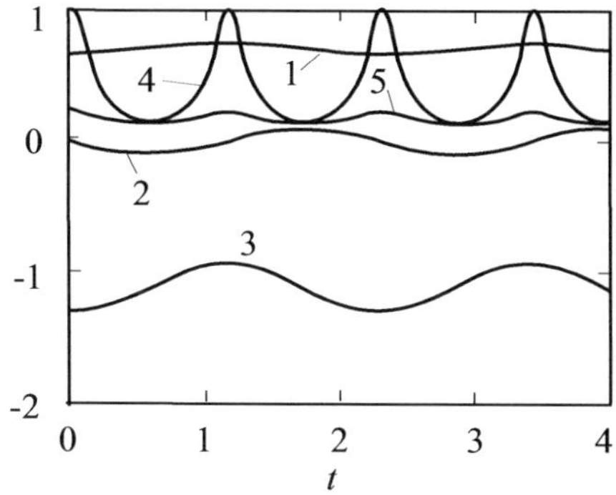

FIGURE 7.7: Plots of the functions $p^+(t)$, $q^+(t)$, $\omega^+(t)$, $\Im\, b^+(t)$, $\Re\, w^+_{max}(t)$ (curves 1, 2, 3, 4, 5, respectively) for the circular cylindrical shell subject to stationary, non-uniformly distribute in the circumferential direction axial forces.

Example 7.5. Consider a simply supported cylindrical shell with an elliptical cross section and with semi-axes $e_1 < e_2$ (see Fig. 6.4). The remaining geometric and physical parameters of the shell are considered constant. It is convenient to go from the coordinate φ to the polar angle θ (see relations (6.145) in Example 6.2).

Let the shell be subject to slowly varying axial forces $T_1 = c_t\, t$, where c_t is a constant (if $c_t < 0$, then the shell is stretched, and for $c_t > 0$, it is compressed).

The generatrix $\theta = 0$ has the least curvature and therefore is the weakest (see Example 4.2). As the initial WP, we choose one of the eigenmodes (4.18) of the shell free of forces T_1, which are localized in the vicinity of the line $\theta = 0$. To this end, we assume in (6.6), (7.57), (7.61):

$$\begin{gathered} a_0 = \sqrt{\sqrt{k(0)}\delta_n - \delta_n^2}, \quad b_0 = i\,\delta_n \frac{\sqrt{k''(0)}}{4a_0}, \\ \tilde{W}_0(x,\varphi;\mu) = w_n^\circ z_n, \quad \tilde{V}_0(x,\varphi;\mu) = v_n^\circ z_n. \end{gathered} \tag{7.90}$$

The parameters a_0, b_0 are found as a solution of Eqs. (6.93), (6.94) at $T_1 = 0$.

Figure 7.8 shows the calculation results for $n = 1$, $e_1 = 1$, $e_2 = 1.5$, $l = 5$, $w_n^\circ = 1$, $v_n^\circ = 1$, $c_t = 1$.

In this case, the value of the critical axial force leading to the shell buckling is $T_1^* \approx 0.889$, therefore, the calculations for curves 2 and 3 were carried out up to the time instants $t^* \approx 2.96$, 1.77, 0.88, respectively. Here, the parameters $p^+ \equiv a_0 \approx 0.155$; $q^+ \equiv 0$, $b^+(t) \equiv b_0 \approx 0.58$ are constant, and the time dependent quantities (for $c_t \neq 0$) are $\omega^+(t)$ and $\Re\, w_{max}(t)$.

It is worth paying attention to the fact that here the variable axial force does not lead to a splitting up of the initial WP which coincides with one of

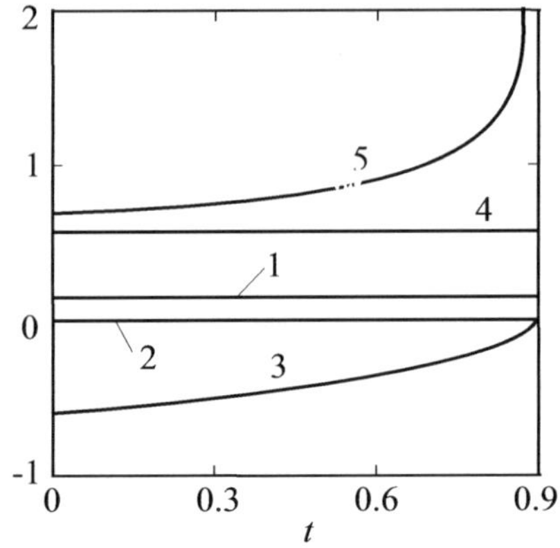

FIGURE 7.8: Plots of the functions $p^+(t)$, $q^+(t)$, $\omega^+(t)$, $\Im b^+(t)$, $\Re w^+_{max}(t)$ (curves 1, 2, 3, 4, 5, respectively) for the elliptic cylindrical shell subject to slowly varying the uniform axial forces.

the eigenmodes with a large number of waves in the both direction. We recall that loading the shell by a growing external pressure leads to the splitting up of the localized eigenmodes with a small number of semi-waves in the axial direction (see Example 7.3) into two unsteady WPs. In this case, the growing tensile force ($c_t < 0$) leads to an increase in the oscillation frequency and slow decrease in the amplitudes of the WP with a center on the generatrix $\theta = 0$. In contrast, the slowly growing compressive force ($c_t > 0$) leads to a decrease in the oscillation frequency and an increase in their amplitudes. It should be noted that for $t \to t^*$ (that is, for axial forces close to the critical value T_1^*), the oscillation amplitudes increase unboundedly. This indicates that in the linear statement of the problem it is possible the dynamic loss of stability at values of $T_1(t)$ is less than the critical forceT_1^*.

Chapter 8

Wave Packets in Long Shells of Revolution Travelling in the Axial Direction

In this chapter, wave packets travelling in the axial direction in an infinitely long shell of revolution are studied. The linear equations written in terms of displacements and based on the Kirchhoff–Love hypotheses are used as the governing ones. Three different wave forms of motion are considered: a) axisymmetric WPs; b) non-axisymmetric WPs with small wave number $m_0 \sim 1$ in the circumferential direction; c) non-axisymmetric WPs with $m_0 \sim h_*^{-1/2}$, where h_* is the relative thickness of a shell.

At first, using the asymptotic approach by Gol'denveizer [65, 67] based on introducing indices of variation for unknown functions, the integrals of the governing equations are classified (both for axisymmetric and for non-axisymmetric WPs), and the next, the simplified systems of differential equations describing predominantly bending, longitudinal and torsional waves are derived.

Then, utilizing the method developed in Chapter 6, solutions of the governing equations are constructed in the form of a superposition of localized families of longitudinal, torsional, and bending waves running in the axial direction. The 2D initial-boundary-value problem is reduced to the sequence of coupled algebraic equations with coefficients being calculated at the moving parallels which are the centers of travelling WPs. The compatibility conditions of inhomogeneous equations arising in the first- and second-order approximations generate the Hamiltonian systems, the Riccati equations and the amplitude equations specifying the dynamic characteristics of WPs (instantaneous frequency, wave parameter, group velocity and amplitude). The effect of geometric and physical parameters of the shell of revolution on the dynamic characteristics of travelling WPs is analyzed. The influence of the material inhomogeneity as well as the thickness variation on the running packets of bending and tangential waves in a cylindrical shell is also studied.

8.1 Governing equations in terms of displacements. Setting the problem

Consider an infinite elastic shell of revolution of thickness h. Let $x = R\,s$ be the arc length of the generatrix ($-\infty < x < +\infty$), and φ be the angle counted in the circumferential direction. Here $R = r^*(0)$ is the characteristic size of the shell, where $r^*(s) = R\,r(s)$ is the distance to the axis of rotation (see Fig. 8.1).

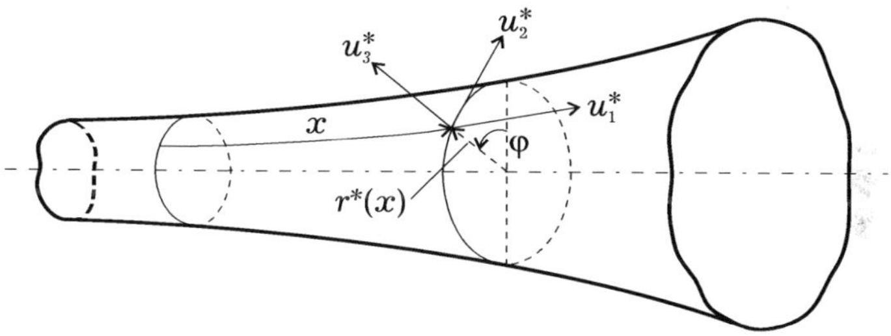

FIGURE 8.1: Shell of revolution.

In the introduced coordinate system on the mid-surface, the first quadratic form is $R^2\left(d\,s^2 + r^2\,d\varphi^2\right)$. The principal radii of curvature R_1, R_2 satisfy the relations

$$R_1 = -R\frac{\sqrt{1-r'^2}}{r''}, \qquad R_2 = R\frac{r}{\sqrt{1-r'^2}}. \tag{8.1}$$

Here and below, the prime denotes differentiation with respect to s.

We shall study axisymmetric and non-axisymmetric forms of motion with a number m_0 of waves along the parallel. The linear differential equations based on the Kirchhoff – Love hypotheses [64] are used as the governing ones:

$$\left(\mathbf{L} + \frac{\partial^2}{\partial t^2}\right)\mathbf{U}^T = \mathbf{0}. \tag{8.2}$$

Here $\mathbf{L}$ is the (3×3)-matrix differential operator and $\mathbf{U} = (u_1,\, u_2,\, u_3)$ is the 3D vector, where

$$u_1^* = Ru_1(s,\,t)\cos\left(m_0\varphi\right), \quad u_2^* = Ru_2(s,\,t)\sin\left(m_0\varphi\right) \tag{8.3}$$

and

$$u_3^* = Ru_3(s,\,t)\cos\left(m_0\varphi\right) \tag{8.4}$$

are the tangential and normal displacements, respectively, of a point at the shell mid-surface, t is dimensionless time (time related to the characteristic time $t_c' = R\left[(1-\nu^2)\rho/E\right]^{1/2}$), ρ, ν, E are the material density, Poisson's ratio, and Young's modulus, respectively. The elements of the matrix operator **L** are as follows [64]:

$$
\begin{aligned}
L_{11}z &= -\frac{\partial}{\partial s}\frac{1}{r}\frac{\partial}{\partial s}(r\,z) + (1-\nu)\left(\frac{m_0^2}{2r^2} - k_1\,k_2\right)z,\\
L_{12}z &= -m_0\frac{\partial}{\partial s}\left(\frac{z}{r}\right) + \frac{(1-\nu)\,m_0}{2\,r^2}\frac{\partial}{\partial s}(r\,z),\\
L_{13}z &= \frac{\partial}{\partial s}[(k_1+k_2)\,z] - (1-\nu)\,k_2\frac{\partial z}{\partial s},\\
L_{21}z &= \frac{m_0}{r^2}\frac{\partial}{\partial s}(rz) - \frac{(1-\nu)\,m_0}{2}\frac{\partial}{\partial s}\left(\frac{z}{r}\right),\\
L_{22}z &= -\frac{1-\nu}{2}\frac{\partial}{\partial s}\frac{1}{r}\frac{\partial}{\partial s}(r\,z) + \left[\frac{m_0}{r^2} - (1-\nu)k_1\,k_2\right]z,\\
L_{23}z &= -\frac{m_0}{r}(k_2+\nu\,k_1)\,z \quad l_{32}z = l_{23}z,\\
L_{31}z &= -\frac{1}{r}(k_1+k_2)\frac{\partial}{\partial s}(rz) + \frac{1-\nu}{r}\frac{\partial}{\partial s}(k_2rz),\\
L_{33}z &= \mu_1^4\left\{\left[\frac{1}{r}\left(\frac{\partial}{\partial s}r\frac{\partial}{\partial s} - \frac{m_0^2}{r}\right)\right]^2 z - \frac{1}{(1+\nu)\,r}\left(\frac{\partial}{\partial s}r''\frac{\partial}{\partial s}z\right)\right.\\
&\left. + \frac{m_0^2r''}{r^2}z\right\} + \left(k_1^2 + 2\nu\,k_1k_2 + k_2^2\right)z,
\end{aligned}
\tag{8.5}
$$

where $\mu_1^4 = h^2/\left(12R^2\right)$ is the small parameter, and $k_1 = R/R_1$, $k_2 = R/R_2$ are the dimensionless curvatures.

The coupled Eqs. (8.2) were deduced in [64] by separating variables in equations of the classical thin shell theory [67]. When deriving them, some terms were omitted, they do not affect the accuracy of the leading approximation of the asymptotic solution constructed below.

We consider the following initial conditions:

$$
\begin{aligned}
u_j|_{t=0} &= u_j^\circ(\zeta,\mu_1)\exp\{\mathrm{i}\mu_1^{-1}S_0(s)\}, \quad j=1,2,3,\\
\dot{u}_j|_{t=0} &= \mathrm{i}\epsilon_j\vartheta_j^\circ(\zeta,\mu_1)\exp\{\mathrm{i}\mu_1^{-1}S_0(s)\}
\end{aligned}
\tag{8.6}
$$

with

$$
\begin{aligned}
&S_0(s) = a_0s + 1/2b_0s^2, \quad \operatorname{Im} b_0 > 0, \quad a_0 > 0,\\
&u_j^\circ = \sum_{k=0}^{\infty}\mu_1^{k/2}u_{jk}^\circ(\zeta), \quad \vartheta_j^\circ = \sum_{k=0}^{\infty}\mu_1^{k/2}\vartheta_{jk}^\circ(\zeta),\\
&\mathrm{i} = \sqrt{-1}, \quad \zeta = \mu_1^{-1/2}s, \quad \epsilon_1 = \epsilon_2 = \mu_1^{-1}, \quad \epsilon_3 = 1,
\end{aligned}
\tag{8.7}
$$

and $u^{\circ}_{jk}(\zeta)$, $\vartheta^{\circ}_{jk}(\zeta)$ are polynomials in ζ with complex coefficients. In particular,

$$u^{\circ}_{j0} = \sum_{i=0}^{M_j} c^{\circ}_{ji}\zeta^i, \quad \vartheta^{\circ}_{j0} = \sum_{i=0}^{M_j} d^{\circ}_{ji}\zeta^i, \quad |c^{\circ}_{ji}|, \, |d^{\circ}_{ji}| \sim 1. \tag{8.8}$$

In (8.8) and below, the dot means differentiation with respect to dimensionless time t.

Functions (8.6), (8.7) define on the shell surface the initial packets of bending, longitudinal and torsional waves with the center on the parallel $s = 0$. In Section 5.5, it was shown that such forms of vibrations can arise as a result of parametric excitation in a long shell compressed in the axial direction in the presence of a small pulsating component of the axial force or pulsating pressure. Formulas (8.6) can also approximate the shell response to a short-term dynamic load applied along the parallel $s = 0$.

Below, the following cases will be considered:

a) $m_0 = 0$; b) $m_0 \sim 1$; c) $m_0 = \mu^{-1}m$, where $m \sim 1$.

8.2 Classification of integrals of the dynamic equations

Following [63, 67], we introduce for the unknown functions u_j the indices of intensity, υ_j, and dynamism, α_t:

$$u_j \sim h_*^{\upsilon_j}, \quad \partial u_j/\partial t \sim h_*^{-\alpha_t} u_j, \quad j = 1, 2, 3, \tag{8.9}$$

where $h_* = h/R \sim \mu_1^2$.

From the initial conditions (8.6), it follows that the index of variation of the integrals of system (8.2) at least for some initial time interval is equal to 1/2:

$$\partial u_j/\partial s \sim h_*^{-1/2} u_j, \quad j = 1, 2, 3. \tag{8.10}$$

The problem is to find consistent values of the indices υ_j, α_t. The consistency criterion [67] is the equality of at least two basic terms in each of equations composing system (8.2). Depending on which terms are assumed as the main ones, various the coupled equations are obtained that have a solution with one or another property.

Among all possible solutions in each of cases a), b) and c), we consider two types corresponding to tangential and bending waves.

8.2.1 Axisymmetric waves

Let $m_0 = 0$. This case is the simplest, because the original system of equations (8.2) splits into the equation describing *torsional* waves,

$$-\frac{1-\nu}{2}\frac{\partial}{\partial s}\frac{1}{r}\frac{\partial}{\partial s}(ru_2) - (1-\nu)k_1k_2u_2 + \frac{\partial^2 u_2}{\partial t^2} = 0, \tag{8.11}$$

and the system of two equations with respect to the longitudinal and normal displacements, u_1, u_3:

$$\left(\mathbf{L}^\circ + \partial^2/\partial t^2\right)\mathbf{V}^T = 0, \tag{8.12}$$

where $\mathbf{V} = (u_1,\, u_3)$ is the 2D-vector, and $\mathbf{L}^\circ$ is the square matrix with elements

$$\begin{aligned}
L^\circ_{11}z &= -\frac{\partial}{\partial s}\frac{1}{r}\frac{\partial}{\partial s}(rz) - (1-\nu)k_1k_2z,\\
L^\circ_{12}z &= \frac{\partial}{\partial s}[(k_1+k_2)z] - (1-\nu)k_2\frac{\partial z}{\partial s},
\end{aligned}$$

$$L^\circ_{21}z = -\frac{1}{r}(k_1+k_2)\frac{\partial}{\partial s}(rz) + \frac{1-\nu}{r}\frac{\partial}{\partial s}(k_2rz), \tag{8.13}$$

$$L^\circ_{22}z = \mu_1^4\left\{\left[\frac{1}{r}\left(\frac{\partial}{\partial s}r\frac{\partial}{\partial s}\right)\right]^2 z - \frac{1}{(1+\nu)r}\left(\frac{\partial}{\partial s}r''\frac{\partial}{\partial s}z\right)\right\}$$

$$+(k_1^2 + 2\nu k_1k_2 + k_2^2)z.$$

In turn, equations (8.12) can be replaced by two simpler equations describing the propagation of axisymmetric *longitudinal* and *flexural (bending)* waves. Using the example of Eqs. (8.12), we consider the method of classification of integrals.

8.2.1.1 Longitudinal waves

We write down the main terms in each equation of system (8.12), requiring that in the first equation the terms containing the largest derivative with respect to s, as well as the inertia term, be of the same order of smallness and be contained in the main part of the resolving system, providing wave nature of the integrals:

$$\text{(the 1-st equation)} \qquad -\frac{\partial}{\partial s}\frac{1}{r}\frac{\partial}{\partial s}(ru_1), \quad -\frac{\partial^2 u_1}{\partial t^2},$$

$$\text{(the 2-nd equation)} \qquad -\frac{k_1+k_2}{r}\frac{\partial}{\partial s}(ru_1) + \frac{1-\nu}{r}\frac{\partial}{\partial s}(k_2ru_1), \quad \frac{\partial^2 u_3}{\partial t^2}. \tag{8.14}$$

In view of (8.9), (8.10), the orders of the main terms in (8.14) are respectively equal to:

$$\begin{aligned}&\text{(the 1-st equation)} \qquad \upsilon_1 - 1, \quad \upsilon_1 - 2\alpha_t,\\ &\text{(the 2-nd equation)} \qquad \upsilon_1 - 1/2, \quad \upsilon_3 - 2\alpha_t.\end{aligned} \tag{8.15}$$

Since the intensity indices of the initial displacements (8.6) are equal to zero (see (8.8)), for the longitudinal waves we set $\upsilon_1 = 0$. Equating the orders in (8.15), we find $\alpha_t = \upsilon_3 = 1/2$. Then for the longitudinal and normal displacements, one can assume

$$u_1 = u_s, \quad u_3 = \mu_1 \, w_s, \tag{8.16}$$

where $u_s, w_s = O(1)$ as $\mu_1 \to 0$. As a result, the coupled equations (8.12) read

$$\left(\mathbf{L}^\circ + \partial^2/\partial t^2\right)\left(\mathbf{E}_{01}\mathbf{V}_s^T\right) = 0 \tag{8.17}$$

with

$$\mathbf{V}_s = (u_s, \, w_s)\,, \quad \mathbf{E}_{01} = \begin{pmatrix} 1 & 0 \\ 0 & \mu_1 \end{pmatrix}. \tag{8.18}$$

Obviously, if the vector $\mathbf{V}_s$ satisfies equation (8.17), then functions (8.16) satisfy the system of equations (8.12). A characteristic feature of this solution is that the function u_1 in the case under consideration ($m_0 = 0$) is the main one in the asymptotic sense. Here the normal deflection u_3 is a quantity of the order of $\sim \mu_1$ and is "generated", as will be shown below, by the tangential displacement u_s. In view of the above, Eqs. (8.12) will be called the equations describing axisymmetric *longitudinal* waves.

8.2.1.2 Bending waves

To describe bending waves, we must require that in the second equation of system (8.12) the inertial term as well as the terms containing the highest derivatives by s be of the same order of smallness and be contained in the asymptotically leading approximation. Equating as before the orders of the main terms, we find $\upsilon_1 = 1/2$, $\upsilon_3 = 0$, $\alpha_t = 0$. Then, to study the bending waves, one needs to assume

$$u_1 = \mu \, u_n, \quad u_3 = w_n, \tag{8.19}$$

where $u_n, w_n = O(1)$. Let $t_1 = \mu_1 \, t$ be the "slow"time that is introduced in order to emphasize the different indexes of dynamism for the in-plane longitudinal waves and bending ones. As a result, to describe *bending* waves, system (8.12) can be replaced by the following system of equations

$$\left(\mathbf{L}^\circ + \mu_1^2 \, \partial^2/\partial t_1^2\right)\left(\mathbf{E}_{10}\mathbf{V}_n^T\right) = 0, \tag{8.20}$$

where

$$\mathbf{V}_n = (u_n, \, w_n)\,, \quad \mathbf{E}_{10} = \begin{pmatrix} \mu_1 & 0 \\ 0 & 1 \end{pmatrix}. \tag{8.21}$$

In this case, the normal displacement u_3 is asymptotically the main one, and the longitudinal displacement $u_1 \sim \mu_1$ is the secondary one "generated" by tu_3. Note that here, in contrast to the notation adopted in Chapter 6, the subscript n means that an unknown function belongs to the family of bending waves.

8.2.2 Non-axisymmetric waves with low variability along a parallel

Let $m_0 \sim 1$. Omitting the analysis of Eqs. (8.2) similar to that which was done for the axisymmetric case, we write down two systems describing the in-plane and out-of-plane motions of the shell (respectively, tangential and bending waves).

8.2.2.1 Tangential waves

Here

$$u_1 = u_\tau, \quad u_2 = v_\tau, \quad u_3 = \mu_1 w_\tau, \tag{8.22}$$

and the corresponding equations read

$$\left(\mathbf{L} + \partial^2/\partial t^2\right)\left(\mathbf{E}_{001}\mathbf{U}_\tau^T\right) = 0 \tag{8.23}$$

with

$$\mathbf{U}_\tau = (u_\tau,\, v_\tau,\, w_\tau)\,, \quad u_\tau,\, v_\tau,\, w_\tau = O(1) \quad \text{as} \quad \mu_1 \to 0,$$

$$\mathbf{E}_{001} = \begin{pmatrix} 1 & 0 & 0 \\ 0 & 1 & 0 \\ 0 & 0 & \mu_1 \end{pmatrix}. \tag{8.24}$$

8.2.2.2 Bending waves

In this case,

$$u_1 = \mu_1 u_n, \quad u_2 = \mu_1^2 v_n, \quad u_3 = w_n, \tag{8.25}$$

where $u_n, v_n, w_n = O(1)$ for $\mu_1 \to 0$, and the equations are as follows:

$$\left(\mathbf{L} + \mu_1^2\partial^2/\partial t_1^2\right)\left(\mathbf{E}_{120}\mathbf{U}_n^T\right) = 0. \tag{8.26}$$

Here,

$$\mathbf{U}_n = (u_n,\, v_n,\, w_n)\,, \quad \mathbf{E}_{120} = \begin{pmatrix} \mu_1 & 0 & 0 \\ 0 & \mu_1^2 & 0 \\ 0 & 0 & 1 \end{pmatrix}. \tag{8.27}$$

Relations (8.25) indicate that the non-axisymmetric bending waves, which have small variability in the circumferential direction, are characterized by small tangential components $u_1 \sim \mu_1$, $u_2 \sim \mu_1^2$, which are "generated" by the normal displacement $u_3 \sim 1$.

8.2.3 Non-axisymmetric waves with large variability along a parallel

Let now $m_0 = \mu_1^{-1} m$, where $m \sim 1$. In this case, the coupled equations (8.2) splits into two systems that describe the non-axisymmetric in-plane and out-of-plane motions of the shell with large number of waves along a parallel.

8.2.3.1 Tangential waves

Here

$$u_1 = u_\tau, \quad u_2 = v_\tau, \quad u_3 = \mu_1 w_\tau, \tag{8.28}$$

$$\left(\mathbf{L}_{-1} + \partial^2/\partial t^2\right)\left(\mathbf{E}_{001}\,\mathbf{U}_\tau^T\right) = 0, \tag{8.29}$$

where u_τ, v_τ, $w_\tau = O(1)$ as $\mu_1 \to 0$,

$$\mathbf{U}_\tau = (u_\tau,\, v_\tau,\, w_\tau)\,, \quad \mathbf{E}_{001} = \begin{pmatrix} 1 & 0 & 0 \\ 0 & 1 & 0 \\ 0 & 0 & \mu_1 \end{pmatrix}, \tag{8.30}$$

and $\mathbf{L}_{-1}$ denotes the matrix $\mathbf{L}$ in which $m_0 = \mu_1^{-1} m$.

8.2.3.2 Bending waves

For this type of waves

$$u_1 = \mu_1 u_n, \quad u_2 = \mu_1 v_n, \quad u_3 = w_n, \tag{8.31}$$

$$\left(\mathbf{L}_{-1} + \mu_1^2 \partial^2/\partial t_1^2\right)\left(\mathbf{E}_{110}\mathbf{U}_n^T\right) = 0, \tag{8.32}$$

where $u_n, v_n, w_n = O(1)$ as $\mu_1 \to 0$, and

$$\mathbf{E}_{110} = \begin{pmatrix} \mu_1 & 0 & 0 \\ 0 & \mu_1 & 0 \\ 0 & 0 & 1 \end{pmatrix}. \tag{8.33}$$

8.3 Asymptotic solution as superposition of packets of axisymmetric flexural and tangential waves

The method of constructing asymptotic solutions in the form of WPs travelling in the axial direction is not much different from the method developed in Chapter 6. At the same time, it has some specific features related to the asymptotic splitting of the original system (8.2) carried out above into simplified equations corresponding to the tangential and bending motion of a shell.

In this section, using the example of Eqs. (8.12), the algorithm for constructing solutions in the form of a superposition of travelling packets of longitudinal and bending waves, which are the integrals of Eqs. (8.17) and (8.20), respectively, is given. Although the method is presented for a special case of axisymmetric shell motion, the asymptotic solutions constructed here, after appropriate changes, may be used to study non-axisymmetric WPs.

It should be noted that to find a solution both of Eqs. (8.12) and of other simplified equations in the form of running WPs, as an alternative method can be used Maslov's method presented in Section 6.4. However, its use is difficult for constructing an asymptotic solution of the original Eqs. (8.2) in the form of a superposition of integrals corresponding to packets of bending, longitudinal and torsional waves.

8.3.1 Axisymmetric packets of bending waves

We shall give a solution of Eqs. (8.20) in the form of the WP travelling in the axial direction of the shell [148]. Let $s = q_n(t_1)$ be the center of this packet, where $q_n(t_1)$ is a twice differentiable function such that $q_n(0) = 0$. We pass in Eqs. (8.20) to a new coordinate system by the formula

$$s = q_n(t_1) + \mu_1^{1/2}\,\xi_n. \tag{8.34}$$

We expand all the functions appearing in Eqs. (8.20) into the Taylor series in powers of $\mu_1^{1/2}\xi_n$ in the neighborhood of the moving parallel $s = q_n(t_1)$. We will construct the solution on a certain time interval $[0,\, t_n']$, where for any $t_1 \in [0, t_n']$ the values of all functions and their derivatives from (8.20) at the point $q_n(t_1)$ are quantities of order $O(1)$ for $\mu_1 \to 0$.

The asymptotic solution of Eqs. (8.20) is sought in the form:

$$\begin{gathered}
\mathbf{V}_n = \sum_{k=0}^{\infty} \mu_1^{k/2}\mathbf{V}_{n,k} \exp\left[\mathrm{i}\mu_1^{-1} S_n\left(\xi_n, t_1, \mu_1\right)\right], \\
S_n = \int_0^{t_1} \omega_n(\tau)\mathrm{d}\tau + \mu_1^{1/2} p_n(t_1)\xi_n + \tfrac{1}{2}\mu_1 b_n(t_1)\xi_n^2, \\
\mathbf{V}_{n,k} = \left(u_{n,k},\, w_{n,k}\right),
\end{gathered} \tag{8.35}$$

where $u_{n,k}\left(\xi_n, t_1\right),\, w_{n,k}\left(\xi_n, t_1\right)$ are polynomials in ξ_n, $\Im b_n\left(t_1\right) > 0$ for any $t_1 \in [0,\, t_n']$. We require also that all unknown functions in ansatz (8.35) satisfy conditions of type (6.20):

$$\left\{\omega_n,\, p_n,\, b_n,\, \dot{\omega}_n,\, \dot{p}_n,\, \dot{b}_n,\, \dot{q}_n,\, \mathbf{V}_{n,k},\, \frac{\mathbf{V}_{n,k}}{\partial \xi_n},\, \frac{\mathbf{V}_{n,k}}{\partial t_1}\right\} = O(1) \quad \text{as} \quad \mu_1 \to 0. \tag{8.36}$$

The substitution of expansion (8.35) into Eqs. (8.20) results in the sequence of algebraic equations:

$$\sum_{j=0}^{k} \mathbf{L}_{n,j}\mathbf{V}^{\mathrm{T}}_{n,k-j} = \mathbf{0}, \quad k = 0, 1, 2, \ldots \tag{8.37}$$

Here $\mathbf{L}_{n,0}$ is the (2×2)-matrix with elements

$$\begin{aligned} &l_{n,11} = -p_n^2, \quad l_{n,12} = \mathrm{i}p_n\left[k_1(q_n) + \nu k_2(q_n)\right], \quad l_{n,21} = -l_{n,12}, \\ &l_{n,22} = p_n^4 + k_1^2(q_n) + 2\nu k_1(q_n)k_2(q_n) + k_2^2(q_n) - (\omega_n - \dot{q}_n p_n)^2. \end{aligned} \tag{8.38}$$

The matrix operators $\mathbf{L}_{n,j}(j \geq 1)$ are defined by formulas (6.23) in which the differential operator $\mathbf{L}$ should be replaced with the matrix operator $\mathbf{L}_{n,0}$, and the operator $\mathbf{N}$ is to be substituted by the (2×2)-matrix $\mathbf{N}_n$ with the elements

$$\begin{aligned} n_{n,11} &= r'r^{-1}p_n, \quad n_{n,12} = \mathrm{i}\left(k_1' + k_2'\right), \\ n_{n,21} &= -\mathrm{i}r'r^{-1}(k_1 + k_2) + \mathrm{i}(1-\nu)(k_2 r)'r^{-1}, \\ n_{n,22} &= 2r'r^{-1}p_n^3 + \ddot{q}_n p_n. \end{aligned} \tag{8.39}$$

The dot in (8.38), (8.39) means differentiation with respect to "slow time" t_1. Values of the functions r, k_1, k_2 and their derivatives are calculated on the moving parallel $s = q_n(t_1)$.

The algorithm for finding unknown functions in expansion (8.35) remains the same. Its peculiarity consists only in the fact that a problem arising at each step of the iteration is a system of algebraic equations.

In the leading approximation (for $k = 0$), one arrives at the homogeneous system of algebraic equations

$$\mathbf{L}_{n,0}\mathbf{V}^{\mathrm{T}}_{n,0} = \mathbf{0}. \tag{8.40}$$

It has the non-trivial solution

$$\mathbf{V}_{n,0} = P_{n,0}(\xi_n, t_1)\mathbf{Y}_n \tag{8.41}$$

when the condition

$$\omega_n(t) = \dot{q}_n(t)p_n(t) \mp H_n\left[p_n(t_1), q_n(t_1)\right] \tag{8.42}$$

holds, where

$$H_n(p, q) = \sqrt{(1-\nu^2)k_2^2(q) + p^4} \tag{8.43}$$

is the Hamiltonian function corresponding to the axisymmetric bending waves, $P_{n,0}(\xi_n, t_1)$ is the polynomial in ξ_n with coefficients depending on t_1, and $\mathbf{Y}_n = (y_{n,1}, y_{n,2})$ is the vector with components

$$y_{n,1} = -\mathrm{i}\{k_1[q_n(t_1)] + \nu k_2[q_n(t_1)]\}p_n^{-1}(t_1), \quad y_{n,2} = 1. \tag{8.44}$$

It follows from (8.44) that $p_n(t_1) \neq 0$ on the time interval $[0,\, t_n']$. Moreover, we accept $p_n(t_1) \sim 1$, because for $p_n \to 0$ one of the conditions (8.36) that guarantees asymptotic nature of the constructed solution does not hold.

For $k = 1$, we get the inhomogeneous system of algebraic equations

$$\mathbf{L}_{n,0}\mathbf{V}_{n,1}^{\mathrm{T}} = -\mathbf{L}_{n,1}\mathbf{V}_{n,0}^{\mathrm{T}}, \tag{8.45}$$

which is solvable under the condition

$$\bar{\mathbf{Z}}_{n,0}\mathbf{L}_{n,1}\mathbf{V}_{n,0}^{\mathrm{T}} = 0. \tag{8.46}$$

Here $\mathbf{Z}_n$ is any non-trivial solution of the system

$$\mathbf{L}_{n,0}^{*}\mathbf{Z}_n = \mathbf{0}, \tag{8.47}$$

where $\mathbf{L}_{n,0}^{*} = \bar{\mathbf{L}}_{n,0}^{\mathrm{T}}$ is the matrix conjugate to $\mathbf{L}_{n,0}$, and the bar means complex conjugation.

By virtue of (6.23), the matrix $\mathbf{L}_{n,1}$ is expressed in terms of the matrices $\frac{\partial \mathbf{L}_{n,0}}{\partial p_n}$, $\frac{\partial \mathbf{L}_{n,0}}{\partial q_n}$. The latter can be found by differentiating the homogeneous system (8.40) with respect to p_n, q_n. Substituting the matrix elements $\frac{\partial \mathbf{L}_{n,0}}{\partial p_n}$, $\frac{\partial \mathbf{L}_{n,0}}{\partial q_n}$ into (8.46), we arrive at the equation with respect to $P_{n,0}$:

$$\begin{gathered}\bar{\mathbf{Z}}_n\left\{\mathbf{L}_{n,0}\frac{\partial \mathbf{Y}_n^{\mathrm{T}}}{\partial p_n}\left(\xi_n P_{n,0} b_n - \mathrm{i}\frac{\partial P_{n,0}}{\partial \xi_n}\right) + \mathbf{L}_{n,0}\frac{\partial \mathbf{Y}_n^{\mathrm{T}}}{\partial q_n}\xi_n P_{n,0}\right.\\ \left. + \left[\xi_n P_{n,0}\left(b_n\frac{\partial \omega_n}{\partial p_n} + \frac{\partial \omega_n}{\partial q_n} - \dot{p}_n\right) - \mathrm{i}\frac{\partial P_{n,0}}{\partial \xi_n}\frac{\partial \omega_n}{\partial p_n}\right]\frac{\partial \mathbf{L}_{n,0}}{\partial \omega_n}\mathbf{Y}_n^{\mathrm{T}}\right\} = 0.\end{gathered} \tag{8.48}$$

From Eqs. (8.38), (8.44), (8.47), one obtains

$$\bar{\mathbf{Z}}_n\mathbf{L}_{n,0}\frac{\partial \mathbf{Y}_n^{\mathrm{T}}}{\partial p_n} \equiv 0, \quad \bar{\mathbf{Z}}_n\mathbf{L}_{n,0}\frac{\partial \mathbf{Y}_n^{\mathrm{T}}}{\partial q_n} \equiv 0. \tag{8.49}$$

Taking into account the last identities as well as relation (8.42), Eq. (8.48) may be simplified:

$$\begin{gathered}\mathrm{i}\left(\frac{\partial H_n}{\partial p_n} - \dot{q}_n\right)\bar{\mathbf{Z}}_n\frac{\partial \mathbf{L}_{n,0}}{\partial \omega_n}\mathbf{Y}_n^{\mathrm{T}}\frac{\partial P_{n,0}}{\partial \xi_n}\\ -\left[b_n\left(\frac{\partial H_n}{\partial p_n} - \dot{q}_n\right) + \left(\frac{\partial H_n}{\partial q_n} + \dot{p}_n\right)\right]\bar{\mathbf{Z}}_n\frac{\partial \mathbf{L}_{n,0}}{\partial \omega_n}\mathbf{Y}_n^{\mathrm{T}}\xi_n P_{n,0} = 0.\end{gathered} \tag{8.50}$$

It is readily to prove that

$$\bar{\mathbf{Z}}_n\frac{\partial \mathbf{L}_{n,0}}{\partial \omega_n}\mathbf{Y}_n^{\mathrm{T}} \neq 0 \tag{8.51}$$

for any $p_n(t_1)$, $q_n(t_1)$ satisfying the accepted above conditions (8.36). In order for Eq. (8.50) to have a solution in the form of a polynomial in ξ_n, it is necessary that the functions $p_n(t_1)$, $q_n(t_1)$ give a solution to the Hamiltonian system

$$\frac{dq_n}{dt_1} = \frac{\partial H_n}{\partial p_n}, \quad \frac{dp_n}{dt_1} = -\frac{\partial H_n}{\partial q_n}. \tag{8.52}$$

Let $p_n(t_1)$, $q_n(t_1)$ be the solution of system (8.52) with the initial conditions $p_n(0) = a_0$, $q_n(0) = 0$. Returning to Eq. (8.45), we find its solution,

$$\mathbf{V}_{n,1} = P_{n,1}\mathbf{Y}_n + \xi_n P_{n,0}\left(b_n \frac{\partial \mathbf{Y}_n}{\partial p_n} + \frac{\partial \mathbf{Y}_n}{\partial q_n}\right) - \mathrm{i}\frac{\partial P_{n,0}}{\partial \xi_n}\frac{\partial \mathbf{Y}_n}{\partial p_n}, \tag{8.53}$$

where $P_{n,0}, P_{n,1}$ are polynomials in ξ_n to be determined.

In the second-order approximation ($k = 2$), we have again the inhomogeneous system of algebraic equations

$$\mathbf{L}_{n,0}\mathbf{V}_{n,2}^{\mathrm{T}} = -\mathbf{L}_{n,1}\mathbf{V}_{n,1}^{\mathrm{T}} - \mathbf{L}_{n,2}\mathbf{V}_{n,0}^{\mathrm{T}}. \tag{8.54}$$

The compatibility condition of this system is the equality

$$\bar{\mathbf{Z}}_n\left(\mathbf{L}_{n,1}\mathbf{V}_{n,1}^{\mathrm{T}} + \mathbf{L}_{n,2}\mathbf{V}_{n,0}^{\mathrm{T}}\right) = 0. \tag{8.55}$$

After simple transformations, which consist in determining elements of the matrices $\frac{\partial^2 \mathbf{L}_{n,0}}{\partial p_n^2}$, $\frac{\partial^2 \mathbf{L}_{n,0}}{\partial q_n^2}$, $\frac{\partial^2 \mathbf{L}_{n,0}}{\partial p_n \partial q_n}$ included in the operator $\mathbf{L}_{n,2}$, the resolvability condition (8.55), taking into account (8.52), (8.53), will take the form:

$$\begin{aligned}
\bar{\mathbf{Z}}_n \Bigg\{ & \frac{1}{2}\left[b_n^2 \frac{\partial^2 \omega_n}{\partial p_n^2} + b_n \frac{\partial^2 \omega_n}{\partial p_n \partial q_n} + \frac{\partial^2 \omega_n}{\partial q_n^2} - \dot{b}_n\right] \frac{\partial \mathbf{L}_{n,0}}{\partial \omega_n} \mathbf{Y}_n^{\mathrm{T}} \xi_n^2 P_{n,0} \\
& + \mathrm{i}\left[\frac{\mathrm{i}}{2}\frac{\partial^2 \omega_n}{\partial p_n^2}\frac{\partial^2 P_{n,0}}{\partial \xi_n^2} - b_n \frac{\partial^2 \omega_n}{\partial p_n^2}\xi_n \frac{\partial P_{n,0}}{\partial \xi_n} - \frac{\partial^2 \omega_n}{\partial p_n \partial q_n}\xi_n P_{n,0}\right. \\
& \left. + \dot{P}_{n,0} - \frac{1}{2} b_n \frac{\partial^2 \omega_n}{\partial p_n^2} P_{n,0}\right] \frac{\partial \mathbf{L}_{n,0}}{\partial \omega_n}\mathbf{Y}_n^{\mathrm{T}} + \mathrm{i}\left[\frac{1}{2}\dot{\omega}_n \frac{\partial^2 \mathbf{L}_{n,0}}{\partial \omega_n^2}\right. \\
& \left. + \dot{p}_n \frac{\partial^2 \mathbf{L}_{n,0}}{\partial \omega_n \partial p_n} + \mathbf{N}_n\right] \mathbf{Y}_n^{\mathrm{T}} P_{n,0} + \mathrm{i}\frac{\partial \mathbf{L}_{n,0}}{\partial p_n}\frac{\partial \mathbf{Y}_n^{\mathrm{T}}}{\partial q_n} P_{n,0} \Bigg\} = 0.
\end{aligned} \tag{8.56}$$

For the resulting equation (8.56) to have a solution in the form of a polynomial in ξ_n, it is necessary that the coefficient by ξ_n^2 be zero. Hence, taking into account (8.43), we arrive at the Riccati equation with respect to the function $b_n(t_1)$:

$$\frac{db_n}{dt_1} + \frac{\partial^2 H_n}{\partial p_n^2} b_n^2 + 2\frac{\partial^2 H_n}{\partial p_n \partial q_n} b_n + \frac{\partial^2 H_n}{\partial q_n^2} = 0. \tag{8.57}$$

Let $b_n(t_1)$ be the solution of Eq. (8.57) with the initial condition $b_n(0) = b_0$. Then Eq. (8.56) takes the form:

$$h_{n,0}(t_1)\,\frac{\partial^2 P_{n,0}}{\partial \xi_n^2} + h_{n,1}(t_1)\,\xi_n\,\frac{\partial P_{n,0}}{\partial \xi_n} + h_{n,2}(t_1)\,\frac{\partial P_{n,0}}{\partial t_1} + h_{n,3}(t_1)\,P_{n,0} = 0. \tag{8.58}$$

We have arrived at the amplitude equation similar to Eq. (6.62), in which the coefficients $h_{n,j}$ for $j = 0, 1, 2$ are calculated by formulas (6.56), and

$$h_{n,3}(t_1) = \mathrm{i}\left(\bar{\mathbf{Z}}_n\,\frac{\partial \mathbf{L}_{n,0}}{\partial \omega_n}\,\mathbf{Y}_n^{\mathrm{T}}\right)^{-1} \bar{\mathbf{Z}}_n \left[\left(\frac{1}{2}\,b_n\,\frac{\partial^2 H_n}{\partial p_n^2}\,\frac{\partial \mathbf{L}_{n,0}}{\partial \omega_n} + \frac{1}{2}\,\dot{\omega}_n\,\frac{\partial^2 \mathbf{L}_{n,0}}{\partial \omega_n^2}\right.\right.$$
$$\left.\left. +\dot{p}_n\,\frac{\partial^2 \mathbf{L}_{n,0}}{\partial \omega_n \partial p_n} + \mathbf{N}_n\right)\mathbf{Y}_n^{\mathrm{T}} + \frac{\partial \mathbf{L}_{n,0}}{\partial p_n}\,\frac{\partial \mathbf{Y}_n^{\mathrm{T}}}{\partial q_n}\right] = 0. \tag{8.59}$$

The solution of Eq. (8.58) is a polynomial

$$P_{n,0}(\xi_n, t_1) = \sum_{k=0}^{M_n} A_{n,k}(t_1; d_{n,i})\,\xi_n^k \tag{8.60}$$

of degree M_n with complex coefficients $A_{n,k}(t_1; d_{n,i})$ containing arbitrary constants $d_{n,i}$ $(i = 0, 1, 2, \ldots, M_n)$. The functions $A_{n,k}(t_1; d_{n,i})$ are defined by formulas (6.65), in which d_i, Ψ_i, M, h_j should be replaced by $d_{n,i}, \Psi_{n,i}, M_n, h_{n,j}$, respectively $(j = 0, 1, 2, 3)$. Recall in passing that the polynomial $P_{n,0}$ can be constructed in another way, in terms of the Hermite polynomials (see Eq. (6.74)).

The compatibility conditions for inhomogeneous systems of algebraic equations (8.37) for $j \geq 3$ lead to inhomogeneous differential equations of type (6.60) with respect to polynomials $P_{n,k}(\xi_n, t_1)$.

8.3.2 Axisymmetric packets of longitudinal waves

Now we consider Eqs. (8.17). Let $s = q_s(t) + \mu_1^{1/2}\xi_s$, where $q_s(t)$ is a twice differentiable function such that $q_s(0) = 0$. The parallel $s = q_s(t)$ will be called the center of the travelling packet of longitudinal waves.

We seek a solution of Eqs. (8.17) in the same form (8.35). Moreover, in ansatz (8.35) as well as in all subsequent formulas, the indices n should be replaced with s, and the "slow time" t_1 should be substituted by t. It is assumed that for any $t \in [0, t_s']$ the values of all functions and their derivatives from Eqs. (8.17), calculated on the parallels $s = q_s(t)$, are quantities of order $O(1)$ for $\mu_1 \to 0$.

The procedure for finding unknown functions included in the asymptotic representation of the vector $\mathbf{V}_s$ remains the same. We point out the changes that need to be made to the formulas and equations obtained in the previous

Subsection. Now, $\mathbf{L}_{s,0}$ and $\mathbf{N}_{s,0}$ are the 2×2 - matrices with the elements

$$\begin{gathered} l_{s,11} = p_s^2 - (\omega_s - \dot{q}_s p_s)^2, \quad l_{s,12} = 0, \\ l_{s,21} = -\mathrm{i} p_s (k_1 + \nu k_2), \quad l_{s,22} = -(\omega_s - \dot{q}_s p_s)^2, \\ n_{s,11} = r' r^{-1} p_s + p_s \ddot{q}_s, \quad n_{s,12} = 0, \\ n_{s,21} = \mathrm{i} r^{-1} [(1-\nu)(k_2 r)' - (k_1 + k_2) r'], \quad n_{s,22} = p_s \ddot{q}_s, \end{gathered} \tag{8.61}$$

$\mathbf{Y}_s = (y_{s,1}, y_{s,2})$ is the two-component vector, where

$$y_{s,1} = 1, \quad y_{s,2} = -\mathrm{i} p_s^{-1} (k_1 + k_2), \tag{8.62}$$

and the Hamilton function corresponding to axisymmetric longitudinal waves has a simple form,

$$H_s(p) = p. \tag{8.63}$$

Here, the Hamiltonian system and the Riccati equation degenerate:

$$\dot{q}_s = 1, \quad \dot{p}_s = 0, \quad \dot{b}_s = 0. \tag{8.64}$$

Hence, taking into account the initial conditions, $q_s(0) = 0, p_s(0) = a_0, b_s(0) = b_0$, one gets

$$q_s(t) = t, \quad p_s = a_0, \quad b_s = b_0. \tag{8.65}$$

The amplitude equation corresponding to axisymmetric longitudinal motion is also simplified:

$$\frac{\partial P_{s,0}}{\partial t} + \frac{r'(t)}{2r(t)} P_{s,0} = 0. \tag{8.66}$$

Its solution is the M_s degree polynomial

$$P_{s,0}\left(\xi_s, t;\, d_{s,i}\right) = \frac{1}{\sqrt{r(t)}} \sum_{i=0}^{M_s} d_{s,i}\, \xi_s^i \tag{8.67}$$

with arbitrary constants $d_{s,i}$.

In this case, the leading approximation of the asymptotic expansion of the vector $\mathbf{V}_s = (u_s, w_s)$ is written in the explicit form:

$$\mathbf{V}_s = \left[\frac{1}{\sqrt{r(t)}} \sum_{i=0}^{M_s} d_{s,i} \xi_s^i \mathbf{Y}_s + O\left(\mu_1^{1/2}\right) \mathbf{I}_2 \right] \times \exp\left[\mathrm{i} \left(\mu_1^{-1/2} a_0 \xi_s + \frac{1}{2} b_0 \xi_s^2 \right) \right], \tag{8.68}$$

where $\xi_s = \mu_1^{-1/2}(s - t)$, $\mathbf{I}_2 = (1, 1)$.

Remark 8.1. The structure of matrices $\mathbf{L}_{s,0}$, $\mathbf{N}_{s,0}$ (here $\ddot{q}_s = 0$) shows that when constructing the leading approximation (8.68), the coupled equations (8.17) can be split into equations with respect to the longitudinal component u_s and an equation linking the components u_s and w_s. Indeed, disregarding in Eqs. (8.17) terms not appearing in the first three approximations, we obtain

$$-\frac{\partial}{\partial s}\frac{1}{r}\frac{\partial}{\partial s}(r\,u_s) + \frac{\partial^2 u_s}{\partial t^2} = 0, \tag{8.69}$$

$$-\frac{1}{r}(k_1 + k_2)\frac{\partial}{\partial s}(ru_s) + \frac{1-\nu}{r}\frac{\partial}{\partial s}(k_2 r u_s) + \mu\frac{\partial^2 u_s}{\partial t^2} = 0. \tag{8.70}$$

Obviously, the component u_s of the vector $\mathbf{V}_s$ defined in accordance with (8.68) is the asymptotic solution of Eq. (8.69).

8.3.3 Axisymmetric packets of torsional waves

Consider Eq. (8.11) corresponding to the torsional waves. For a solution to have the wave character, one needs to require the estimate, $\partial u_2/\partial t \sim \mu_1^{-1}$. Hence, taking into account the algorithm for constructing travelling WPs, we can conclude that the second term in Eq. (8.11) affects only the higher approximations of the constructed solution and therefore can be discarded when determining the leading approximation. The equation derived in such a way differs from Eq. (8.69) for the longitudinal waves only by the presence of the factor $(1-\nu)/2$ in the first term. Having made the replacement $\tilde{t} = \sqrt{(1-\nu)/2}\,t$, it is readily to pass to Eq. (8.69), the solution of which is known (see Remark 8.1).

Thus, the leading approximation of the solution of Eq. (8.11) in the form of the travelling WP is written in the form

$$\begin{aligned} u_2 &= \left[\frac{1}{\sqrt{r\left(\sqrt{(1-\nu)/2}\;t\right)}}\sum_{i=0}^{M_\varphi} d_{\varphi,i}\,\xi_\varphi^i + O\left(\mu_1^{1/2}\right)\right] \\ &\quad\times \exp\left[\mathrm{i}\left(\mu_1^{-1/2} a_0\,\xi_\varphi + \tfrac{1}{2}\,b_0\,\xi_\varphi^2\right)\right], \\ \xi_\varphi &= \mu_1^{-1/2}\left(s - \sqrt{\tfrac{1-\nu}{2}}\;t\right), \end{aligned} \tag{8.71}$$

where $d_{\varphi,\,i}$ are arbitrary constants.

8.3.4 Superposition of axisymmetric WPs

The above asymptotic expansions for u_2, $\mathbf{V}_n$, $\mathbf{V}_s$ are the positive branches of the constructed solutions. In order to pass to the negative branches, in all equations, the signs at the Hamiltonians and their derivatives should be changed into opposite. In particular, in expansions (8.68), (8.71), this leads to a change in sign in the argument t. We supply all undefined constants as well as defined functions in the expansion of the vector $\mathbf{V}_n$ with indices (+) and (−) corresponding to the positive and negative branches. Compose the vector-function

$$\mathbf{U} = (u_1,\, u_2,\, u_3)\,, \tag{8.72}$$

where

$$u_2 = u_2^+ + u_2^-,$$

$$u_2^\pm = \left[\frac{1}{\sqrt{r\left(\pm\sqrt{(1-\nu)/2}\,t\right)}} \sum_{i=0}^{M_\varphi} d_{\varphi,i}^\pm \left(\xi_\varphi^\pm\right)^i + O\left(\mu_1^{1/2}\right) \right] \times \exp\left[\mathrm{i}\left(\mu_1^{-1/2} a_0\, \xi_\varphi^\pm + \tfrac{1}{2}\, b_0 \left(\xi_\varphi^\pm\right)^2\right)\right], \tag{8.73}$$

and the components u_1, u_3 forming the vector $\mathbf{V} = (u_1, u_3)$ are found as follows:

$$\mathbf{V} = \left(\mathbf{E}_{01}\left(\mathbf{V}_s^+ + \mathbf{V}_s^-\right)^{\mathrm{T}}\right)^{\mathrm{T}} + \left(\mathbf{E}_{10}\left(\mathbf{V}_n^+ + \mathbf{V}_n^-\right)^{\mathrm{T}}\right)^{\mathrm{T}}. \tag{8.74}$$

Here

$$\mathbf{V}_s^\pm = \left(u_s^\pm,\, w_s^\pm\right), \quad \mathbf{V}_n^\pm = \left(u_n^\pm,\, w_n^\pm\right), \tag{8.75}$$

$$\begin{aligned} \mathbf{V}_s^\pm &= \left[\frac{1}{\sqrt{r(\pm t)}} \sum_{i=0}^{M_s} d_{s,i}^\pm \left(\xi_s^\pm\right)^i \mathbf{Y}_s + O\left(\mu_1^{1/2}\right) \mathbf{I}_2 \right] \\ &\quad \times \exp\left[\mathrm{i}\left(\mu_1^{-1/2} a_0\, \xi_s^\pm + \tfrac{1}{2} b_0 \left(\xi_s^\pm\right)^2\right)\right], \\ \mathbf{V}_n^\pm &= \left[P_{n,0}^\pm\left(\xi_n^\pm,\, \mu_1 t;\, d_{n,i}^\pm\right) \mathbf{Y}_n + O\left(\mu_1^{1/2}\right) \mathbf{I}_2\right] \\ &\quad \times \exp\left[\mathrm{i}\mu_1^{-1} S_n^\pm\left(\xi_n^\pm, \mu_1 t,\, \mu_1\right)\right], \end{aligned} \tag{8.76}$$

$$S_n^\pm = \int_0^{\mu_1 t} \omega_n^\pm(\tau) d\tau + \mu_1^{1/2} p_n^\pm(\mu_1 t) \xi_n^\pm + \frac{1}{2} \mu_1 b_n^\pm(\mu_1 t) \left(\xi_n^\pm\right)^2, \tag{8.77}$$

$$\begin{aligned} \xi_\varphi^\pm &= \mu_1^{-1/2}\left(s \mp \sqrt{\tfrac{1-\nu}{2}}\, t\right), \\ \xi_s^\pm &= \mu_1^{-1/2}(s \mp t), \quad \xi_n^\pm = \mu_1^{-1/2}\left[s - q_n^\pm(\mu_1 t)\right], \end{aligned} \tag{8.78}$$

where the polynomials $P_{n,0}^\pm$ are found by formula (8.60) for both branches, and components of the vectors $\mathbf{Y}_n$, $\mathbf{Y}_s$ are determined according to (8.44), (8.62), respectively.

By virtue of the performed constructions, and also taking into account the linearity of the governing equations (8.2), the vector $\mathbf{U}$ is the formal asymptotic solution (the leading approximation) of Eqs. (8.2) for $m_0 = 0$. This solution contains undefined complex constants $d_{\varphi,i}^\pm$, $d_{s,i}^\pm$, $d_{n,i}^\pm$ of integrating.

To find them, we substitute (8.72)–(8.78) into the initial conditions (8.6) and equate the coefficients by the same powers of $\mu_1^{1/2}$. Moreover, we take into account that $\xi_n^{\pm}|_{t=0} = \zeta$. As a result, we obtain

$$d_{\varphi,i}^{\pm} = \frac{1}{2}\left(c_{2i}^{\circ} \mp \frac{d_{2i}^{\circ}}{a_0\sqrt{(1-\nu)/2}}\right), \qquad i = 0, 1, \ldots, M_2, \tag{8.79}$$

and

$$d_{s,i}^{\pm} = \frac{1}{2}\left(c_{1i}^{\circ} \mp \frac{d_{1i}^{\circ}}{a_0}\right),$$
$$d_{n,i}^{\pm} = \frac{c_{3i}^{\circ} H_n^{\circ} \mp d_{3i}^{\circ} \mp \mathrm{i}\,[k_1(0) + \nu k_2(0)]\, a_0^{-1} d_{1i}^{\circ}}{2H_n^{\circ}}, \tag{8.80}$$

where $i = 0, 1, \ldots, M^*$, $M^* = \max\{M_1, M_3\}$, $H_n^{\circ} = H_n(a_0, 0)$. It follows that the orders of the polynomials $P_{\varphi,0}^{\pm}$ and $P_{s,0}^{\pm}$, $P_{n,0}^{\pm}$ are $M_\varphi = M_2$ and $M_s = M_n = M^*$, respectively.

8.3.5 Solution properties. Examples

It can be seen from formulas (8.72)—(8.78) that the solution of problem (8.2), (8.6) in the axisymmetric case is the superposition of packets of torsional, longitudinal and bending waves with the centers on the parallels $s = \pm\sqrt{\frac{1-\nu}{2}}\, t$, $\pm t$, $q_n^{\pm}(\mu_1 t)$, respectively. At that, the torsion waves propagate regardless of the longitudinal and bending waves. For the convenience of analyzing the interaction of the longitudinal and normal components of the constructed solution, we write out the vector $\mathbf{V}$ (see (8.74)) component-wise:

$$u_1 = u_s^+\left(\xi_s^+, t\right) + u_s^-\left(\xi_s^-, t\right) + \mu_1\left[u_n^+\left(\xi_n^+, \mu_1 t\right) + u_n^-\left(\xi_n^-, \mu_1 t\right)\right],$$
$$u_3 = w_n^+\left(\xi_n^+, \mu_1 t\right) + w_n^-\left(\xi_n^-, \mu_1 t\right) + \mu_1\left[w_s^+\left(\xi_s^+, t\right) + w_s^-\left(\xi_s^-, t\right)\right]. \tag{8.81}$$

The longitudinal displacement u_1 consists of the components $u_s^{\pm}$, which are the main ones, and the secondary component $\mu_1 u_n^{\pm}$ "generated" by the functions $w_n^{\pm}$. The normal displacement u_3 is the combination of the main components $w_n^{\pm}$ and the minor ones $\mu_1 w_s^{\pm}$, which in turn are "generated" by the longitudinal displacements $u_s^{\pm}$. In formulas (8.81), when calculating the displacements of points on the shell mid-surface, the minor terms proportional to the small parameter μ_1 can not be taken into account, since their orders exceed the orders of summands which were not taken into account in (8.76). However, under evaluating the velocities of normal displacements, $\partial u_3/\partial t$, it should borne in mind that $\partial w_n^{\pm}/\partial t$ and $\mu_1 \partial w_s^{\pm}/\partial t$ are of the same order $O(1)$ as $\mu_1 \to 0$. This is explained by different indices of the dynamism of the main longitudinal and normal components:

$$\frac{\partial u_s^{\pm}}{\partial t} = O\left(\mu_1^{-1}\right), \quad \frac{\partial w_n^{\pm}}{\partial t} = O(1) \qquad \mu_1 \to 0. \tag{8.82}$$

This is also a reason for the dependence of the amplitude of bending waves on the amplitude of initial velocities of longitudinal displacements (see (8.8) and (8.80)). This circumstance is also the reason for the dependence of the bending waves amplitude on the amplitude of the initial velocities of longitudinal displacements (see Eqs. (8.8) and (8.80)).

The first relation in (8.44) shows that for $p_n^\pm \to 0$ expansions (8.81) become unusable because the terms $\mu_1 u_n^\pm$, which are secondary in solution (8.81), become infinitely large. Thus, if there exists $t_n^* < +\infty$ such that $\lim\limits_{t_1 \to t_n^*} p_n^\pm(t_1) = 0$, then solution (8.81) should be considered for some finite time interval $0 \leq t \leq t_n'/\mu_1 < t_n^*/\mu_1$, where $p_n^\pm(t) \sim 1$. From this, in particular, it follows that the initial parameter $a_0 > 0$ should also be of the order of unity.

Remark 8.2. Let in Eqs. (8.80) the equalities $c_{3,i}^\circ = d_{3i}^\circ = d_{1i}^\circ = 0$ hold for any i, and $c_{1i}^\circ \neq 0$ for some i. Then $w_n^\pm \equiv 0$, $u_n^\pm \equiv 0$ and solution (8.72)-(8.78) defines only the longitudinal and torsional waves. In this case the requirement $a_0 \sim 1$ is redundant. In particular, when assuming $a_0 = 0$, then one obtains a running longitudinal wave of the bell-shaped profile with an amplitude varying in time.

Let us study the effect of shell geometry on the dynamic characteristics of the travelling WPs. The derived in the explicit form formulas for $u_s^\pm$, $u_2^\pm$ show that in the axisymmetric packets of longitudinal and torsional waves the parameters $p_\varphi^\pm = p_s^\pm = a_0$, $b_\varphi^\pm = b_s^\pm = b_0$ as well as the group velocities

$$\dot{q}_\varphi^\pm = \pm\sqrt{(1-\nu)/2}, \quad \dot{q}_s^\pm = \pm 1$$

remain constant, and the amplitudes change proportionally to the factors

$$\frac{1}{\sqrt{r\left(\pm\sqrt{(1-\nu)/2}\; t\right)}}, \quad \frac{1}{\sqrt{r(\pm t)}}$$

for the torsional and longitudinal waves, respectively.

We write out the Hamiltonian system for bending waves:

$$\dot{q}_n^\pm = \pm\frac{2\left(p_n^\pm\right)^3}{H_n^\circ}, \quad \dot{p}_n^\pm = \mp\frac{(1-\nu^2)k_2\left(q_n^\pm\right)k_2'\left(q_n^\pm\right)}{H_n^\circ},$$
$$q_n^\pm(0) = 0, \quad p_n^\pm(0) = a_0. \tag{8.83}$$

Let $k_2'(s) > 0$ for $-\infty < s < +\infty$. From the analysis of system (8.83), we obtain the inequalities

$$\dot{p}_n^- > 0, \quad \dot{q}_n^- < 0, \quad \ddot{q}_n^- < 0 \quad \text{for any} \quad t_1 \geq 0, \tag{8.84}$$

which indicate that the movement of the packet of bending waves in the direction of decreasing the curvature $k_2(s)$ is accompanied by the increase of the wave parameter p_n^- and the group velocity $|\dot{q}_n^-|$.

Let

$$\sup_{0<s<+\infty} k_2^2(s) < k_2^2(0) + \frac{a_0^4}{1-\nu^2}. \tag{8.85}$$

Then the inequalities

$$\dot{p}_n^+ < 0, \quad \dot{q}_n^+ > 0, \quad \ddot{q}_n^+ < 0 \tag{8.86}$$

are valid for any $t_1 \geq 0$. Inequalities (8.86) indicate that the group velocity $\dot{q}_n^+$ as well as the variability of waves in the packet travelling towards increasing the curvature $k_2(s)$ decrease. If condition (8.85) is not satisfied, then inequalities (8.86) hold for $t_1 < t_n^*$, where

$$t_n^* = \frac{H_n^\circ}{2} \int_0^{q_n^*} \left\{ (1-\nu^2)\left[k_2^2(0) - k_2^2(q)\right] + a_0^4 \right\}^{-3/4} dq, \tag{8.87}$$

and the number q_n^* is found from the equation

$$k_2^2(q_n) = k_2^2(0) + \frac{a_0^4}{1-\nu^2}. \tag{8.88}$$

Note that under the above assumptions regarding the function $k_2(s)$, integral (8.87) is convergent. At the time $t_1 = t_n^*$, the function $p_n^+(t_1)$ vanishes, which, by virtue of (8.44), violates the estimates w_n, $u_n = O(1)$. Thus, in the study of the packets of bending waves running towards increasing the curvature $k_2(s)$, solution (8.72)-(8.78) should be considered at some finite time interval $0 \leq t_1 \leq t_n' < t_n^*$, where $p_n^+(t_1) \sim 1$.

Example 8.1. Consider a cylindrical shell for which $r = k_2 = 1$. In this case, the Hamiltonian system and the Riccati equation have the simplest solutions

$$p_n^\pm = a_0, \quad q_n^\pm(t) = \pm H_p^\circ t, \quad b_n^\pm(t) = \frac{b_0}{1 \pm b_0 H_{pp}^\circ t}, \tag{8.89}$$

where H_p° and H_{pp}° denote the first and second derivatives of the Hamiltonian H_n for $p = a_0$, $q = 0$. The coefficients $A_{n,k}^\pm$ of polynomial (8.60) can be found from formulas (6.138), in which it is necessary to make changes in notation:

$$A_k^\pm \rightsquigarrow A_{n,k}^\pm, \quad d_k^\pm \rightsquigarrow d_{n,k}^\pm, \quad \Psi_k^\pm \rightsquigarrow \Psi_{n,k}^\pm, \quad M \rightsquigarrow M^*. \tag{8.90}$$

From Eqs. (8.89), (6.138), it follows that in a cylindrical shell with constant parameters, the packets of bending waves moving in the axial direction "spread" in time over the shell surface.

Example 8.2. Consider a conical shell for which $r(s) = 1 - s/\kappa$. Here, $\kappa = 1/\sin\theta$, $-\infty < s < \kappa$, where 2θ is the opening angle of the cone apex. Let $\Im b_0 \geq 1$, and θ is a small number. Then the center of the initial WP is sufficiently far from the cone top.

We performed calculations for the axisymmetric packets of bending waves at $a_0 = 1$, $b_0 = \mathrm{i}$, $k = 4.8$, $u_{30}^{\circ} = 1$ and zero values of the remaining parameters u_{j0}°, ϑ_{j0}° (see (8.8)). Fig. 8.2 shows solutions of the Hamiltonian system. The computations were performed at the time interval $0 \le t \le 2/\mu_1$, where the estimate $p_n^+(\mu_1 t) \sim 1$ holds. It can be seen that the variability of waves in the packet running to the cone top decreases, and on the contrary, increases in the WP moving in the opposite direction.

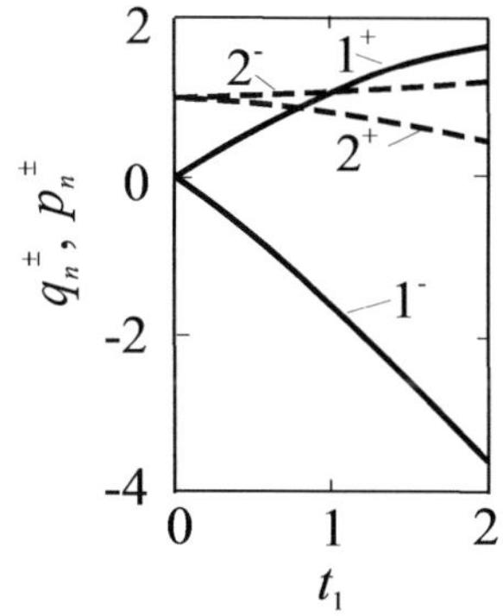

FIGURE 8.2: Plots of the functions $q_n^+(\mu t)$, $q_n^-(\mu t)$ (the solid lines $1^+, 1^-$, respectively) and $p_n^+(\mu t)$, $p_n^-(\mu t)$ (the dashed lines $2^+, 2^-$, respectively).

Plots of the functions $v_n^{\pm} = \dot{q}_n^{\pm}(\mu_1 t)$ and $\omega_n^{\pm}(\mu_1 t)$ are depicted in Fig. 8.3. The curves given are in full agreement with the conclusions of paper [47]

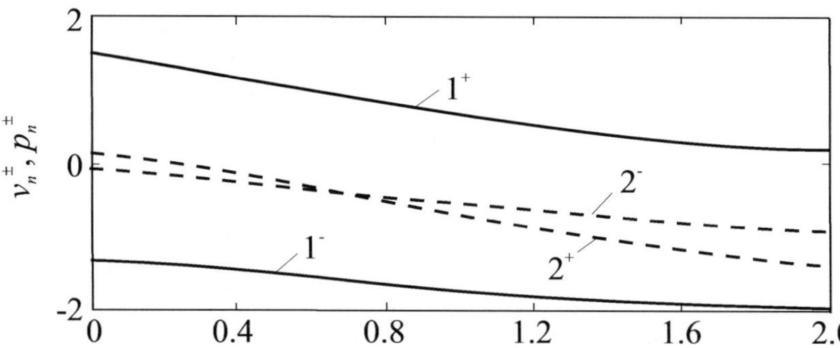

FIGURE 8.3: Plots of the functions $v_n^+(\mu t)$, $v_n^-(\mu t)$ (the solid lines $1^+, 1^-$, respectively) and $\omega_n^+(\mu t)$, $\omega_n^-(\mu t)$ (the dashed lines $2^+, 2^-$, respectively).

regarding the dispersion of bending waves along the cone generatrix. The curves $J^{\pm} = \Im b_n^{\pm}(\mu_1 t)$ and $w_{max}^{\pm} = \max_s |w_n^{\pm}(s, \mu_1 t)|$ shown in Fig. 8.4 indicate that WPs running toward the cone top spread out more slowly than packets travelling in the opposite direction.

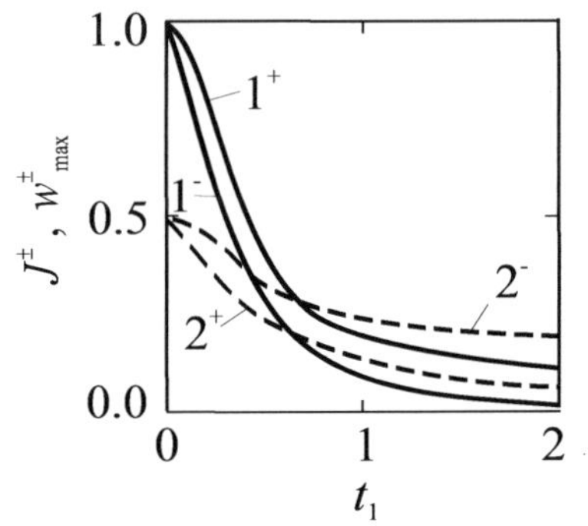

FIGURE 8.4: Plots of the functions $J^+(\mu_1 t)$ and $J^-(\mu_1 t)$ (the solid lines $1^+, 1^-$, respectively) and $w_{max}^+(\mu_1 t)$, $w_{max}^-(\mu_1 t)$ (the dashed lines $2^+, 2^-$, respectively).

8.4 Non-axisymmetric WPs in shells of revolution with a small number of waves in the circumferential direction

Consider the coupled Eqs. (8.2) in the case $m_0 \sim 1$. Their solution can be found in the form of a linear combination of travelling WPs of tangential and bending waves:

$$\mathbf{U} = \left(\mathbf{E}_{001}\mathbf{U}_{\tau}^{\mathrm{T}}\right)^{\mathrm{T}} + \left(\mathbf{E}_{120}\mathbf{U}_{n}^{\mathrm{T}}\right)^{\mathrm{T}}, \tag{8.91}$$

where $\mathbf{U}_{\tau} = (u_{\tau}, v_{\tau}, w_{\tau})$ and $\mathbf{U}_n = (u_n, v_n, w_n)$ are the asymptotic solutions of Eqs. (8.23) and (8.26), respectively. This case is in many respects similar to the case of axisymmetric waves considered in detail above. Therefore, omitting the details of defining the vectors $\mathbf{U}_{\tau}$, $\mathbf{U}_n$, we dwell only on some features of this case.

The matrices $\mathbf{L}_{\tau,0}$, $\mathbf{L}_{n,0}$ appearing in the process of constructing the vectors $\mathbf{U}_\tau$ and $\mathbf{U}_n$ admit splitting of homogeneous algebraic equations (arising in the leading approximation)

$$\mathbf{L}_{\tau,0}\mathbf{U}_{\tau,0}^{\mathrm{T}} = 0, \quad \mathbf{L}_{n,0}\mathbf{U}_{n,0}^{\mathrm{T}} = 0 \tag{8.92}$$

into equations and systems exactly coinciding with those that occurred in the axisymmetric case. Here, the additional equation is only the equation coupling $v_{n,0}$ to the components $u_{n,0}$ and $w_{n,0}$. This circumstance, despite the complex form of the matrices $\mathbf{N}_\tau$, $\mathbf{N}_n$ (depending now on the parameter $m_0 \sim 1$), leads to the fact that in the axisymmetric case the components of the vector (8.92)

$$\begin{aligned}
u_1 &= u_\tau^+\left(\xi_\tau^+, t\right) + u_\tau^-\left(\xi_\tau^-, t\right) + \mu_1\left[u_n^+\left(\xi_n^+, \mu_1 t\right) + u_n^-\left(\xi_n^-, \mu_1 t\right)\right],\\
u_2 &= v_\tau^+\left(\xi_\tau^+, t\right) + v_\tau^-\left(\xi_\tau^-, t\right) + \mu_1^2\left[v_n^+\left(\xi_n^+, \mu_1 t\right) + v_n^-\left(\xi_n^-, \mu_1 t\right)\right],\\
u_3 &= w_n^+\left(\xi_n^+, \mu_1 t\right) + w_n^-\left(\xi_n^-, \mu_1 t\right) + \mu_1\left[w_\tau^+\left(\xi_\tau^+, t\right) + w_\tau^-\left(\xi_\tau^-, t\right)\right]
\end{aligned} \tag{8.93}$$

coincide up to the terms $\mu_1^2\left(v_n^+ + v_n^-\right)$ with the components of vector (8.72). Here $u_\tau^\pm \equiv u_s^\pm$, $v_\tau^\pm \equiv u_2^\pm$, $w_\tau^\pm \equiv w_s^\pm$ and $w_n^\pm, u_n^\pm$ are the same as in the case $m_0 = 0$. Thus, the asymptotically main approximation of the solution of problem (8.2), (8.6) for $m_0 \sim 1$ differs from the similar solution in the axisymmetric case only by an additional component of torsion waves of order $O\left(\mu^2\right)$, generated by the normal component of $w_n^\pm$. In practical calculations, the terms $\mu_1^2 v_n^\pm$ can be discarded, since their order exceeds the error order of the leading approximation.

8.5 Non-axisymmetric WPs in shells of revolution with a large number of waves in the circumferential direction

Now, we assume that $m_0 = \mu_1^{-1}m$, where $m \sim 1$. This case, which was studied in [149], differs in many respects from the above problems at $m_0 = 0$ and $m_0 \sim 1$. Consider it in more detail.

8.5.1 Non-axisymmetric packets of tangential waves

First, we turn to Eqs. (8.29) which describe non-axisymmetric tangential waves. Let $s = q_\tau(t)$ be the center of a travelling WP, where the function $q_\tau(t)$ is twice differentiable and $q_\tau(0) = 0$. We pass in (8.29) to the new variable $\xi_\tau = mu_1^{-1/2}[s - q_\tau(t)]$. The solution of Eqs. (8.29) is sought in the form:

$$\mathbf{U}_\tau = \sum_{k=0}^{\infty} \mu_1^{k/2}\mathbf{U}_{\tau,k} \exp\left[\mathrm{i}\mu_1^{-1} S_\tau\left(\xi_\tau, t, \mu_1\right)\right],$$

$$S_\tau = \int_0^t \omega_\tau\left(\tilde{t}\right) d\tilde{t} + \mu_1^{1/2} p_\tau(t)\xi_\tau + \tfrac{1}{2}\mu_1 b_\tau(t)\xi_\tau^2,$$

$$\mathbf{U}_{\tau,k} = \left(u_{\tau,k},\, v_{\tau,k},\, w_{\tau,k}\right), \tag{8.94}$$

$$\{\omega_\tau,\, p_\tau,\, b_\tau,\, \dot{\omega}_\tau,\, \dot{p}_\tau,\, \dot{b}_\tau,\, \dot{q}_\tau\} = O(1),$$

$$\left\{\mathbf{U}_{\tau,k},\, \dot{\mathbf{U}}_{\tau,k},\, \tfrac{\partial \mathbf{U}_{\tau,k}}{\partial \xi_\tau}\right\} = O(1)\,\mathbf{I}_3,$$

where $\Im\, b_\tau(t) > 0$ for any $t \in \left[0, t_\tau'\right]$, $\mathbf{I}_3 = (1, 1, 1)$, and components of the vector $\mathbf{U}_{\tau,k}$ are polynomials in ξ_τ.

It is assumed that for any $t \in \left[0, t_\tau'\right]$ the values of the functions $r(s), k_1(s), k_2(s)$ and their derivatives are quantities of order $O(1)$ as $\mu_1 \to 0$.

Substituting (8.94) into Eqs. (8.29), we arrive at the sequence of systems of algebraic equations

$$\sum_{j=0}^{k} \mathbf{L}_{\tau,j}\mathbf{U}_{\tau,\,k-j}^{\mathrm{T}} = \mathbf{0}, \quad k = 0, 1, 2, \ldots \tag{8.95}$$

Here, $\mathbf{L}_{\tau,0}$ is the (3×3)-matrix with elements

$$\begin{gathered}
l_{\tau,11} = p_\tau^2 + \frac{(1-\nu)m^2}{2r^2} - \left(\omega_\tau - \dot{q}_\tau p_\tau\right)^2, \quad l_{\tau,12} = -\frac{\mathrm{i}m(1+\nu)p_\tau}{2r}, \\
l_{\tau,21} = -l_{\tau,12}, \quad l_{\tau,22} = \frac{1-\nu}{2}p_\tau^2 + \frac{m^2}{r^2} - \left(\omega_\tau - \dot{q}_\tau p_\tau\right)^2, \\
l_{\tau,31} = -\mathrm{i}p_\tau\left(k_1 + \nu k_2\right), \quad l_{\tau,32} = -\frac{m}{r}\left(k_2 + \nu k_1\right), \\
l_{\tau,33} = -\left(\omega_\tau - \dot{q}_\tau p_\tau\right)^2, \quad l_{\tau,13} = l_{\tau,23} = 0.
\end{gathered} \tag{8.96}$$

The matrix operators $\mathbf{L}_{\tau,j}$ are defined by formulas (6.23) in which the differential operator $\mathbf{L}$ should be replaced by the matrix operator $\mathbf{L}_{\tau,0}$ and the operator $\mathbf{N}$ is changed by the (3×3)-matrix $\mathbf{N}_\tau$ with elements

$$
\begin{aligned}
&n_{\tau,11} = \frac{r'}{r} p_\tau + \ddot{q}_\tau p_\tau, \quad n_{\tau,12} = \frac{\mathrm{i}(3-\nu)mr'}{2r^2}, \\
&n_{\tau,21} = n_{\tau,12}, \quad n_{\tau,22} = \frac{(1-\nu)r'}{2r} p_\tau + \ddot{q}_\tau p_\tau, \\
&n_{\tau,31} = -\frac{\mathrm{i}}{r}\left[(1-\nu)(k_2 r)' - (k_1 + k_2) r'\right], \\
&n_{\tau,13} = n_{\tau,23} = n_{\tau,32} = 0, \quad n_{\tau,33} = \ddot{q}_\tau p_\tau.
\end{aligned}
\tag{8.97}
$$

The procedure for determining the unknown functions contained in expansion (8.94) remains the same up to indices (see Section 8.3) and the dimensions of vectors and matrices. However, now the condition for the existence of a nontrivial solution of system (8.95) for $k = 0$ leads to four relations coupling the parameters ω_τ, p_τ, q_τ:

$$
\omega_\tau^{(\pm \jmath)}(t) = \dot{q}_\tau^{(\pm \jmath)} p_\tau^{(\pm \jmath)}(t) - H_\tau^{(\pm \jmath)}\left[p_\tau^{(\pm \jmath)},\, q_\tau^{(\pm \jmath)}(t)\right], \quad \jmath = 1, 2, \tag{8.98}
$$

where

$$
H_\tau^{(\pm \jmath)}(p, q) = \pm \delta^{(\jmath)} \sqrt{p^2 + \frac{m^2}{r^2(q)}}, \quad \delta^{(1)} = 1, \quad \delta^{(2)} = \sqrt{\frac{1-\nu}{2}}. \tag{8.99}
$$

Here, $H_\tau^{(+1)}$, $H_\tau^{(-1)}$ are the Hamiltonian functions corresponding to the two branches (positive and negative) of longitudinal waves, and $H_\tau^{(+2)}$, $H_\tau^{(-2)}$ corresponds to the two branches of torsional waves. If we take $m = 0$ in (8.99), then one obtains the Hamiltonian functions for the axisymmetric motion of the shell.

Under condition (8.98) for the vector $\mathbf{U}_{\tau,0}$, we arrive at the four relations

$$
\mathbf{U}_{\tau,0}^{(\pm \jmath)} = P_{\tau,0}^{(\pm \jmath)} \mathbf{Y}_\tau^{(\pm \jmath)}, \tag{8.100}
$$

where $P_{\tau,0}^{(\pm \jmath)}$ are polynomial in $\xi_\tau^{(\pm \jmath)} = \mu_1^{-1/2}[s - q_\tau^{(\pm \jmath)}(t)]$, and

$$
\mathbf{Y}_\tau^{(\pm \jmath)} = \left(y_{\tau,1}^{(\pm \jmath)},\, y_{\tau,2}^{(\pm \jmath)},\, y_{\tau,3}^{(\pm \jmath)}\right) \tag{8.101}
$$

are any non-trivial solution of Eqs. (8.95) in the leading approximation ($k = 0$). Here

$$
\begin{aligned}
&y_{\tau,1}^{(\pm1)} = 1, \quad y_{\tau,2}^{(\pm1)} = -\frac{l_{\tau,11}^{(\pm1)}}{l_{\tau,12}^{(\pm1)}}, \quad y_{\tau,3}^{(\pm1)} = \frac{l_{\tau,32}^{(\pm1)} l_{\tau,11}^{(\pm1)} - l_{\tau,31}^{(\pm1)} l_{\tau,12}^{(\pm1)}}{l_{\tau,12}^{(\pm1)} l_{\tau,33}^{(\pm1)}}, \\
&y_{\tau,2}^{(\pm2)} = 1, \quad y_{\tau,1}^{(\pm2)} = -\frac{l_{\tau,22}^{(\pm2)}}{l_{\tau,21}^{(\pm2)}}, \quad y_{\tau,3}^{(\pm2)} = \frac{l_{\tau,31}^{(\pm2)} l_{\tau,22}^{(\pm2)} - l_{\tau,32}^{(\pm2)} l_{\tau,21}^{(\pm2)}}{l_{\tau,21}^{(\pm2)} l_{\tau,33}^{(\pm2)}}.
\end{aligned}
\tag{8.102}
$$

In (8.102), the superscripts $(\pm 1), (\pm 2)$ mean that the elements of matrix (8.96) are calculated for the solutions $p_\tau = p_\tau^{(\pm \jmath)}(t)$, $q_\tau = q_\tau^{(\pm \jmath)}$ of the Hamiltonian system with the Hamiltonians $H_\tau^{(\pm \jmath)}$ at $\jmath = 1, 2$.

Formulas (8.99), (8.100) show that the case under consideration fundamentally differs from the cases considered above at $m_0 = 0$ and $m_0 \sim 1$. Now, the WPs of longitudinal and torsional waves propagate together, forming "bundles"corresponding to the functions $H_\tau^{(\pm \jmath)}$ (more on this below).

8.5.2 Non-axisymmetric packets of bending waves

Consider the coupled Eqs. (8.32). Their solution is sought at some interval $0 \le t_1 \le t_n'$ in the form of a travelling WP of bending waves with the center on the parallel $s = q_n(t_1)$. As the ansatz of a solution, function (8.94) is considered, where the subscript τ and the dimensionless time t are replaced by n and t_1, respectively:

$$
\begin{aligned}
&\mathbf{U}_n = \sum_{k=0}^{\infty} \mu_1^{k/2} \mathbf{U}_{n,k} \exp\left[\mathrm{i}\mu_1^{-1} S_n\left(\xi_n, t_1, \mu_1\right)\right], \\
&S_n = \int_0^{t_1} \omega_n\left(\tilde{t}\right) d\tilde{t} + \mu_1^{1/2} p_n(t_1)\xi_n + \tfrac{1}{2}\mu_1 b_n(t_1)\xi_n^2, \\
&\mathbf{U}_{n,k} = \left(u_{n,k},\, v_{n,k},\, w_{n,k}\right), \\
&\omega_n,\, p_n,\, b_n,\, \dot{\omega}_n,\, \dot{p}_n,\, \dot{b}_n,\, \dot{q}_n = O(1), \\
&\mathbf{U}_{n,k},\, \dot{\mathbf{U}}_{n,k},\, \partial \mathbf{U}_{n,k}/\partial \xi_n = O(1)\, \mathbf{I}_3.
\end{aligned}
\tag{8.103}
$$

In this case, the dot means the differentiation with respect to the dimensionless "slow time"t_1.

We write out the elements of the (3×3)– matrices, $\mathbf{L}_{n,0}$ and $\mathbf{N}_n$, appearing in the algorithm for constructing vector (8.103):

$$l_{n,11} = p_n^2 + \frac{(1-\nu)m^2}{2r^2}, \quad l_{n,12} = -\frac{\mathrm{i}(1+\nu)mp_n}{2r},$$

$$l_{n,13} = \mathrm{i}p_n(k_1+\nu k_2), \quad l_{n,21} = -l_{n,12}, \quad l_{n,22} = \frac{(1-\nu)p_n^2}{2} + \frac{m^2}{r^2},$$

$$l_{n,23} = -(k_2+\nu k_1)\frac{m}{r}, \quad l_{n,31} = -l_{n,13}, \quad l_{n,32} = l_{n,23},$$

$$l_{n,33} = \left(p_n^2 + \frac{m^2}{r^2}\right)^2 + \left(k_1^2 + 2\nu k_1 k_2 + k_2^2\right) - \left(\omega_n - \dot{q}_n p_n\right)^2,$$

$$n_{n,11} = \frac{r'}{r}p_n, \quad n_{n,12} = \frac{\mathrm{i}(3-\nu)mr'}{2r^2}, \quad n_{n,13} = \mathrm{i}\left(k_1' + k_2'\right), \tag{8.104}$$

$$n_{n,21} = n_{n,12}, \quad n_{n,22} = \frac{(1-\nu)r'}{2r}p_n, \quad n_{n,23} = 0,$$

$$n_{n,31} = -\frac{\mathrm{i}}{r}\left[r'(k_1+k_2)(1-\nu)(\kappa_2 r)'\right],$$

$$n_{n,32} = 0, \quad n_{n,33} = 2\frac{r'}{r}\left(p_n^3 - m^2 p_n\right) + \ddot{q}_n p_n.$$

The relations coupling the parameters ω_n, p_n, q_n have the same form (8.42) as in the case of axisymmetric waves. However, the Hamiltonian function is much more complicated:

$$H_n(p,q) = \left\{G(p,q) + \frac{1}{G(p,q)}\left[g_1(q)p^4 + \frac{m^2 g_2(q)p^2}{r^2(q)} + \frac{m^4 g_3(q)}{r^4(q)}\right] + g_4(q)\right\}^{1/2}. \tag{8.105}$$

Here,

$$G = \left[p^2 + \frac{m^2}{r^2(q)}\right]^2, \quad g_1(q) = -\left[k_1(q) + \nu k_2(q)\right]^2,$$

$$g_3(q) = -\frac{\left[k_2(q)+\nu k_1(q)\right]^2}{1-\nu}, \quad g_4(q) = k_1^2(q) + 2\nu k_1(q)k_2(q) + k_2^2(q). \tag{8.106}$$

If we assume $m = 0$ in (8.105), then we get the Hamiltonian function (8.43) for axisymmetric bending waves.

8.5.3 Superposition of solutions

Consider the vector-function

$$\mathbf{U} = \left(\mathbf{E}_{001}\mathbf{U}_\tau^{\mathrm{T}}\right)^{\mathrm{T}} + \left(\mathbf{E}_{110}\mathbf{U}_n^{\mathrm{T}}\right)^{\mathrm{T}}, \tag{8.107}$$

where

$$\mathbf{U}_\tau = \sum\nolimits^{\pm} \sum_{\jmath=1,2} \left[P_{\tau,0}^{(\pm\jmath)}\left(\xi_\tau^{(\pm\jmath)}, t, d_{\tau,i^{(\pm\jmath)}}\right)\mathbf{Y}_\tau^{(\pm\jmath)} + O\left(\mu_1^{1/2}\right)\mathbf{I}_3\right]$$
$$\times \exp\left\{\mathrm{i}\left[\mu_1^{-1}\int_0^t \omega_\tau^{(\pm\jmath)}\left(\tilde{t}\right)d\tilde{t} + \mu_1^{-1/2}p_\tau^{(\pm\jmath)}(t)\xi_\tau^{(\pm\jmath)} + \tfrac{1}{2}b_\tau^{(\pm\jmath)}(t)\left(\xi_\tau^{(\pm\jmath)}\right)^2\right]\right\}, \tag{8.108}$$

$$\mathbf{U}_n = \sum\nolimits^{\pm}\left[P_{n,0}^{\pm}\left(\xi_n^{\pm}, \mu_1 t, d_{n,i^{\pm}}\right)\mathbf{Y}_n^{\pm} + O\left(\mu_1^{1/2}\right)\mathbf{I}_3\right]$$
$$\times \exp\left\{\mathrm{i}\left[\mu_1^{-1}\int_0^{\mu_1 t}\omega_n^{\pm}\left(\tilde{t}\right)d\tilde{t} + \mu_1^{-1/2}p_n^{\pm}(\mu_1 t)\xi_n^{\pm} + \tfrac{1}{2}b_n^{\pm}(\mu_1 t)\left(\xi_n^{\pm}\right)^2\right]\right\},$$

$$\xi_\tau^{(\pm\jmath)} = \mu_1^{-1/2}\left[s - q_\tau^{(\pm\jmath)}(t)\right], \quad \jmath = 1,2,$$
$$\xi_n^{\pm} = \mu_1^{-1/2}\left[s - q_n^{\pm}(\mu_1 t)\right]. \tag{8.109}$$

and $\mathbf{E}_{001}, \mathbf{E}_{110}$ are the matrices introduced in Section 8.2. In (8.108), the superscripts $(+\jmath)$ and $(-\jmath)$ designate the functions corresponding to the positive and negative branches of the longitudinal ($\jmath = 1$) and torsional ($\jmath = 2$) waves, while in relation (8.109), signs " + " and " − " identify the quantities corresponding to the positive and negative branches of the bending waves; the symbol $\sum^{\pm}$ means summing the expressions following it and supplied with the upper signs " + " and " − "; the vectors $\mathbf{Y}_n^+$, $\mathbf{Y}_n^-$ denote any nonzero solutions of the system

$$\mathbf{L}_{n,0}\mathbf{U}_{n,0}^{\mathrm{T}} = \mathbf{0}$$

with coefficients computed on the positive and negative branches. Polynomials $P_{\tau,0}^{(\pm\jmath)}$, $P_{n,0}^{\pm}$ are found up to constants $d_{\tau,i}^{(\pm\jmath)}$, $d_{n,i}^{\pm}$ by Eqs. (6.63), (6.74). To determine the constants $d_{\tau,i}^{(\pm\jmath)}$, $d_{n,i}^{\pm}$, we substitute (8.107)-(8.109) into the initial conditions (8.6) and equate coefficients by the same powers of $\mu_1^{1/2}$. Taking into account the equalities $\xi_\tau^{(\pm\jmath)}|_{t=0} = \xi_n^{\pm}|_{t=0} = \zeta$, we arrive at the system consisting of the $4(1+M^*)$ equations,

$$\begin{aligned}
&\sum_{\jmath=1,2} y_{\tau,1}^{(\jmath)}\left(d_{\tau,i}^{(+\jmath)}+d_{\tau,i}^{(-\jmath)}\right)=c_{1i}^{\circ},\\
&\sum_{\jmath=1,2} y_{\tau,2}^{(\jmath)}\left(d_{\tau,i}^{(+\jmath)}+d_{\tau,i}^{(-\jmath)}\right)=c_{2i}^{\circ},\\
&\sum_{\jmath=1,2} \delta^{(\jmath)} y_{\tau,1}^{(\jmath)}\left(d_{\tau,i}^{(+\jmath)}-d_{\tau,i}^{(-\jmath)}\right)=-\frac{d_{1i}^{\circ}}{H_\tau^{\circ}},\\
&\sum_{\jmath=1,2} \delta^{(\jmath)} y_{\tau,2}^{(\jmath)}\left(d_{\tau,i}^{(+\jmath)}-d_{\tau,i}^{(-\jmath)}\right)=-\frac{d_{2i}^{\circ}}{H_\tau^{\circ}},
\end{aligned} \tag{8.110}$$

with respect to $d_{\tau,i}^{(\pm\jmath)} (\jmath = 1,2;\ i = 0,1,\ldots, M^*)$, and at the system of the $2(1+M^*)$ equations

$$\begin{aligned}
&\left(d_{n,i}^{+}+d_{n,i}^{-}\right)=c_{3i}^{\circ},\\
&\left(d_{n,i}^{+}-d_{n,i}^{-}\right)=-\frac{1}{H_n^{\circ}}\left[d_{3i}^{\circ}+H_\tau^{\circ}\sum_{\jmath=1,2}\delta^{(\jmath)}y_{\tau,3}^{(\jmath)}\left(d_{\tau,i}^{(+\jmath)}-d_{\tau,i}^{(-\jmath)}\right)\right],
\end{aligned} \tag{8.111}$$

for the constants $d_{n,i}^{\pm}$. Here

$$y_{\tau,\ell}^{(\jmath)}=y_{\tau,\ell}^{(\pm\jmath)}\Big|_{t=0},\quad \ell=1,2,3,\quad \jmath=1,2,$$

$$H_\tau^{\circ}=H_\tau^{(\pm1)}(a_0,0),\quad H_n^{\circ}=H_n^{\pm}(a_0,0),\quad M^*=\max\{M_1,\,M_2,\,M_3\},$$

where M_ℓ are the degrees of polynomials (8.8). If for some $\ell = 1,2,3$ it turns out that $M_\ell < M^*$, then in (8.110), (8.111) one should take $c_{\ell i}^{\circ} = d_{\ell i}^{\circ} = 0$ for $i = M_\ell + 1, \ldots, M^*$. Thus, the degrees of all the polynomials $P_{\tau,0}^{(\pm\jmath)}$, $P_{n,0}^{\pm}$ do not exceed M^*.

Equations (8.110) and (8.111) are solved in stages: at first, one needs to find the constants $d_{\tau,i}^{(\pm\jmath)}$ from (8.110), and then $d_{n,i}^{\pm}$ are determined from Eqs. (8.111). As can be seen from (8.111), at $m_0 \sim \mu_1^{-1}$ the amplitudes of bending waves depend on the initial tangential displacements and velocities of points on the shell mid-surface.

The constructed vector (8.107), (8.108) is the leading approximation of a formal asymptotic solution of Eqs. (8.2) with the initial conditions (8.6) for

$m_0 \sim \mu_1^{-1}$. In the scalar form, it can be written as

$$u_1 = \sum^{\pm} \left\{ u_\tau^{(\pm 1)} \left(\xi_\tau^{(\pm 1)}, t \right) + u_\tau^{(\pm 2)} \left(\xi_\tau^{(\pm 2)}, t \right) + \mu_1 u_n^{\pm} \left(\xi_n^{\pm}, \mu_1 t \right) \right\},$$

$$u_2 = \sum^{\pm} \left\{ v_\tau^{(\pm 1)} \left(\xi_\tau^{(\pm 1)}, t \right) + v_\tau^{(\pm 2)} \left(\xi_\tau^{(\pm 2)}, t \right) + \mu_1 v_n^{\pm} \left(\xi_n^{\pm}, \mu_1 t \right) \right\},$$

$$u_3 = \sum^{\pm} \left\{ w_n^{\pm} \left(\xi_n^{\pm}, \mu_1 t \right) + \mu_1 w_\tau^{(\pm 1)} \left(\xi_\tau^{(\pm 1)}, t \right) + \mu_1 w_\tau^{(\pm 2)} \left(\xi_\tau^{(\pm 2)}, t \right) \right\}. \tag{8.112}$$

When disregarding the terms of order $O(\mu_1)$, solution (8.112) is the superposition of two the packets of bending waves and four the packets of longitudinal and torsional waves with centers on the parallels $s = q_n^{\pm}(\mu_1 t)$ and $s = q_\tau^{(\pm \jmath)}(t)$ $(\jmath = 1, 2)$, respectively. Moreover, the (± 1)-st packets of longitudinal waves (the functions $u_\tau^{(+1)}$ and $u_\tau^{(-1)}$) corresponding to the Hamiltonians $H_\tau^{(+1)}$ and $H_\tau^{(-1)}$ propagate in "bundles"with the (± 1)-st packets of torsion waves generated by them (the functions $v_\tau^{(+1)}$ and $v_\tau^{(-1)}$), and the (± 2)-nd packets of torsional waves ($v_\tau^{(+2)}$ and $v_\tau^{(-2)}$) with the Hamiltonians $H_\tau^{(+2)}$ and $H_\tau^{(-2)}$ generate two the packets of longitudinal waves ($u_\tau^{(+2)}$ and $u_\tau^{(-2)}$). The terms of order $O(\mu_1)$ in (8.112) should be kept if terms of the same order are taken into account in expansions (8.94), (8.103). However, when calculating the velocities of the transverse displacements, the functions $\mu_1 w_\tau^{(\pm \jmath)}$ (as in the axisymmetric case) should be taken into account, because

$$\partial w_n^{\pm} / \partial t, \; \mu_1 \partial w_\tau^{(\pm \jmath)} / \partial t = O(1).$$

The formal asymptotic solution (8.107), (8.108) constructed above is uniformly valid on some time interval $0 \le t \le t' = \min\{t'_\tau, \mu_1^{-1} t'_n\}$ if on this interval, estimates (8.94) and (8.103) are simultaneously satisfied for the functions appearing in (8.108). The fulfilment of these estimates depends on the correlation of parameters included in the problem. In particular, formulas (8.102) show that if any of the wave parameters $p_\tau^{(\pm \jmath)}$ becomes small over time, then conditions (8.94) for the amplitudes are violated. Thus, as in the axisymmetric case, the restrictions imposed on the wave parameters must be stronger: $p_\tau^{(\pm \jmath)}, \, p_n^{\pm} \sim 1$.

8.5.4 The effect of the shell geometry on dynamics of WPs of tangential waves

A simple form of the Hamiltonian function (8.99) for tangential waves with a large number of waves in the circumferential direction allows us to establish some properties of the WPs of longitudinal and torsional waves depending on the properties of the function $r(s)$.

Let $r'(s) > 0$ for $-\infty < s < +\infty$. The Hamiltonian system

$$\dot{q}_{\tau}^{(\pm\jmath)} = \pm\frac{p_{\tau}^{(\pm\jmath)}(t)}{H_{\tau}^{(\pm\jmath)}}, \quad q_{\tau}^{(\pm\jmath)}(0) = 0,$$

$$\dot{q}_{\tau}^{(\pm\jmath)} = \pm\frac{m^2 r_q'\left[q_{\tau}^{(\pm\jmath)}\right]}{r^3\left[q_{\tau}^{(\pm\jmath)}\right] H_{\tau}^{(\pm\jmath)}}, \quad p_{\tau}^{(\pm\jmath)}(0) = a_0 \tag{8.113}$$

obviously implies the inequalities

$$\dot{p}_{\tau}^{(+\jmath)} > 0, \quad \dot{q}_{\tau}^{(+\jmath)} > 0, \quad \ddot{q}_{\tau}^{(+\jmath)} > 0, \tag{8.114}$$

which are valid for any $t \geq 0$.

Inequalities (8.114) indicate that the packet motion of tangent waves in the direction of increasing the function $r(s)$ (see Fig. 8.1) is accompanied by the increase in both the group velocity of the WP and the variability of waves in the axial direction.

Let

$$\inf_{-\infty<s<+\infty} r(s) > \sqrt{\frac{m^2}{a_0^2 + m^2}}. \tag{8.115}$$

Then

$$\dot{p}_{\tau}^{(-\jmath)} < 0, \quad \dot{q}_{\tau}^{(-\jmath)} < 0, \quad \frac{d|\dot{q}_{\tau}^{(-\jmath)}|}{dt} < 0 \tag{8.116}$$

for any $t \geq 0$, which indicates a decrease in the wave number and group velocity of the WP running towards decreasing the function $r(s)$. If condition (8.115) does not hold, then inequality (8.116) is valid in the interval $0 \leq t < t_{\tau}^*$, where

$$t_{\tau}^* = H_{\tau}^{\circ} \int_0^{q_{\tau}^*} \frac{r(q)dq}{\sqrt{\left(H_{\tau}^{\circ}\right)^2 r^2(q) - m^2}}, \tag{8.117}$$

and a number q_{τ}^* is determined from the equation

$$r^2(q_{\tau}) = \frac{m^2}{a_0^2 + m^2}. \tag{8.118}$$

For $t = t_{\tau}^*$, the function $p_{\tau}^{(-\jmath)}(t)\,(\jmath = 1, 2)$ vanishes, but this contradicts the asymptotic estimates (8.103) by virtue of (8.102). Therefore, if inequality (8.116) does not hold, then the constructed asymptotic solution should be considered for $t < t_{\tau}^*$.

We note that if $r(s)$ is an infinitely differentiable function and $r'(s) > 0$ for $-\infty < s < +\infty$, then integral (8.117) is convergent.

It follows from the analysis that the WPs of tangential waves with large number of waves in the circumferential direction reveal a greater dependence on the shell geometry than axisymmetric packets of longitudinal and torsional waves.

8.6 Wave packets in long cylindrical shells with variable parameters

In this section, we consider the problem on travelling packets of the tangential (in-plane) and bending waves in an infinite cylindrical shell with the thickness and material parameters being functions of the axial coordinate.

Let $h(s), E(s), \rho(s)$ be the dimensionless variable thickness, Young modulus and mass density (the dimensional quantities related to their characteristic values h_0, E_0, ρ_0, respectively). As the characteristic values, h_0, E_0, ρ_0, the numerical values of the correspondent dimensional parameters at the origin $s = 0$ may be taken. Poisson's ratio is assumed to be constant. The coordinate system s, φ is introduced here as in Sect. 8.1.

After separation of the circumferential coordinate φ by formulas (8.3), (8.4), the equations of non-axisymmetric motion of a cylindrical shell with variable characteristics can be written as follows:

$$\left(\mathbf{L} + \gamma \frac{\partial^2}{\partial t^2}\right) \mathbf{U}^T = \mathbf{0}. \tag{8.119}$$

Here, $\gamma = \rho(s)h(s)$ is the surface dimensionless density of a material, $t = t^*/t_c$ is the dimensionless time, $t_c = R\sqrt{\rho_0/E_0}$ is the characteristic time, $\mathbf{U}$ is the three-component vector introduced in Sect. 8.1, and the elements of the matrix $\mathbf{L}$ are the following differential operators:

$$\begin{gathered}
L_{11}z = -\frac{\partial}{\partial s}\left(g\frac{\partial}{\partial s}\right) + \frac{1}{2}(1-\nu)m_0^2\, g\, z, \\
L_{12}z = -m_0\frac{\partial}{\partial s}\left(\nu\, g\, z\right) - \frac{1}{2}m_0(1-\nu)g\frac{\partial z}{\partial s}, \; L_{13}z = \frac{\partial}{\partial s}\left(\nu\, g\, z\right), \\
L_{21}z = \frac{1}{2}(1+\nu)m_0 g\frac{\partial z}{\partial s}, \; L_{22}z = -\frac{1}{2}(1-\nu)g\,\frac{\partial^2 z}{\partial s^2} + m_0^2 g\, z, \\
L_{23}z = -m_0\, g\, z, \; L_{31}z = -\nu g\frac{\partial z}{\partial s}, \; L_{32}z = L_{23}z, \\
L_{33}z = \mu_1^4\left\{\frac{\partial^2}{\partial s^2}\left[d\left(\frac{\partial^2 z}{\partial s^2} - m_0^2\nu z\right)\right] + d m_0^2\left(m_0^2 z - \nu\frac{\partial^2 z}{\partial s^2}\right)\right. \\
\left. -\, 2m_0^2\frac{\partial}{\partial s}\left[(1-\nu)\, d\,\frac{\partial z}{\partial s}\right]\right\} + g\, z,
\end{gathered} \tag{8.120}$$

where $g(s) = Eh/\left(1-\nu^2\right)^{-1}$, $d(s) = Eh^3\left(1-\nu^2\right)^{-1}$ are variable dimensionless in-plane and bending stiffness, respectively, and $\mu_1^4 = h_0^2/\left(12R^2\right)$. As the initial conditions, we consider WP (8.6)–(8.8) centered on the parallel $s = 0$.

As in Section 8.1, we distinguish here three cases: a) $m_0 = 0$, b) $m_0 \sim 1$, c) $m_0 \sim \mu_1^{-1}$. In each of these cases, the classification of the integrals of Eqs. (8.2) leads to governing equations that, up to the matrix operators $\mathbf{L}^\circ$ and $\mathbf{L}$ (for axisymmetric and non-axisymmetric motions, respectively), coincide with equations obtained in Section 8.2. Thus, for $m_0 = 0$, the solution of Eqs. (8.2) can be constructed in the form of separately running WPs of torsion waves and the superposition of packets of longitudinal and bending waves. Case b) is not considered here, because, as above, the solutions for $m_0 = 0$ and $m_0 \sim 1$ coincide in the leading approximation (see Sect. 8.4). For $m_0 \sim \mu_1^{-1}$, the solution of Eqs. (8.2) is representable as superposition (8.107) of tangential and bending waves. In all cases, the algorithm for construction of propagating WPs remains the same. Considering the above, below in cases when it is impossible to explicitly write the solution, we shall indicate only those changes that should be made in the process of its construction, paying more attention to the study of properties of solutions under various assumptions regarding the functions E, ν, ρ, h.

8.6.1 Axisymmetric packets of bending waves

Consider the system of equations

$$\left(\mathbf{L}^\circ + \gamma\, \mu_1^2\, \partial^2/\partial t_1^2\right)\left(\mathbf{E}_{10}\mathbf{V}_n^T\right) = 0, \tag{8.121}$$

describing the axisymmetric bending motion of the shell. Here, $\mathbf{V}_n = (u_n, w_n)$, where the components u_n, w_n coupled to the magnitudes u_1, u_3 by relations (8.19), and $\mathbf{L}^\circ$ is the square matrix with the operator elements

$$\begin{aligned} \mathbf{L}_{11}^\circ z &= -\frac{\partial}{\partial s}\left(g\frac{\partial z}{\partial s}\right), \quad \mathbf{L}_{11}^\circ z = \frac{\partial}{\partial s}\left(\nu\, g\, z\right), \\ \mathbf{L}_{21}^\circ &= -\nu\, g\frac{\partial z}{\partial s}, \quad \mathbf{L}_{22}^\circ z = \mu_1^4 \frac{\partial^2}{\partial s^2}\left(d\frac{\partial^2 z}{\partial s^2}\right) + g\, z. \end{aligned} \tag{8.122}$$

We seek a solution of Eqs. (8.121) in the form of travelling WPs (8.35) of bending waves with the center on parallel $s = q_n(t_1)$. Here the matrices $\mathbf{L}_{n,0}, \mathbf{N}_n$ have the elements

$$\begin{aligned} l_{n,11} &= g(q_n)p_n^2, \quad l_{n,12} = \mathrm{i}p_n\nu(q_n)g(q_n), \\ l_{n,21} &= -l_{n,12}, \quad l_{n,22} = d(q_n)p_n^4 + g(q_n) - \gamma(q_n)\left(\omega_n - \dot q_n p_n\right)^2, \\ n_{n,11} &= g'(q_n)p_n, \quad n_{n,12} = \mathrm{i}\left[\nu'(q_n)g(q_n) + \nu(q_n)g'(q_n)\right], \\ n_{n,21} &= 0, \quad n_{n,22} = 2d'(q_n)p_n^3 + \gamma(q_n)\ddot q_n p_n, \end{aligned} \tag{8.123}$$

and the Hamiltonian function is

$$H_n(p,q) = f(q)\sqrt{1 - \nu^2(q) + h^2(q)p^4} \tag{8.124}$$

with

$$f(q) = \sqrt{E(q)\left[1-\nu^2(q)\right]^{-1}\rho^{-1}(q)}.$$

8.6.1.1 Effect of the material inhomogeneity on the WP dynamics

Let $h = 1$, ν be constants, and $f(s)$ be a positive sufficiently differentiable function such that $f' > 0$ for $-\infty < s < +\infty$. Then the Hamiltonian systems corresponding to the positive and negative branches of the constructed solution take the form

$$\begin{aligned}\dot{q}_n^{\pm} &= \pm\frac{2\left(p_n^{\pm}\right)^3 f^2\left(q_n^{\pm}\right)}{H_n^{\circ}}, \quad q_n^{\pm}(0) = 0,\\ \dot{p}_n^{\pm} &= \mp\frac{f'\left(q_n^{\pm}\right)}{f\left(q_n^{\pm}\right)}H_n^{\circ}, \quad p_n^{\pm}(0) = a_0,\end{aligned} \tag{8.125}$$

where $H_n^{\circ} = H_n(a_0, 0)$. Equations (8.125) obviously imply the inequality

$$\dot{p}_n^- > 0, \quad \dot{q}_n^- < 0, \quad \frac{d|\dot{q}_n^-|}{dt_1} > 0, \tag{8.126}$$

which are valid for any $t \geq 0$.

Let

$$\sup_{-\infty<s<+\infty} f^2(s) < \frac{a_0^4 + 1 - \nu^2}{(1-\nu^2)^2}. \tag{8.127}$$

Then

$$\dot{p}_n^+ < 0, \quad \dot{q}_n^+ > 0, \quad \ddot{q}_n^+ < 0. \tag{8.128}$$

for any $t_1 \geq 0$. If inequality (8.127) does not hold, then solution (8.35) is to be considered at some finite interval $0 \leq t_1 \leq t_n' < t_n^*$ at which inequalities (8.128) are valid. Here,

$$t_n^* = \frac{H_n^{\circ}(1-\nu^2)^{3/4}}{2}\int_0^{q_n^*}\frac{dq}{\sqrt{f(q)}\sqrt[4]{\left[a_0^4 - (1-\nu^2) - (1-\nu^2)^2 f^2(q)\right]^3}}, \tag{8.129}$$

where the number q_n^* is found from the equation

$$a_0^4 + 1 - \nu^2 = (1-\nu^2)^2 f^2(q). \tag{8.130}$$

As $t \to t_n^*$, one obtains $\dot{q}_n^+ \to 0$, $p_n^+ \to 0$. The last limit relation means the existence of a singularity at $t = t_n^*$ for the vector function $\mathbf{Y}_n$, which is a solution of the homogeneous system of equations $\mathbf{L}_{n,0}\,\mathbf{Y}_n = \mathbf{0}$ (see elements (8.123) of the matrix $\mathbf{L}_{n,0}$). The obtained inequalities (8.126), (8.128) allow concluding: the motion of the WP of bending waves in the direction of decreasing the function $f(s)$ is accompanied by an increase in the group velocity and variability of waves, and when the WP moves in the opposite direction, these magnitudes decrease.

In order to analyze the effect of the function $f(s)$ on the WP width and the wave amplitude, we consider

Example 8.3. Let

$$h = 1, \quad \rho = 1, \quad f(s) = \sqrt{E/(1-\nu^2)} = e^{\alpha s},$$

where $\alpha > 0$. Then

$$p_n^{\pm}(t_1) = a_0 \mp \alpha\sqrt{1-\nu^2+\alpha_0^4}\, t_1,$$

$$q_n^{\pm}(t_1) = \frac{1}{2} \ln \sqrt{\frac{1-\nu^2+\alpha_0^4}{1-\nu^2+\left[p_n^{\pm}(t_1)\right]^4}}. \tag{8.131}$$

The solution of the Riccati equation can be written in the form (6.86)

$$b_n^{\pm}(t_1) = B_n^{\pm}(t_1)/C_n^{\pm}(t_1), \tag{8.132}$$

where $B_n^{\pm}(t_1), C_n^{\pm}(t_1)$ are the solutions of the system in variations (6.85) with the initial conditions

$$B_n^{\pm}(0) = b_0,\ C_n^{\pm}(0) = 1. \tag{8.133}$$

The considered distribution of Young's modulus along the cylinder axis makes it easy to integrate system (6.85):

$$B_n^{\pm}(t_1) = \beta_2 + \beta_1\, p_n^{\pm}(t_1),$$

$$C_n^{\pm}(t_1) = \frac{1}{\alpha}\left\{\beta_1 - \frac{2\left[p_n^{\pm}(t_1)\right]^3\left[\beta_2+\beta_1\, p_n^{\pm}(t_1)\right]}{1-\nu^2+\left[p_n^{\pm}(t_1)\right]^4}\right\}, \tag{8.134}$$

$$\beta_1 = \alpha + \frac{2a_0^3 b_0}{1-\nu^2+a_0^4}, \quad \beta_2 = \frac{b_0\left(1-\nu^2-a_0^4\right)}{1-\nu^2+a_0^4} - \alpha\, a_0.$$

For simplicity, suppose that in the initial conditions (8.6), (8.7) $u_3^\circ = c_{30}^\circ$ is a constant, and $u_1^\circ = u_2^\circ = \vartheta_j^\circ = 0$, $j = 1, 2, 3$. Then the polynomial $P_{n,0}^{\pm}$ which is determined according to (8.60), can be represented by only one term

$$A_{n,0}^{\pm}\left(t_1; d_{n,0}^{\pm}\right) = \frac{d_{n,0}^{\pm}}{\sqrt{C_n^{\pm}(t_1)}}, \tag{8.135}$$

where $d_{n,0}^{\pm}$ are arbitrary constants.

It is of interest to note that

$$\lim_{t_1\to\infty} \Im\, b_n^-(t_1) = b_\infty^- > 0, \quad \lim_{t_1\to\infty} |A_{n,0}^-(t_1; d_{n,0}^-)| = A_\infty^- > 0, \tag{8.136}$$

that is, the packet of bending waves running in the direction of decreasing the function $f(s)$ does not "spread out"at infinity, the width of the packet and its amplitude tending to some finite positive values. If we pass in (8.131)–(8.135) to the limit as $\alpha \to 0$, then we can obtain the corresponding formulas (8.89) for the shell with constant parameters.

Figures 8.5 and 8.6 display plots of the functions $\Im\, b_n^{\pm}(t_1)$ and $w_{max}^{\pm}(t_1) = \max_s |w_n^{\pm}(t_1)|$ for $\alpha = 0.2, 0.5, 1$ (curves 1, 2, 3, respectively), $h/R = 4 \cdot 10^{-3}$; $\nu = 0.3$, $a_0 = 1$, $b_0 = \mathrm{i}$; $c^{\circ}_{3,0} = 1$. The solid lines correspond to the WPs

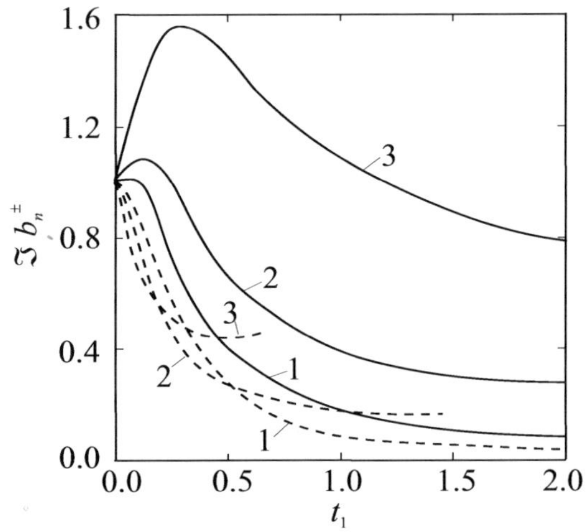

FIGURE 8.5: Plots of the functions $\Im b_n^{\pm}(t_1)$ for $\alpha = 0.2, 0.5, 1$ (curves 1, 2, 3, respectively).

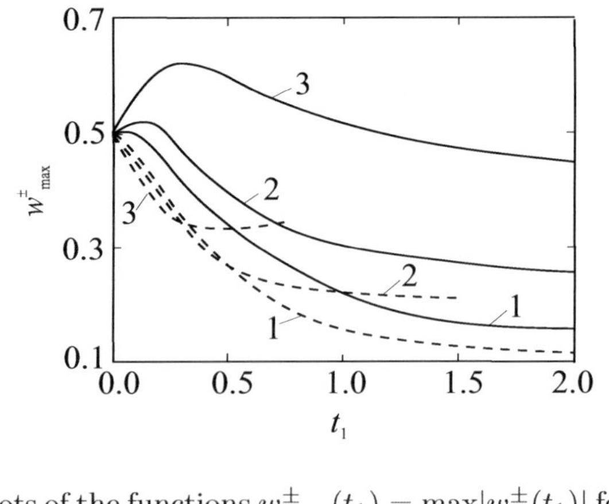

FIGURE 8.6: Plots of the functions $w_{max}^{\pm}(t_1) = \max_s |w_n^{\pm}(t_1)|$ for $\alpha = 0.2, 0.5, 1$ (curves 1, 2, 3, respectively).

propagating towards decreasing the function $f(s)$, and the dashed curves do to packets travelling in the direction of increasing this function. In the latter case, the calculations were performed for

$$0 \leq t_1 < \frac{a_0}{\alpha(\delta + a_0^4)^{1/2}}.$$

The plots of the functions $\Im b_n^-$, $w_{\max}^-$ (solid lines) indicate a possible focusing of the WPs running in the direction of decreasing $f(s)$; moreover, the faster the function $f(s)$ decreases, the stronger focusing is. The calculations of the functions $\Im b_n^-$, $w_{\max}^-$ for large values of t_1 verify the existence of limits (8.136).

8.6.1.2 Effect of the thickness variation on the WP dynamics

Let $E = 1, \rho = 1, \nu$ be constants, and $h(s)$ is the positive function such that $h'(s) > 0$ for any s. Here the analysis of the Hamiltonian system results in the inequalities

$$\begin{gathered} \dot{p}_n^- > 0, \quad \dot{q}_n^- < 0, \quad \frac{d|\dot{q}_n^-|}{dt_1} > 0, \\ \dot{p}_n^+ < 0, \quad \dot{q}_n^+ > 0, \quad \ddot{q}_n^+ < 0, \end{gathered} \tag{8.137}$$

which, in the contrast to the previous case, hold at any finite time interval $0 \leq t \leq t_n'$. It follows from inequalities (8.137) that the WP group velocity and the wave variability increase if the WP runs in the direction of decreasing the shell thickness, and vice versa, decreases when the WP moves in the opposite direction.

Example 8.4. The outcomes of numerical solutions of the Riccati and amplitude equations for $h(s) = 1 + \tanh(\alpha s)$, $E = \rho = 1$ are depicted in Figs. 8.7 and 8.8. Here, $\tanh(x)$ denotes the hyperbolic tangent of an argument x. The values of the remaining parameters are taken from the previous example. The solid lines correspond to the WP running in the direction of the shell thickness reduction, and the dashed lines do to the packet running in the direction of increasing thickness. It can be seen that an increase in the thickness assists the rapid spreading the WP of flexural waves, and its decrease leads to focusing. Numerical calculations at a large values of t_1 indicate an unlimited increase in the wave amplitude in the WP running in the direction of decreasing $h(s)$. We also note that the greater the parameter α characterizing the change rate in the shell thickness is, the greater the degree of focusing WPs and the growth rate of the waves amplitude are.

8.6.2 Axisymmetric packets of longitudinal waves

Consider the system of Eqs. (8.17)

$$\left(\mathbf{L}^\circ + \partial^2/\partial t^2\right)\left(\mathbf{E}_{01}\mathbf{V}_s^T\right) = 0, \tag{8.138}$$

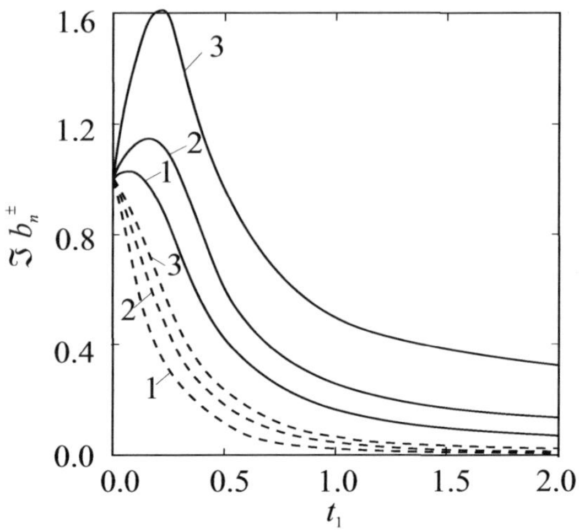

FIGURE 8.7: Plots of the functions $\Im b_n^{\pm}(t_1)$ for $\alpha = 0.25, 0.5, 1$ (curves 1, 2, 3, respectively).

for the axisymmetric longitudinal waves, where $\mathbf{L}^{\circ}$ is the matrix operator introduced in accordance with (8.122), $\mathbf{V}_s = (u_s, w_s)$, and the components u_s, w_s are coupled with the displacements by relations (8.16). We seek the solution to Eqs. (8.138) in the form of WP (8.35) centered on the parallel $s = q_s(t)$, where the indices n should be replaced by s, and t_1 by t. Here, the elements of the matrices $\mathbf{L}_{s,0}$ and $\mathbf{N}_s$ appearing in the algorithm for constructing the solution are

$$
\begin{gathered}
l_{s,11} = g(q_s)p_s^2 - \gamma(q_s)(\omega_s - \dot{q}_s p_s)^2, \quad l_{s,12} = 0, \\
l_{s,21} = -\mathrm{i}\nu(q_s)g(q_s)p_s, \quad l_{s,22} = -\gamma(q_s)(\omega_s - \dot{q}_s p_s)^2; \\
n_{s,11} = g'(q_s)p_s + \gamma(q_s)p_s\ddot{q}_s, \quad n_{s,12} = 0, \\
n_{s,21} = 0, \quad n_{s,22} = \gamma(q_s)p_s\ddot{q}_s,
\end{gathered}
\tag{8.139}
$$

and the Hamiltonian function is

$$
H_s(p, q) = f(q)p, \quad f(q) = \sqrt{\frac{g(q)}{\gamma(q)}}. \tag{8.140}
$$

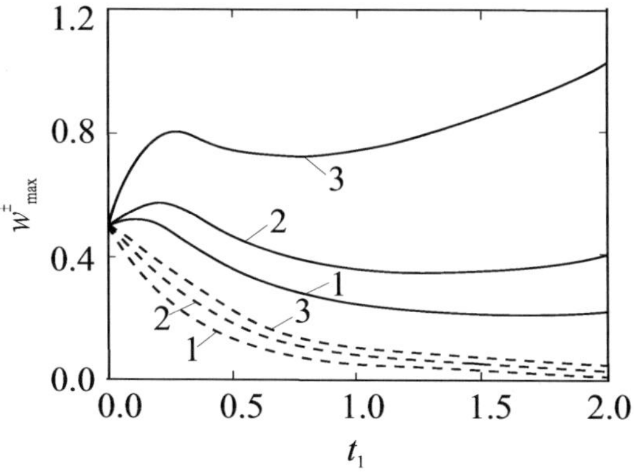

FIGURE 8.8: Plots of the functions $w^{\pm}_{max}(t_1) = \max_s |w^{\pm}_n(t_1)|$ for $\alpha = 0.25, 0.5, 1$ (curves 1, 2, 3, respectively).

A simple form of the Hamiltonian allows one to find

$$p_s^{\pm}(t) = \frac{a_0}{f\left[q_s^{\pm}(t)\right]}, \qquad t = \pm\int\limits_0^{q_s^{\pm}} \frac{dq}{f(q)}, \qquad \omega_s^{\pm} = 0,$$
$$b_s^{\pm}(t) = \frac{b_0 + a_0\left\{f'(0) - f'\left[q_s^{\pm}(t)\right]\right\}}{f^2\left[q_s^{\pm}(t)\right]}. \tag{8.141}$$

Then, omitting the indices + and −, we construct the positive branch of a solution. The equation for amplitudes (8.58) is simplified here:

$$h_{s,1}(t)\xi_s\frac{\partial P_{s,0}}{\partial \xi_s} + h_{s,2}(t)\frac{\partial P_{s,0}}{\partial t} + h_{s,3}(t)P_{3,0} = 0. \tag{8.142}$$

Its solution is the polynomial

$$P_{s,0} = \sum_{k=0}^{M_s} A_{s,k}(t;\, d_{s,k})\, \xi_s^k \tag{8.143}$$

of degree M_s with the coefficients

$$A_{s,k} = d_{s,k}\exp\left\{-\int\left[kh_{s,1}(t) + h_{s,3}(t)\right]h_{s,2}^{-1}(t)dt\right\}$$
$$= \frac{d_{s,k}f^{-k}[q_s(t)]}{\sqrt[4]{\gamma[q_s(t)]\,g[q_s(t)]}}, \tag{8.144}$$

where $d_{s,k}$ are arbitrary constants, and the functions $h_{s,j}$ are defined by relations (8.59) and (6.56) for $j = 3$ and $j = 1, 2$, respectively, with replacing the subscript n by s.

Finally, the leading approximation of the positive branch of the solution to Eqs. (8.138) is as follows:

$$\mathbf{V}_s = \left\{ \frac{1}{\sqrt[4]{\gamma[q_s(t)]\, g[q_s(t)]}} \sum_{k=0}^{M_s} \frac{d_{s,k}[s - q_s(t)]^k \mathbf{Y}_s}{\mu_1^{k/2} f^k[q_s(t)]} + O\left(\mu_1^{1/2}\right) \mathbf{I}_2 \right\}$$

$$\times \exp\left\{ \frac{\mathrm{i}}{\mu_1} \left[p_s(t)[s - q_s(t)] + \frac{1}{2} b_s(t)[s - q_s(t)]^2 \right] \right\}, \tag{8.145}$$

$$\mathbf{Y}_s = (1, y_{s,2}), \quad y_{s,2} = -\frac{\mathrm{i}\nu[q_s(t)]}{a_0} \sqrt{\frac{E[q_s(t)]\{1 - \nu^2[q_s(t)]\}}{\rho[q_s(t)]}}.$$

To pass to the negative branch, one needs, in view of (8.141), to substitute in (8.145) the functions $p_s^+ \equiv p_s, q_s^+ \equiv q_s, b_s^+ \equiv b_s$ by the functions p_s^-, q_s^-, b_s^-, respectively.

The explicitly obtained solution (8.145) as well as relations (8.140), (8.141) allow one to study the influence of the variables h, E, ρ, ν on the dynamics of axisymmetric WPs of longitudinal waves. It can be seen from (8.141) that the wave parameter p_s, the group velocity $\dot{q}_s$ as well as the parameter $\Im\, b_s$ characterizing the width of the WP are independent of $h(s)$ and can be determined from the properties of $f(s)$. In particular, the WP travelling towards the direction of decreasing the function $f(s)$ is contracted. In Eq. (8.145), the ratio

$$(\gamma g)^{-1/4} = \sqrt[4]{\frac{1 - \nu^2[q_s(t)]}{E[q_s(t)]\rho[q_s(t)]h^2[q_s(t)]}} \tag{8.146}$$

shows that the thickness variation affects only the amplitude of axisymmetric packets of longitudinal waves, namely, with decreasing thickness the amplitude of waves increases.

8.6.3 Axisymmetric packets of torsional waves

Consider the equation

$$-\frac{\partial}{\partial s}\left(\frac{1 - \nu}{2} g \frac{\partial u_2}{\partial s}\right) + \gamma \frac{\partial^2 u_2}{\partial t^2} = 0 \tag{8.147}$$

governing the axisymmetric torsional waves. Before proceeding to its integration, we again turn to the system of Eqs. (8.138). If we disregard in the first equation from (8.138) the terms not involved in the construction of the leading approximation of solution (8.145), then we obtain the equation

$$-\frac{\partial}{\partial s}\left(g\frac{\partial u_1}{\partial s}\right)+\gamma\frac{\partial^2 u_1}{\partial t^2}=0 \tag{8.148}$$

for the longitudinal waves. Its solution is the first component $u_s \equiv u_1$ of the vector $\mathbf{V}_s$ defined by (8.145). Let

$$g_1(s)=\frac{1-\nu(s)}{2}g(s),\quad f_1(s)=\sqrt{\frac{1-\nu(s)}{2}}\,f(s). \tag{8.149}$$

Now Eq. (8.147) takes form (8.148). Since the component u_s of vector (8.145) is invariant with respect to the form of the functions $g(s), f(s)$, it follows that under performing the formal replacements of g, f with g_1, f_1, respectively, we find

$$\begin{aligned} u_2 = &\left\{\frac{1}{\sqrt[4]{\gamma[q_\varphi(t)]\,g_1[q_\varphi(t)]}}\sum_{k=0}^{M_\varphi}\frac{d_{\varphi,k}[s-q_\varphi(t)]^k}{\mu_1^{k/2}f_1^k[q_\varphi(t)]}+O\left(\mu_1^{1/2}\right)\right\} \\ &\times\exp\left\{\frac{\mathrm{i}}{\mu_1}\left[p_\varphi(t)[s-q_\varphi(t)]+\frac{1}{2}b_\varphi(t)[s-q_\varphi(t)]^2\right]\right\}. \end{aligned} \tag{8.150}$$

Here $p_\varphi(t), q_\varphi(t), b_\varphi(t)$ are determined by formulas (8.141) with the replacement of f by f_1.

It can be seen from (8.150) that the maximum wave amplitude in the axisymmetric WP of torsion waves is (compare with (8.146))

$$(\gamma g_1)^{-1/4}=\sqrt[4]{\frac{1+\nu[q_\varphi(t)]}{E[q_\varphi(t)]\rho[q_\varphi(t)]h^2[q_\varphi(t)]}}\,. \tag{8.151}$$

8.6.4 Non-axisymmetric WPs with a large number of waves in the circumferential direction

Consider system (8.119) in the case when $m_0=\mu_1^{-1}m$, $m\sim 1$. Its solution with the initial conditions (8.6) can be represented in the form of superposition (8.107) of solutions to Eqs. (8.29) and (8.32) in which $\mathbf{L}_{-1}$ is the matrix operator with elements (8.120) at $m_0=\mu_1^{-1}m$. All the unknown functions contained in (8.107), (8.108) can be found according to the scheme described in Section 8.5. It remains only to specify the matrix operators and Hamiltonians necessary for constructing vector (8.107). For the tangential waves,

$$
\begin{gathered}
l_{\tau,11} = gp_\tau^2 + \frac{1-\nu}{2} m^2 g - \gamma\left(\omega_\tau - \dot{q}_\tau p_\tau\right)^2, \\
l_{\tau,12} = -\mathrm{i}\frac{1+\nu}{2} mgp_\tau, \\
l_{\tau,22} = \frac{1-\nu}{2} gp_\tau^2 + m^2 g - \gamma\left(\omega_\tau - \dot{q}_\tau p_\tau\right)^2, \\
l_{\tau,31} = -\mathrm{i}\nu g p_\tau, \quad l_{\tau,32} = -mg, \quad l_{\tau,33} = -\gamma\left(\omega_\tau - \dot{q}_\tau p_\tau\right)^2, \\
l_{\tau,21} = -l_{\tau,12}, \quad l_{\tau,13} = l_{\tau,23} = 0; \\
n_{\tau,11} = g' p_\tau + \gamma \ddot{q}_\tau p_\tau, \quad n_{\tau,21} = -\mathrm{i}m(\nu g)', \\
n_{\tau,21} = \mathrm{i}m\left(\frac{1-\nu}{2} g\right)', \quad n_{\tau,22} = \left(\frac{1-\nu}{2} g\right)' p_\tau + \gamma \ddot{q}_\tau p_\tau, \\
n_{\tau,13} = n_{\tau,23} = n_{\tau,31} = n_{\tau,32} = 0, \quad n_{\tau,33} = \gamma \ddot{q}_\tau p_\tau
\end{gathered}
\tag{8.152}
$$

with the Hamiltonians

$$
\begin{aligned}
H_\tau^{(\pm 1)}(p,q) &= \pm f(q)\sqrt{p^2+m^2}, \\
H_\tau^{(\pm 2)}(p,q) &= \pm f_1(q)\sqrt{p^2+m^2},
\end{aligned}
\tag{8.153}
$$

where

$$
f(q) = \sqrt{\frac{g(q)}{\gamma(q)}}, \quad f_1(q) = \sqrt{\frac{1-\nu(q)}{2}}\, f(q), \tag{8.154}
$$

and for the bending waves, we get

$$
\begin{gathered}
l_{n,11} = gp_n^2 + \frac{1-\nu}{2} m^2 g, \quad l_{n,12} = -\mathrm{i}\frac{1+\nu}{2} mgp_n, \quad l_{n,13} = \mathrm{i}\nu g p_n, \\
l_{n,21} = -l_{n,12}, \quad l_{n,22} = \frac{1-\nu}{2} gp_n^2 + m^2 g, \quad l_{n,23} = -mg, \\
l_{n,31} = -l_{n,13}, \quad l_{n,32} = l_{n,23}, \\
l_{n,33} = d\left(p_n^2 + m^2\right)^2 + g - \gamma\left(\omega_n - \dot{q}_n p_n\right)^2, n_{n,11} = g' p_n, \\
n_{n,12} = -\mathrm{i}m(\nu g)', \quad n_{n,13} = \mathrm{i}(\nu g)', \\
n_{n,21} = \mathrm{i}m\left(\frac{1-\nu}{2} g\right)', \quad n_{n,22} = \left(\frac{1-\nu}{2} g\right)' p_n, \quad n_{n,23} = 0, \\
n_{n,31} = 0, \quad n_{n,32} = 0, n_{n,33} = 2d'\left(p_n^3 - m^2 p_n\right) + \gamma \ddot{q}_n p_n
\end{gathered}
\tag{8.155}
$$

with the Hamilton functions

$$H_n^{\pm} = \pm f(q)\sqrt{h^2(q)\,(p^2+m^2)^2 + \frac{[1-\nu^2(q)]p^4}{(p^2+m^2)^2}}. \tag{8.156}$$

Here $H_\tau^{(\pm 1)}$ and $H_\tau^{(\pm 2)}$ correspond to the longitudinal and torsional waves, respectively.

Thus, in contrast to the axisymmetric case, the tangential non-axisymmetric wave forms of the shell motion with a large number of waves along a parallel are four the "bundles"of travelling WPs of longitudinal and torsional waves.

When assuming $m = 0$ in (8.153), (8.156), one obtains Hamiltonians (8.140) and (8.124) for axisymmetric waves.

Here we consider in more detail the tangential motion under the assumption that $E = \rho = 1$, ν are constants, and $h(s)$ is a function. Then $f = (1-\nu)^{-1/2}$ and the Hamiltonian system and the Riccati equation give

$$\begin{gathered} p_\tau^{(\pm l)} = a_0, \quad q_\tau^{(\pm l)}(t) = \pm\frac{a_0\delta^{(l)}}{(1-\nu^2)H_\tau^\circ}\,t, \\ \omega_\tau^{(\pm l)} = \mp\frac{\delta^{(l)}m^2}{(1-\nu^2)H_\tau^\circ}, \quad b_\tau^{(\pm l)}(t) = \frac{b_0}{1\pm\delta^{(l)}b_0\,H_{pp}^\circ\,t}, \quad l=1,2, \\ H_\tau^\circ = H_\tau^{(+1)}(a_0,0), \quad H_{pp}^\circ = \left.\frac{\partial^2 H_\tau^{(+1)}}{\partial p^2}\right|_{\substack{q=0\\ p=a_0}}, \\ \delta^{(1)} = 1, \quad \delta^{(1)} = \sqrt{\frac{1-\nu}{2}}. \end{gathered} \tag{8.157}$$

Below, omitting the superscripts $(\pm l)$, we perform all calculations for the functions corresponding to the Hamiltonian $H_\tau^{(1)}$.

We turn to the equation for amplitudes

$$h_{\tau,0}(t)\,\frac{\partial^2 P_{\tau,0}}{\partial \xi_n^2} + h_{\tau,1}(t)\,\xi_\tau\,\frac{\partial P_{\tau,0}}{\partial \xi_\tau} + h_{\tau,2}(t)\,\frac{\partial P_{\tau,0}}{\partial t} + h_{\tau,3}(t)\,P_{\tau,0} = 0, \tag{8.158}$$

where $\xi_\tau = \mu_1^{-1/2}[s - q_\tau(t)]$, and the variable coefficients $h_{\tau,i}(t)$ are specified as

$$\begin{gathered} h_{\tau,0} = \frac{1}{2}H_{pp}^\circ, \quad h_{\tau,1}(t) = \mathrm{i}\frac{\dot C_\tau}{C_\tau}, \quad h_{\tau,2} = \mathrm{i}, \\ h_{\tau,3}(t) = \left\{\frac{1}{2}\frac{C_\tau}{C_\tau} + \frac{p_\tau h'[q_\tau(t)]}{2(1-\nu^2)H_\tau^\circ h[q_\tau(t)]}\right\}. \end{gathered} \tag{8.159}$$

Here the function

$$C_\tau(t) = 1 + b_0 H^\circ_{pp}\, t \tag{8.160}$$

is the solution of the system in variations (6.85) with the Hamiltonian $H_\tau^{(+1)}$. When defining functions (8.159), we used the identity

$$b_\tau H_{pp} \equiv \frac{\dot C_\tau}{C_\tau} - H_{pq}. \tag{8.161}$$

Equation (8.158) has a solution in the form of a polynomial

$$P_{\tau,0}\left(\xi_\tau, t; d_{\tau,i}\right) = \sum_{k=0}^{M} A_{\tau,k}\left(t;\, d_{\tau,i}\right) \xi_\tau^k \tag{8.162}$$

of degree M, where the complex-valued functions $A_{\tau,k}$ are found from the recurrence relations (6.65). We substitute (8.159) into (6.65) and, taking into account the differential relation

$$dt = \frac{(1-\nu^2)H^\circ_\tau}{a_0} dq_\tau,$$

perform where necessary the integration over the variable q_τ. Using induction, we again obtain formulas like (6.138):

$$A_{\tau,M}(t) = d_{\tau,M}\Psi_{\tau,M}(t), \quad A_{\tau,M-1}(t) = d_{\tau,M-1}\Psi_{\tau,M-1}(t),$$

$$A_{\tau,M-r}(t) = \Psi_{\tau,M-r}(t) d_{\tau,M-r} + \Psi_{\tau,M-r}(t) \times$$

$$\sum_{j=1}^{N} \frac{\mathrm{i}^j (M-r+1)(M-r+2)\dots(M-r+2j) d^{\pm}_{\tau,M-r+2j} \left[C_\tau(t)-1\right]^j}{2^j j! \left[b_0 C_\tau(t)\right]^j}. \tag{8.163}$$

However, in this case

$$\Psi_{\tau,j}(t) = \frac{1}{\sqrt{h[q_\tau(t)]\,\left[C_\tau(t)\right]^{2j+1}}}, \qquad j = 0, 1, 2, \dots, M. \tag{8.164}$$

In order to derive similar relations corresponding to the Hamiltonians $H_\tau^{(\pm l)}$, it is necessary in Eqs. (8.163), (8.164) to provide all functions and constants with the superscript $(\pm l)$, at that $C_\tau^{(\pm l)}t) = 1 + \delta^{(l)} b_0 H^\circ_{pp} t$.

Finally, we find the tangential (in-plane) components u_1, u_2 of vector (8.107):

$$u_j = \sum_{l=1,2}^{\pm} \left\{ \sum_{k=0}^{M_\tau} \mu_1^{-k/2} A_{\tau,k}^{(\pm l)}(t) \left[s \pm \frac{\delta^{(l)} a_0 t}{(1-\nu^2) H_\tau^\circ} \right]^k y_j^{(\pm l)} + O\left(\mu_1^{1/2}\right) \right\}$$

$$\times \exp\left\{ \mathrm{i}\mu_1^{-1} \left[\mp \frac{\delta^{(l)} m^2 t}{(1-\nu^2) H_\tau^\circ} + a_0 \left(s \pm \frac{\delta^{(l)} a_0 t}{(1-\nu^2) H_\tau^\circ} \right) \right.\right.$$

$$\left.\left. + \frac{b_0}{2\left(1 \pm \delta^{(l)} b_0 H_{pp}^\circ t\right)} \left(s \pm \frac{\delta^{(l)} a_0 t}{(1-\nu^2) H_\tau^\circ} \right)^2 \right] \right\}, \quad j = 1, 2,$$

$$y_1^{(\pm 1)} = y_2^{(\pm 2)} = 1, \quad y_1^{(\pm 2)} = y_2^{(\pm 1)} = \frac{\mathrm{i}m}{a_0}, \tag{8.165}$$

where the functions $A_{\tau,k}^{(\pm l)}(t)$ are calculated by formulas (8.163).

We note that now, by virtue of (8.157), solution (8.165) is valid for any $t \geq 0$ while the similar solution (8.108), which was previously constructed for an infinite shell of revolution, makes sense only at the time interval where $p_\tau^{(\pm l)} \sim 1$.

It follows from (8.165) that, just as in the axisymmetric case, the thickness variability of a shell affects only the waves amplitude. However, now the amplitude of waves in the WP running in the direction of decreasing thickness grows more slowly; with a decrease in the parameter m, the growth rate of the waves amplitude increases. For $m = 0$, expansion (8.165) degenerates into the asymptotic solutions constructed above for axisymmetric waves.

Note that the effect of internal inhomogeneous pressure on travelling in the axial direction packets of non-axisymmetric waves was studied in [154]. The Flügge type equations were used as the governing ones, the solutions of which were constructed in the form of a superposition of bending, longitudinal, and torsional waves. It was detected a strong dependence of the dynamic characteristics of packets of bending waves and, to a lesser extent, longitudinal waves on the law of variation of internal pressure.

Chapter 9

Two-dimensional Wave Packets in Shells of Arbitrary Shape

In this chapter, the method developed in Chapters 6-8 is generalized to the case of two-dimensional (2D) wave packets. The object of the study is a shell of an arbitrary shape with variable physical and geometric characteristics, subject to external forces. Assuming a short wavelength in both directions on the shell mid-surface, the semi-momentless equations of shallow shells are used as the governing ones. Solutions of the equations are constructed in the form of travelling bending waves localized in the vicinity of some moving points on the shell surface. The idea of an asymptotic approach is the introduction of a local coordinate system with the origin in the moving point coinciding with the center of a WP. The asymptotic procedure applied to the equations written in the new coordinate system results in: (i) the dispersion-like equation relating the frequency, the group velocity and the wave number, (ii) the Hamiltonian system of four differential equations, (iii) the matrix Riccati equation and (iv) the amplitude equation. The effect of the curvature and thickness variability on the travelling 2D wave packets is studied. In particular, it is observed that the thickness variability in a shell of revolution may cause a complicated trajectory of 2D packets of bending waves and result in their multiple reflection from regions with high bending stiffness.

9.1 Statement of problem

Consider a thin elastic shell of arbitrary shape. We introduce on its reference (middle) surface the curvilinear coordinates $\alpha = R\alpha_1, \beta = R\alpha_2$ as shown in Fig. 2.1. The system axes coincide with the curvature lines. The first quadratic form of the reference surface is

$$d\sigma^2 = R^2 \left(A_1^2 d\alpha_1^2 + A_2^2 d\alpha_2^2 \right),$$

where R is the characteristic size of the shell, and A_1, A_2 are Lame's coefficients for the reference surface. We denote by R_1, R_2 the dimensionless radii of curvature (the principal radii of curvature divided by R). Let the thickness $h(\alpha_1, \alpha_2)$, Young's modulus $E(\alpha_1, \alpha_2)$, Poisson's ratio $\nu(\alpha_1, \alpha_2)$, the material density $\rho(\alpha_1, \alpha_2)$, and the dimensionless radii $R_1(\alpha_1, \alpha_2), R_2(\alpha_1, \alpha_2)$ be functions of α_1, α_2. We introduce also the characteristic values h_0, E_0, ν_0 and ρ_0 for $h(\alpha_1, \alpha_2)$, $E(\alpha_1, \alpha_2)$, $\nu(\alpha_1, \alpha_2)$ and $\rho(\alpha_1, \alpha_2)$, respectively.

The shell is assumed to be under external slowly varying forces which cause a momentless (in the general case, unsteady) stress state. Let this state be characterized by the membrane (in-plane) stress resultants $T_i^\circ(\alpha_1, \alpha_2, t) = \mu^2 E_0 h_0 T_i(\alpha_1, \alpha_2, t)$ satisfying the conditions

$$T_i = O(1) \quad \text{as} \quad \mu \to 0, \tag{9.1}$$

where

$$\mu^4 = \frac{h_0^2}{12\left(1 - \nu_0^2\right) R^2} \tag{9.2}$$

is the small parameter. In addition, it is assumed that the forces T_i do not reach their critical values at which there is a loss of stability of the initial momentless stress state.

Assuming a large variability of waves in both directions on the shell surface, we use the system of differential equations for shallow shells written in the dimensionless form:

$$\begin{aligned} &\mu^2 \Delta d \Delta W - \mathrm{K}[(1-\nu)d, W] + \Delta_k \Phi + \Delta_T W + \gamma \frac{\partial^2 W}{\partial t^2} = 0, \\ &\mu^2 \Delta g \Delta \Phi - \mathrm{K}[(1+\nu)g, \Phi] - \Delta_k W = 0, \end{aligned} \tag{9.3}$$

where

$$\Delta z = \frac{1}{A_1 A_2} \sum_i \frac{\partial}{\partial \alpha_i} \left(\frac{A_j}{A_i} \frac{\partial z}{\partial \alpha_i} \right), \quad \Delta_k z = \frac{1}{A_1 A_2} \sum_i \frac{\partial}{\partial \alpha_i} \left(\frac{A_j}{A_i R_j} \frac{\partial z}{\partial \alpha_i} \right),$$

$$\Delta_T z = \frac{1}{A_1 A_2} \sum_i \frac{\partial}{\partial \alpha_i} \left(\frac{A_j T_i}{A_i} \frac{\partial z}{\partial \alpha_i} + T_3 \frac{\partial z}{\partial \alpha_j} \right), \quad i, j = 1, 2, \quad i \neq j,$$

$$d(\alpha_k) = \frac{E h^3 (1 - \nu_0^2)}{E_0 h_0^3 (1 - \nu^2)}, \quad g(\alpha_k) = \frac{E h}{E_0 h_0}, \quad \gamma(\alpha_k) = \frac{\rho h}{\rho_0 h_0}. \tag{9.4}$$

Here t, W, Φ are the dimensionless time, normal deflection and stress function introduced in accordance with (7.54), and $\mathrm{K}[\chi, z]$ is a differential operator whose contribution to the asymptotic solution constructed below is of order $O(\mu)$ and should be taken into account only when constructing higher approximations.

It is assumed that all the coefficients in Eqs. (9.3) are a sufficient number of times differentiable with respect to the coordinates α_i.

We set the initial conditions for the normal displacements and velocities of points on the shell mid-surface:

$$\begin{gathered} W|_{t=0} = w^\circ(\zeta_1, \zeta_2) \exp \left[\mathrm{i}\mu^{-1} S^\circ(\alpha_1, \alpha_2) \right], \\ \dot{W}|_{t=0} = \mathrm{i}\mu^{-1} \vartheta^\circ(\zeta_1, \zeta_2) \exp \left[\mathrm{i}\mu^{-1} S^\circ(\alpha_1, \alpha_2) \right], \\ S^\circ = \mathbf{p}^\circ \mathbf{\Lambda}^{\mathrm{T}} + \tfrac{1}{2} \mathbf{\Lambda} \mathbf{B}^\circ \mathbf{\Lambda}^{\mathrm{T}}, \\ w^\circ = \sum_{j=0}^{\infty} \mu^{j/2} w_j^\circ(\zeta_1, \zeta_2), \quad \vartheta^\circ = \sum_{j=0}^{\infty} \mu^{j/2} \vartheta_j^\circ(\zeta_1, \zeta_2), \\ \zeta_k = \mu^{-1/2} \alpha_k, \quad k = 1, 2, \end{gathered} \tag{9.5}$$

where $\mathbf{\Lambda} = (\alpha_1, \alpha_2)$, $\mathbf{p}^\circ = (p_1^\circ, p_2^\circ)$ are the 2D vectors, w_k°, ζ_k° are the polynomials of degrees M_k of the arguments ζ_1, ζ_2 with complex coefficients, $\mathbf{B}^\circ$ is the symmetric 2×2 complex matrix with the positively definite imaginary part $\Im \mathbf{B}^\circ$. The symmetry of the matrix $\mathbf{B}^\circ$ does not detract from the generality. The real and imaginary parts of functions (9.5) define on the shell surface a pair of two-dimensional WPs with the center at the point $(0, 0)$. The assumption of positive definiteness of $\Im \mathbf{B}^\circ$ guarantees the localization of the initial conditions in the neighborhood of the origin. In (9.5) and below, the T icon means the transposition.

Consider the problem on the shell response to the initial displacements and velocities given by relations (9.5). It is assumed that the center of the initial WP is located far from the shell boundaries. Therefore, the boundary conditions and the effects of reflection from the boundaries are not considered here. The solution to this problem without taking into account the membrane forces T_i was constructed in [152].

9.2 The algorithm for constructing two-dimensional WPs

The solution to problem (9.3), (9.5) will be sought in the form of a running two-dimensional WP with a center in the point $Q(q_1, q_2)$, where $q_1(t), q_2(t)$ are twice differentiable functions of time t such that $q_1(0) = q_2(0) = 0$. If the shell has edges, then it is assumed that they are far enough from the center of $(0,0)$, so that their influence can be neglected for some finite period of time. We pass to the new coordinate system with the origin in the moving point $Q(q_1, q_2)$:

$$\alpha_k = q_k(t) + \mu^{1/2}\xi_k, \quad k = 1, 2. \tag{9.6}$$

The ansatz of an asymptotic solution is assumed in the form:

$$\begin{aligned} W &= \sum_{j=0}^{\infty} \mu^{j/2} w_j\left(\xi_1, \xi_2, t\right) \exp\left[\mathrm{i}\mu^{-1} S\left(\xi_1, \xi_2, t, \mu\right)\right], \\ \Phi &= \sum_{j=0}^{\infty} \mu^{j/2} \Phi_j\left(\xi_1, \xi_2, t\right) \exp\left[\mathrm{i}\mu^{-1} S\left(\xi_1, \xi_2, t, \mu\right)\right] \end{aligned} \tag{9.7}$$

with

$$\begin{aligned} S(\xi_1, \xi_2, t, \mu) &= \int_0^t \omega(\tau) d\tau + \mu^{1/2} \mathbf{p}(t) \Xi + \tfrac{1}{2} \mu \Xi^{\mathrm{T}} \mathbf{B}(t) \Xi, \\ \mathbf{p}(t) &= \left(p_1(t), p_2(t)\right), \quad \Xi = \left(\xi_1, \xi_2\right)^{\mathrm{T}}. \end{aligned} \tag{9.8}$$

Here $\mathbf{B}(t)$ is a symmetric complex matrix with positively definite imaginary part $\Im \mathbf{B}(t)$ for any $t \geq 0$, $p_1(t), p_2(t)$ are real functions of time t, w_k, Φ_k are polynomials in ξ_1, ξ_2 with coefficients depending on t. It is assumed that all unknown functions contained in (9.7), (9.8), together with their derivatives, are quantities of order $O(1)$. To determine them, we substitute (9.6)-(9.8) into the governing Eqs. (9.3) previously expanding all the coefficients in (9.3) into the Taylor series in the neighborhood of the moving point $Q(q_1, q_2)$.

9.2.1 The leading approximation

Equating in Eqs. (9.3) coefficients of μ^{-2}, one obtains the homogeneous system of algebraic equations with respect to w_0, Φ_0:

$$\mathbf{L}_0 \mathbf{X}_0 = \mathbf{0}, \tag{9.9}$$

where $\mathbf{X}_0 = (w_0, \Phi_0)^{\mathrm{T}}$ is the two-component vector, and $\mathbf{L}_0$ is the (2×2)-matrix with the elements

$$l_{0,11} = d\left(\frac{p_1^2}{A_1^2} + \frac{p_2^2}{A_2^2}\right) - T_1\frac{p_1^2}{A_1^2} - 2T_3\frac{p_1}{A_1}\frac{p_2}{A_2}$$

$$-T_2\frac{p_2^2}{A_2^2} - \gamma\left(\omega - \dot{q}_1 p_1 - \dot{q}_2 p_2\right)^2, \tag{9.10}$$

$$l_{0,12} = -l_{0,21} = -\left(\frac{p_1^2}{A_1^2 R_2} + \frac{p_2^2}{A_2^2 R_1}\right), \quad l_{0,22} = g\left(\frac{p_1^2}{A_1^2} + \frac{p_2^2}{A_2^2}\right).$$

In (9.10) and below, the values of the functions $g, d, g, A_k, R_k, T_k, T_3 (k = 1, 2)$ are calculated at the point $Q(q_1, q_2)$. The condition for the existence of a non-trivial solution of Eqs. (9.9) gives

$$\omega = \dot{q}_1 p_1 + \dot{q}_2 p_2 \mp H\left(p_1, p_2, q_1, q_2, t\right) \tag{9.11}$$

with

$$H = \sqrt{\frac{d}{\gamma}\left(\frac{p_1^2}{A_1^2} + \frac{p_2^2}{A_2^2}\right)^2 + \frac{\left(\dfrac{p_1^2}{A_1^2 R_2} + \dfrac{p_2^2}{A_2^2 R_1}\right)^2}{\gamma g\left(\dfrac{p_1^2}{A_1^2} + \dfrac{p_2^2}{A_2^2}\right)^2} - \frac{N}{\gamma}}, \tag{9.12}$$

$$N(p_1, p_2, q_1, q_2, t) = T_1\frac{p_1^2}{A_1^2} + 2T_3\frac{p_1}{A_1}\frac{p_2}{A_2} + T_2\frac{p_2^2}{A_2^2}.$$

Here $H\left(p_1, p_2, q_1, q_2, t\right)$ is the Hamiltonian function. The function N depends on the membrane forces (in the general case, unsteady) arising in the middle surface of the shell. Obviously, the radicand in (9.12) is positive because, by assumption, the membrane forces do not reach their critical values.

Performing the asymptotic constructions, we will assume that $p_1^2(t) + p_2^2(t) \neq 0$ at some time interval $t \in [0, T]$.

Let $w_0 = P_0\left(\xi_1, \xi_2, t\right)$ be a polynomial in ξ_1, ξ_2. Then

$$\Phi_0 = \lambda P_0, \quad \lambda = -\frac{l_{0,11}}{l_{0,12}}. \tag{9.13}$$

9.2.2 The first-order approximation

In the first-order approximation (by $\mu^{-3/2}$), one arrives at the non-homogeneous system of algebraic equations

$$\mathbf{L}_0\mathbf{X}_1 = -\mathbf{L}_1\mathbf{X}_0, \tag{9.14}$$

where $\mathbf{X}_1 = (w_1, \Phi_1)^{\mathrm{T}}$, and $\mathbf{L}_1$ is the matrix whose elements $l_{1,kl}$ $(k, l = 1, 2)$ are the operators

$$l_{1,kl} = \left(\frac{\partial l_{0,kl}}{\partial \mathbf{p}} \mathbf{B} + \frac{\partial l_{0,kl}}{\partial \mathbf{q}} + \dot{\mathbf{p}} \frac{\partial l_{0,kl}}{\partial \omega} \right) \Xi - \mathrm{i} \frac{\partial l_{0,kl}}{\partial \mathbf{p}} \frac{\partial}{\partial \Xi}. \tag{9.15}$$

In (9.15) and below,

$$\frac{\partial}{\partial \mathbf{p}} = \left(\frac{\partial}{\partial p_1}, \frac{\partial}{\partial p_2} \right), \quad \frac{\partial}{\partial \mathbf{q}} = \left(\frac{\partial}{\partial q_1}, \frac{\partial}{\partial q_2} \right), \quad \frac{\partial}{\partial \Xi} = \left(\frac{\partial}{\partial \xi_1}, \frac{\partial}{\partial \xi_2} \right)^{\mathrm{T}} \tag{9.16}$$

are the vector operators, and $\mathbf{q} = (q_1, q_2)$.

The compatibility condition for the non-homogeneous Eqs. (9.14) results in a differential equation with respect to P_0. This equation has a polynomial solution if the vector functions $\mathbf{q}(t)$, $\mathbf{p}(t)$ satisfy the Hamiltonian system

$$\dot{\mathbf{q}} = \frac{\partial H}{\partial \mathbf{p}}, \quad \dot{\mathbf{p}} = -\frac{\partial H}{\partial \mathbf{q}}. \tag{9.17}$$

consisting of four scalar equations. Equations (9.17) and all further constructions correspond to the case when in (9.11) the Hamiltonian function is taken with a minus sign (the positive branch).

Comparing (9.5) and (9.6)–(9.8), we obtain the initial conditions for the Hamiltonian systems:

$$\mathbf{q}(0) = \mathbf{0}, \quad \mathbf{p}(0) = \mathbf{p}^\circ. \tag{9.18}$$

Let $\mathbf{q}(t)$, $\mathbf{p}(t)$ be a solution of the Cauchy problem (9.17)–(9.18) for the Hamiltonian system. Then the general solution of the inhomogeneous system (9.14) is as follows:

$$w_1 = P_1, \quad \Phi_1 = \lambda P_1 + \frac{\partial \lambda}{\partial \mathbf{p}} \mathbf{B}\, \Xi P_0 + \frac{\partial \lambda}{\partial \mathbf{q}} \Xi\, P_0 - \mathrm{i} \frac{\partial \lambda}{\partial \mathbf{p}} \frac{\partial P_0}{\partial \Xi}, \tag{9.19}$$

where $P_1(\xi_1, \xi_2, t)$ is unknown polynomial in ξ_1, ξ_2.

9.2.3 The second- and higher-order approximations

Consider the inhomogeneous system of algebraic equations

$$\mathbf{L}_0 \mathbf{X}_2 = -\mathbf{L}_1 \mathbf{X}_1 - \mathbf{L}_2 \mathbf{X}_0, \tag{9.20}$$

arising in the second-order approximation (by μ^{-1}). Here, $\mathbf{L}_2$ is the matrix with elements

$$l_{2,kl} = \frac{1}{2}\Xi\left(\mathbf{B}\frac{\partial^2 l_{0,kl}}{\partial \mathbf{p}^2}\mathbf{B} + 2\frac{\partial^2 l_{0,kl}}{\partial \mathbf{q}\partial \mathbf{p}}\mathbf{B} + \frac{\partial^2 l_{0,kl}}{\partial \mathbf{q}^2} + \dot{\mathbf{p}}^{\mathrm{T}}\frac{\partial^2 l_{0,kl}}{\partial \omega^2}\dot{\mathbf{p}} + 2\dot{\mathbf{p}}\frac{\partial^2 l_{0,kl}}{\partial \mathbf{q}\partial \omega}\right.$$

$$\left. + 2\dot{\mathbf{p}}\frac{\partial^2 l_{0,kl}}{\partial \mathbf{p}\partial \omega}\mathbf{B}\right)\Xi - \mathrm{i}\Xi^{\mathrm{T}}\left(\mathbf{B}\frac{\partial^2 l_{0,kl}}{\partial \mathbf{p}^2} + \frac{\partial^2 l_{0,kl}}{\partial \mathbf{q}\partial \mathbf{p}} + \dot{\mathbf{p}}\frac{\partial^2 l_{0,kl}}{\partial \mathbf{p}\partial \omega}\right)\frac{\partial}{\partial \Xi}$$

$$-\frac{\mathrm{i}}{2}\mathrm{tr}\left(\frac{\partial^2 l_{0,kl}}{\partial \mathbf{p}^2}\mathbf{B}\right) - \frac{1}{2}\mathrm{tr}\left(\frac{\partial^2 l_{0,kl}}{\partial \mathbf{p}^2}\frac{\partial^2}{\partial \Xi^2}\right) - \frac{\mathrm{i}}{2}\left(\dot{\omega}\frac{\partial^2 l_{0,kl}}{\partial \omega^2} + 2\dot{\mathbf{p}}^{\mathrm{T}}\frac{\partial^2 l_{0,kl}}{\partial \mathbf{p}\partial \omega}\right)$$

$$-\mathrm{i}\frac{\partial l_{0,kl}}{\partial \omega}\frac{\partial}{\partial t} + \frac{1}{2}\frac{\partial l_{0,kl}}{\partial \omega}\Xi^{\mathrm{T}}\dot{\mathbf{B}}\Xi + \mathrm{i}n_{kl}, \quad k,l = 1,2, \tag{9.21}$$

where

$$n_{11} = -\sum_k \left\{\frac{2}{A_1A_2}\left[\frac{\partial}{\partial \alpha_k}\left(\frac{dA_j}{A_k^3}\right)p_k^3 + \frac{\partial}{\partial \alpha_k}\left(\frac{d}{A_1A_2}\right)p_kp_j^2\right] + \gamma\ddot{q}_kp_k\right\},$$

$$n_{12} = \frac{1}{A_1A_2}\sum_k \frac{\partial}{\partial \alpha_k}\left(\frac{A_j}{A_kR_j}\right)p_k^3,$$

$$n_{22} = -\frac{2}{A_1A_2}\sum_k \left\{\frac{\partial}{\partial \alpha_k}\left(\frac{gA_j}{A_k^3}\right)p_k^3 + \frac{\partial}{\partial \alpha_k}\left(\frac{g}{A_kA_j}\right)p_kp_j^2\right\},$$

$$n_{21} = -n_{12}, \quad k,j = 1,2, \quad k \neq j \tag{9.22}$$

In (9.21) and below, the notation tr means the trace of a matrix, and $\frac{\partial^2}{\partial \mathbf{p}^2}, \frac{\partial^2}{\partial \mathbf{p}\partial \mathbf{q}}, \frac{\partial^2}{\partial \mathbf{q}^2}, \frac{\partial^2}{\partial \Xi^2}$ denote the 2×2 matrix operators. In particular,

$$\frac{\partial^2}{\partial \mathbf{p}^2} = \begin{pmatrix} \frac{\partial^2}{\partial p_1^2} & \frac{\partial^2}{\partial p_1\partial p_2} \\ \frac{\partial^2}{\partial p_2\partial p_1} & \frac{\partial^2}{\partial p_2^2} \end{pmatrix}, \quad \frac{\partial^2}{\partial \mathbf{p}\partial \mathbf{q}} = \begin{pmatrix} \frac{\partial^2}{\partial q_1\partial p_1} & \frac{\partial^2}{\partial q_1\partial p_2} \\ \frac{\partial^2}{\partial q_2\partial p_1} & \frac{\partial^2}{\partial q_2\partial p_2} \end{pmatrix}. \tag{9.23}$$

The condition for the existence of a solution of the non-homogeneous system of algebraic equations (9.20), taking into account the Hamilton system (9.17), leads to the Riccati matrix equation

$$\dot{\mathbf{B}} + \mathbf{B}H_{\mathbf{pp}}\mathbf{B} + H_{\mathbf{qp}}\mathbf{B} + \mathbf{B}H_{\mathbf{qp}}^{\mathrm{T}} + H_{\mathbf{qq}} = \mathbf{0}, \tag{9.24}$$

with respect to $\mathbf{B}$, where $H_{\mathbf{pp}}$, $H_{\mathbf{qp}}$, $H_{\mathbf{qq}}$ denote the matrices being obtained by the differentiation of the Hamiltonian H by the vectors $\mathbf{p}, \mathbf{q}$ in accordance with (9.23). Then the function P_0 can be found from the equation

$$-\frac{\mathrm{i}}{2}tr\left(H_{\mathbf{pp}}\frac{\partial^2 P_0}{\partial \Xi^2}\right)+\Xi\left(H_{\mathbf{qp}}\frac{\partial P_0}{\partial \Xi}\right)$$
$$+(\mathbf{B}\Xi)\left(H_{\mathbf{pp}}\frac{\partial P_0}{\partial \Xi}\right)+\frac{\partial P_0}{\partial t}+\mathbf{G}P_0=0, \tag{9.25}$$

where

$$\mathbf{G}=\frac{1}{2}tr(H_{\mathbf{pp}}\mathbf{B})-\frac{1}{2H}\dot{\omega}-\frac{1}{H}\left(H_{q_1}H_{p_1}+H_{q_2}H_{p_2}\right)$$
$$+\frac{l_{0,22}}{2\gamma H}\frac{\partial\lambda}{\partial\mathbf{p}}\frac{\partial\lambda}{\partial\mathbf{q}}+\frac{1}{2\gamma H l_{0,22}}\left(2l_{0,12}n_{21}-l_{0,11}n_{22}-l_{0,22}n_{11}\right). \tag{9.26}$$

Equation (9.25) is an analogue of the amplitude Eqs. (6.62), (8.58) derived when constructing solutions in the form of one-dimensional WPs.

Comparing (9.5) with (9.7), (9.8), one obtains the initial condition

$$\mathbf{B}(\mathbf{0})=\mathbf{B}^{\circ} \tag{9.27}$$

for Eq. (9.25). It can be proved [16] that on some interval $[0,T]$, problem (9.25), (9.27) has a unique and continuous solution $\mathbf{B}(t)$, such that $\Im\mathbf{B}(t)$ is the positively definite matrix for any $t\in[0,T]$.

Equation (9.25) has a solution in the form of a polynomial

$$P_0=\sum_{s=0}^{N}\sum_{k+l=s}A_{kl}(t;c_j)\xi_1^k\xi_2^l \tag{9.28}$$

of any degree $N\geq 0$ with coefficients A_{kl} containing unknown integration constants c_j. A number N depend on the degrees M_k of the polynomials $w_k^{\circ},\zeta_k^{\circ}$ appearing in the initial conditions (9.5).

The compatibility conditions for inhomogeneous algebraic equations arising in higher approximations (by $\mu^{-1+k/2}$, where $k\geq 1$) lead to inhomogeneous differential equations for the functions $P_k(\xi_1,\xi_2,t)$ the left-hand side of which coincide with the left-hand sides of the amplitude equation (9.25), and the right-hand sides are some known polynomials in ξ_1,ξ_2.

9.2.4 Solution of the initial problem (9.3), (9.5)

Consider the linear combination

$$\mathbf{X}=\left[\mathbf{X}_0^+\left(\xi_j^+,t;c_j^+\right)+O\left(\mu^{1/2}\right)\right]\exp\{\mathrm{i}\mu^{-1}S^+\left(\xi_j^+,t,\mu\right)\}$$
$$+\left[\mathbf{X}_0^-\left(\xi_j^-,t;c_j^-\right)+O\left(\mu^{1/2}\right)\right]\exp\{\mathrm{i}\mu^{-1}S^-\left(\xi_j^-,t,\mu\right)\}, \tag{9.29}$$

where

$$\mathbf{X} = (w, \Phi)^{\mathrm{T}}, \quad \mathbf{X}_0^{\pm} = P_0^{\pm}\left(\xi_j^{\pm}, t; c_j^{\pm}\right)\left(1, \lambda^{\pm}\right)^{\mathrm{T}}$$

$$S^{\pm} = \int_0^t \omega^{\pm}(\tau)d\tau + \mu^{1/2}\mathbf{p}^{\pm}(t)\Xi^{\pm} + \tfrac{1}{2}\mu(\Xi^{\pm})^{\mathrm{T}}\mathbf{B}^{\pm}(t)\Xi^{\pm}, \tag{9.30}$$

$$\Xi^{\pm} = \left(\xi_1^{\pm}, \xi_2^{\pm}\right)^{\mathrm{T}}, \quad \xi_j^{\pm} = \mu^{-1/2}\left[\alpha_j - q_j^{\pm}(t)\right].$$

Here, as above, the superscripts $-$ and $+$ correspond to the negative and positive branches of the constructed solutions. The function $\lambda^{\pm}(t)$ is defined by formula (9.13) for $\mathbf{q} = \mathbf{q}^{\pm}(t)$, $\mathbf{p} = \mathbf{p}^{\pm}(t)$. The vector function (9.29) is the formal asymptotic solution (the leading approximation) of Eq. (9.3). Indefinite constants $c_j^{\pm}$ entering into it can be found from the initial conditions. The substitution of (9.29), (9.30) into (9.5) leads to the system of equations

$$\left(P_0^{+} + P_0^{-}\right)|_{t=0} = w^{\circ}(\zeta_1, \zeta_2), \quad \left(P_0^{+} - P_0^{-}\right)|_{t=0} = -\frac{\vartheta^{\circ}(\zeta_1, \zeta_2)}{H^{\circ}}, \tag{9.31}$$

where $H^{\circ} = H\left(p_1^{\circ}, p_2^{\circ}, 0, 0\right)$, and $w^{\circ}, \vartheta^{\circ}$ are the known polynomials of degree M_0. From the identity $\xi_j^{\pm} \equiv \zeta_j$, which holds for $t = 0$, it follows that the left-hand sides of the above equations are also the M_0-degree polynomials of the arguments ζ_1, ζ_2. Equating the coefficients by the same degrees of $z_1^n \zeta_2^m$, we get a system consisting of $(M_0 + 1)(M_0 + 2)$ algebraic equations with respect to complex constants $c_j^{\pm}$, where $j = 0, 1, \ldots \frac{1}{2}\left(M_0 + 1\right)\left(M_0 + 2\right)$.

If the matrix $\Im\mathbf{B}$ is positively definite, the level lines of the function $(\Xi)^{\mathrm{T}}\mathbf{B}(t)\Xi$ are concentric ellipses centered at the moving point $Q = (q_1(t), q_2(t))$. Thus, the constructed asymptotic solution (9.29) represents a pair of WPs with diameters of order

$$\sim \mu^{1/2}\left\{\mathrm{tr}(\mathrm{Im}\,\mathbf{B}^{\pm}) - \sqrt{(\mathrm{tr}(\mathrm{Im}\,\mathbf{B}^{\pm}))^2 - 4\mathrm{det}(\mathrm{Im}\,\mathbf{B}^{\pm})}\right\}^{-1/2} \tag{9.32}$$

and centers in the points $\left(q_1^{\pm}(t), q_2^{\pm}(t)\right)$, which propagate with the group velocities $\mathbf{v}_g^{\pm} = \left(\pm H_{p_1}^{\pm}, \pm H_{p_2}^{\pm}\right)$.

9.3 Effect of shell geometry on dynamic characteristics of two-dimensional WPs

Consider examples illustrating the influence of the shell geometry on two-dimensional WPs.

Example 9.1. Let a shell have the shape of a paraboloid of revolution. Its middle surface obtained by rotating the curve $r = \sqrt{z}$ around the Oz-axis is shown in Fig. 9.1. Let $\alpha_1 = z - 1$, $\alpha_2 = \theta$ be the curvilinear coordinate at

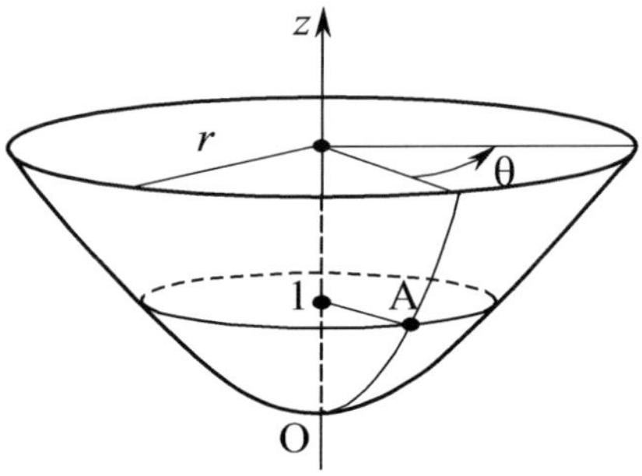

FIGURE 9.1: Paraboloid of revolution.

the middle surface as shown in Fig. 9.1. We place the initial WP at the point A with the coordinates $\alpha_1 = 0, \alpha_2 = 0$ and consider the following functions and parameters from (9.3), (9.5):

$$d = g = \gamma = 1, \quad p_2(0) = p^\circ = 1, \quad \mathbf{B}^\circ = \begin{pmatrix} 2\mathrm{i} & 1 \\ 1 & 3\mathrm{i} \end{pmatrix}. \tag{9.33}$$

The shell is free of an external load.

As it turned out, in this case, the wave parameter p_1° characterizing the variability of the initial WP in the direction of the coordinate α_1 has a decisive influence on the nature of the WP propagation. Calculations were performed for three different values, $p_1^\circ = -1, 0, 1$. As shown in Fig. 9.2, at $p_1^\circ = 0$ or $p_1^\circ = 1$ the WP moves away from the top of the paraboloid with increasing the wave parameter $p_1(t)$. For $p_1^\circ = -1$, the trajectory of the WP is more complicated: first, it is observed a movement to the paraboloid top with a decrease in the parameter $p_1(t)$, then there is the reflection from some parallel, its subsequent movement is similar to the WP at $p_1^\circ = 0$ or $p_1^\circ = 1$. The detected reflection is explained by the fact that the region near the top of the shell is more rigid, since the curvature increases as one approaches the top. The movement in the circular direction is unlimited with a constant number $p_2 = p_2^\circ = 1$ for all three cases. Numerical integration of the Riccati equation as well as the amplitude equation shows that for all the considered values of p_1°, there is spreading out the wave packets with a decrease in the wave amplitude.

Example 9.2. Consider the paraboloid of revolution from the previous example, whose thickness changes with distance from its top according to the law $h(z) = (1 + z^2)$. The coordinate system here is the same as above.

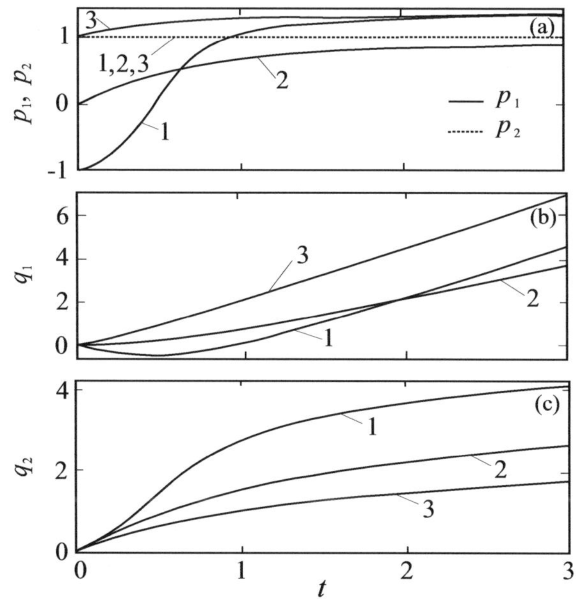

FIGURE 9.2: Solution of the Hamiltonian system for $p_1^\circ = -1, 0, 1$ (curves 1, 2, 3, respectively) for the shell in the form of a paraboloid of revolution with constant thickness.

The center of the initial WP is in the point A (see Fig. 9.1). The calculations were performed for

$$\begin{gathered} d(z) = \frac{h^3(z)}{1-\nu^2}, \quad g(z) = \frac{1}{h(z)}, \quad m(z) = h(z), \quad T_i = 0, \quad \nu = 0.3, \\ p_2(0) = p_2^\circ = 2, \quad \mathbf{B}^\circ = \begin{pmatrix} 1+2\mathrm{i} & 0 \\ 0 & 1+2\mathrm{i} \end{pmatrix}, \quad w^\circ = 1, \quad v^\circ = 0 \end{gathered} \tag{9.34}$$

and different values of $p_1^\circ = -1, 0, 1$.

In Fig. 9.3, solutions of the Hamiltonian system for different p_1° are depicted. It is seen that in the case $p_1^\circ = 0$ the parameters q_1, p_1, p_2 remain constant, and the circumferential coordinate q_2 is variable.

Figure 9.4 demonstrates the projections of the trajectory of the WP center on the plane orthogonal to the axis of rotation of the shell. Point O marks the paraboloid top. The line segment corresponds to the projection of the meridian $\theta = 0$, and the outer circle does to the projections of the paraboloid points with coordinates $z = 2$. In all three cases, the initial "energy" of the packets is not large enough, so that the WP does not go beyond the region defined by the inequality $z < 2$. Moreover, in the case $p_1^\circ = 0$, the WP runs in a circle, and for $p_1^\circ = -1, 1$, it is observed multiple reflections of packets from

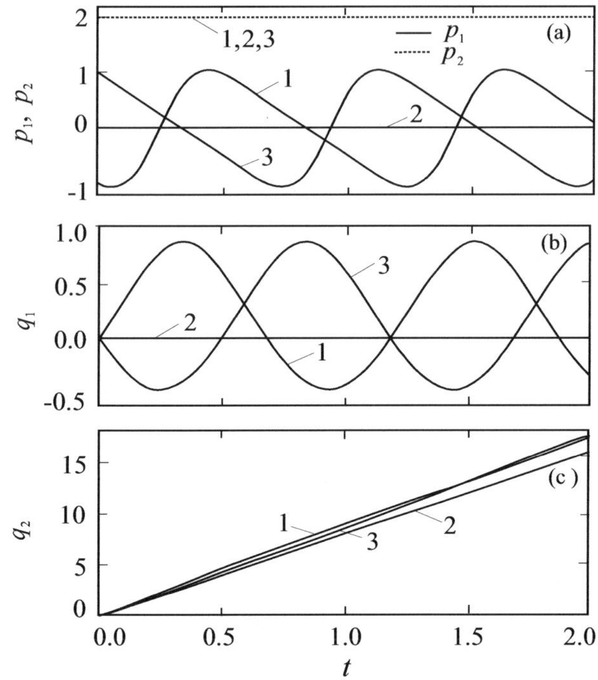

FIGURE 9.3: Solution of the Hamiltonian system for $p_1^\circ = -1,\ 0,\ 1$ (curves 1, 2, 3, respectively) for the shell in the form of a paraboloid of revolution with variable thickness.

some parallels. These reflections, on the one hand, are caused by increasing the paraboloid curvature as one approaches its apex, and on the other hand, by increasing the shell thickness with distance from the apex.

The plots of the instantaneous frequency ω of travelling vibrations, the parameter B_w characterizing the WP width as well as the maximum amplitude w_{max} of waves versus dimensionless time t for different values of $p_1^\circ = -1,\ 0,\ 1$ are shown in Fig. 9.5. For $p_1^\circ = 0$, the instantaneous frequency $\omega \equiv 8$ (its plot is not shown in Fig. 9.5). In this case, the WP width increases uniformly, and the waves amplitude decreases. For $p_1^\circ = -1$ and $p_1^\circ = 1$, at the moments of reflection of the WPs, it is observed noticeable oscillations of both the packets width and the waves amplitude.

It follows from the considered examples that the nature of unsteady bending vibrations of thin shells caused by the initial local perturbations (in the form of normal displacements and velocities) largely depends on the shell geometry, particularly, on changing the curvatures and the thickness. We also point to the example considered in paper [156], where two-dimensional packets of bending waves in a cylindrical shell with variable parameters were studied.

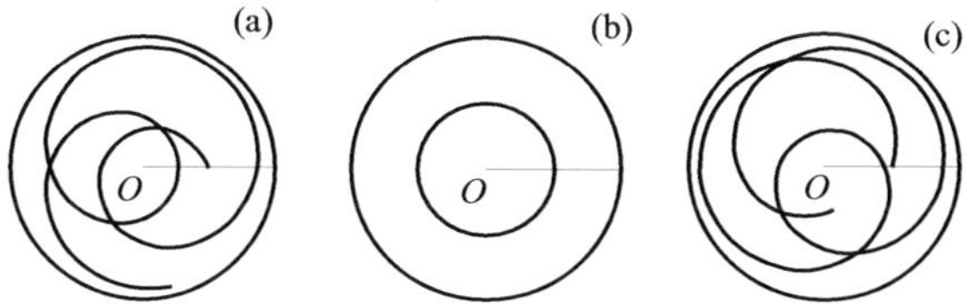

FIGURE 9.4: Trajectories of the WP center for $p_1^\circ = -1$ (a), $p_1^\circ = 0$ (b), $p_1^\circ = 1$ (c).

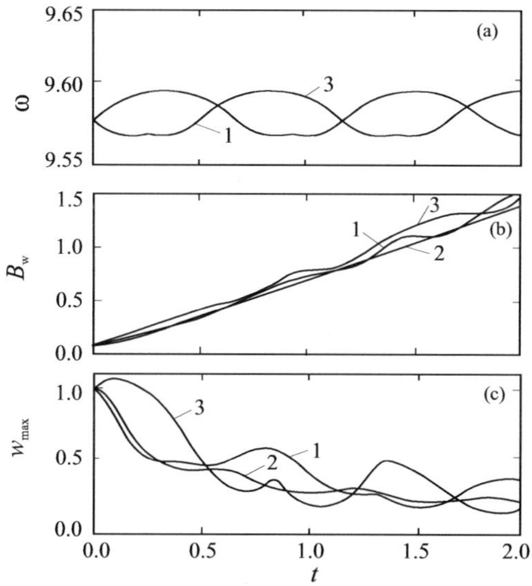

FIGURE 9.5: Plots of the functions $\omega(t)$ (a), $B_w(t)$ (b), $w_{max}(t)$ (c) at $p_1^\circ = -1, 0, 1$ (curves 1, 2, 3, respectively) for the shell in the form of a paraboloid of revolution with variable thickness.

It is revealed that the center of a two-dimensional WP moves along the cylindrical surface in a spiral, in some cases, the effects of reflection of WPs from regions characterized by high bending stiffness being possible.

Bibliography

[1] M. Abramovitz and I.A. Stigun. *Handbook of mathematical functions with formulas, graphs and mathematical tables.* Dover Publications, New York, 1972.

[2] A. A. Al-Jawi, A. G. Ulsoy, and C. Pierre. Vibration localization in dual-span, axially moving beams. *Journal of Sound and Vibration*, 179(2):243–266 (part I), 267–287 (part II), 1995.

[3] N. A. Alfutov. *Fundamentals of Calculation for the Stability of Elastic Systems (in Russ.).* Mashinostroenie, Moscow, 1978.

[4] H. Altenbach. Theories for laminated and sandwich plates. A review. *Mechanics of Composite Materials*, 34(3):243–252, 1998.

[5] H. Altenbach, J. Altenbach, and W. Kissing. *Mechanics of Composite Structural Elements.* Springer, Berlin, Heidelberg, 2nd edition, 2018.

[6] N. A. Alumyae. On the general solution of equations of small axisymmetric stable vibrations of an elastic conic shell of revolution (in Russ.). *Proc. Estonian Acad. Sci.*, 10(1):3–15, 1960.

[7] M. Amabili. A new third-order shear deformation theory with nonlinearities in shear for static and dynamic analysis of laminated doubly curved shells. *Composite Structures*, 128:260 – 273, 2015.

[8] S.A. Ambarcumyan. *Theory of Anisotropic Plates.* Technomic Publishing, Stamford, 1970.

[9] S.A. Ambarcumyan. *Theory of Anisotropic Plates: Strength, Stability, and Vibrations.* Hemispher Publishing, Washington, 1991.

[10] P.W. Anderson. Absence of diffusion in certain random lattices. *Phys. Rev.*, 109(5):1492, 1958.

[11] I.V. Andrianov and J. Awrejcewicz. Edge-localized effects in buckling and vibrations of a shell with free in circumferential direction ends. *Acta Mechanica*, 173:41–47, 2004.

[12] H. Aron. Das Gleichgewicht und die Bewegung einer unendlich dünnen, beliebig gekrümmten elastischen Schale. *Journal für die reine und angewandte Mathematik*, 78:136–174, 1874.

[13] A.G. Aslanyan and V.B. Lidsky. *Distribution of Free Frequencies of Thin Elastic Shells (in Russ.)*. Nauka, Moscow, 1974.

[14] B.A. Auld and E.M. Tsao. Variational analysis of edge resonance in a semi-infinite plate. *IEEE Transactions on Sonics and Ultrasonics*, 24:317–326, 1977.

[15] I.V. Avdoshka and G.I. Mikhasev. Wave packets in a thin cylindrical shell under a non-uniform axial load. *Journal of Applied Mathematics and Mechanics*, 65(2):301–309, 2001.

[16] V. M. Babich, V. S. Buldyrev, and I.A. Molotkov. *Space-time ray method. Linear and Non-linear Waves (in Russ.)*. St. Petersburg Univ. Press, St. Petersburg, 1985.

[17] V.M. Babich and V.S. Buldyrev. *Short-wavelength Diffraction Theory: Asymptotic Methods (English transl.)*. Springer, Berlin, New York, 1991.

[18] V.M. Babich and A.P. Kiselev. *Elastic Waves: High Frequency Theory*. CRC Press, Boca Raton, London, New York, 2018.

[19] V.M. Babich and V.V. Ulin. Complex space-time ray method and quasi-fotons. *Zap. Nauchn. Sem. LOMI*, 117:5–12, 1981.

[20] R.A. Bagdasaryan, M.V. Belubekyan, and K. B Kazaryan. Rayleigh type waves in a semi-infinite closed cylindrical shell. In A. I. Vesnitskii and V. I. Erofeev, editors, *Wave Problems in Mechanics (In Russ.)*, pages 87–93. Nijnii Novgorod Branch of Mechanical Engineering Institute, Nijnii Novgorod, 1992.

[21] M. Baruch. Parametric instability of stiffened cylindrical shells. *Israel J. Tech.*, 7:297–301, 1969.

[22] S. M. Bauer, S. B. Filippov, A.L. Smirnov, and P. E. Tovstik. *Asymptotic Methods in the Mechanics of Thin-walled Structures (In Russ.)*. St. Petersburg University Press, St. Petersburg, 1995.

[23] S. M. Bauer, S. B. Filippov, A.L. Smirnov, and P. E. Tovstik. *Asymptotic Methods in Examples and Problems (In Russ.)*. St. Petersburg University Press, St. Petersburg, 1997.

[24] S. M. Bauer, S. B. Filippov, A.L. Smirnov, and P. E. Tovstik. *Asymptotic Methods in Mechanics of Solids (in Russ.)*. Regulyarnaya i khaotich-eskaya dinamika, Moscow, Izhevsk, 2007.

[25] S. M. Bauer, S. B. Filippov, A.L. Smirnov, P. E. Tovstik, and Vaillancourt. *Asymptotic methods in mechanics of solids*, volume 167 of *International Series of Numerical Mathematics*. Birkhauser Basel, Switzerland, 2015.

[26] S. V. Bauer, S. B. Filippov, B. N. Semenov, P. E. Tovstik, and Y. L. Vorokhovsky. The effect of the temperature on the vibration frequencies of plane-convex plate of quartz resonators. In *Proc.1992 IEEE Frequency Control Symp. Proe.*, pages 128–134, Pensilvania, USA, 1992. Springer.

[27] C. W. Bert. Simplified analysis of static shear correction factors for beams of nonhomogeneous cross section. *Journal of Composite Materials*, 7:525–529, 1973.

[28] C.W. Bert and V. Birman. Parametric instability of thick, orthotropic, circular cylindrical shells. *Acta Mechanica*, 71:61–76, 1988.

[29] I. A. Birger and Ya. G. Panovko. *Strength. Stability. Vibrations: Directory in 3 vol (in Russ.)*. Mashinostroenie, Moskow, 1978.

[30] V. Birman and C.W. Bert. Dynamic stability of reinforced composite cylindrical shells in thermal field. *Journal of Sound and Vibration*, 142(2):183–190, 1990.

[31] A.E. Bogdanovich. Dynamic stability of a viscoelastic orthotropic cylindrical shell. *Polymer Mechanics*, 9(4):626–632, 1973.

[32] A.E. Bogdanovich. Dynamic stability of an elastic orthotropic cylindrical shell with allowance for transverse shears. *Polymer Mechanics*, 9(2):268–274, 1973.

[33] A.E. Bogdanovich and E.G. Feldmane. Nonlinear parametric vibrations of viscoelastic orthotropic cylindrical shells. *Sov. Appl. Mech.*, 16(4):305–309, 1980.

[34] V. V. Bolotin. *Dynamic Stability of Elastic Systems (in Russ.)*. Gostekhizdat, Moscow, 1956.

[35] V. V. Bolotin. On the density of natural frequencies of thin elastic shells. *Appl. Math. Mech.*, 27(2):362–364, 1963.

[36] A.A. Bondarenko and P.I. Galaka. Parametric instability of glass-plastic cylindrical shells. *Sov. Appl. Mech.*, 13(4):411–414, 1977.

[37] M.G. Botogova and G.I. Mikhasev. Free vibrations of non-uniformly heated viscoelastic cylindrical shell. *Technische Mechanik*, 16(3):251–257, 1996.

[38] M. Bresse. *Cours de Mechanique Applique. P.1.* Mallet-Bachelier Imprimeur-Libraire du Bureau des Longitudes, Paris, 1859.

[39] M.F. Caliri, A.J. Ferreira, and V. Tita. A review on plate and shell theories for laminated and sandwich structures highlighting the finite element method. *Composite Structures*, 156:63–77, 2016.

[40] E. Carrera. Theories and finite elements for multilayered, anisotropic, composite plates and shells. *Arch Comput Methods Eng*, 9(2):87–140, 2002.

[41] E. Carrera. Historical review of zig-zag theories for multilayered plates and shells. *Appl Mech Rev*, 56(3):287–308, 2003.

[42] A.P. Chattopadhyay and R.C. Batra. Free and forced vibrations of monolithic and composite rectangular plates with interior constrained points. *Journal of Vibration and Acoustics*, 141(1):011018–1– 011018–15, 2019.

[43] V. N. Chelomei. *Dynamic Stability of Elements of Avia-structures (in Russ.)*. Aeroflot Press, Moscow, 1939.

[44] K. F. Chernykh. *Linear Theory of Shells. Part 1. (in Russ.)*. Leningrad University Press, Leningrad, 1962.

[45] K. F. Chernykh. *Linear Theory of Shells. Part 2. (in Russ.)*. Leningrad University Press, Leningrad, 1964.

[46] T. S. Chow. On the propagation of flexural waves in an orthotropic laminated plate and its response to an impulse load. *Journal of Composite Materials*, 5:306–319, 1971.

[47] H. Cohen and A.B. Berkal. Wave propagation in conical shells . *Trans. ASME*, E 39(4):1166–1168, 1972.

[48] P. J. Cornwell and O. Bendiksen. Localization of vibrations in large space reflectors. *AIAA Journal*, 27(2):219–226, 1989.

[49] R.V. Craster and J. Kaplunov (eds). *Dynamic Localization Phenomena in Elasticity, Acoustics and Electromagnetism.* Springer, Wien, Heidelberg, New York, Dordrecht, London, 2013.

[50] L. H. Donnell. *Beams, Plates and Shells.* McGraw-Hill, Inc New York, New York, 1976.

[51] V.A. Eremeyev and L.M. Zubov. *Mechanics of Elastic Shells (In Russ.)*. Nauka, Moscow, 2008.

[52] Z. G. Ershova and P. E. Tovstik. To the problem of free vibrations of a cylindrical panel (in Russ.). *Vestnik St. Petersburg Univ. Mathematics*, (4):82–89, 1998.

[53] Z. G. Ershova and P. E. Tovstik. A cylindrical plate with a weakly fixed curvilinear edge made of a transversely isotropic material (English transl.). *Vestnik St.Petersburg. Univ. Mathematics*, (1):27–37, 2011.

[54] Z. G. Ershova and P.E. Tovstik. Vibrations and stability of cylindrical panels with a weakly supported rectilinear edge (in Russ.). *Applied Mechanics. Dynamics and stability of mechanical systems*, pages 215–222, 1995.

[55] S. B. Filippov. *Theory of Coupled and Supported Shells (in Russ.)*. St. Petersburg Univ. Press, St. Petersburg, 1999.

[56] S. B. Filippov and Naumova N. V. Axisymmetric vibrations of thin shells of revolution joint by a small angle. *Technische Mechanik*, 18(4):285–291, 1998.

[57] W. Flügge. *Statik und Dynamik der Schalen (in Germ.)*. Springer, Berlin, 1934.

[58] D.G. Gazis and R.D. Mindlin. Extensional vibrations and waves in a circular disk and a semi-infinite plate. *Journal of Applied Mechanics*, 27:541–547, 1960.

[59] I.P. Gertman and O.N. Lisitskii. Reflection and transmission of sound waves through the interfacial boundary of two joined elastic half-strips(in Russ.). *Journal of Appl. Math. And Mech. 1*, 52(6):1044–1048, 1988.

[60] G.R. Ghulgazaryan, Melqonyan A.T., and Sahakyan S.A. About free interfacial and boundary vibrations of thin elastic plate-strip and rectangular plate with free opposite edges (in Russ.). *Armenian State pedagogical university after Kh. Abovyan. Scientific News.*, 2-3(24-25):54–68, 2015.

[61] G.R. Ghulghazaryan, L.G. Ghulghazaryan, and G.I. Mikhasev. Free interfacial and boundary vibrations of thin elastic circular cylindrical shells with free ends (in Russ.). *Proceedings of the National Academy of Sciences of Armenia. Mechanics*, 71(1):61–77, 2018.

[62] G.R. Ghulghazaryan, R.G. Ghulghazaryan, and Mikhasev G.I. On free boundary and interfacial vibrations of thin elastic semi-infinite plates with free end (in Russ.). *Mechanics of Machines, Mechanisms and Materials*, 2(31):29–36, 2015.

[63] A. L. Gol'denveizer. Classification of integrals of the dynamic equations of the linear two-dimensional theory of shells. *Appl. Math. & Mech.*, 37(4):591–603, 1973.

[64] A. L. Gol'denveizer, V. B. Lidsky, and P. E. Tovstik. *Free Vibrations of Thin Elastik Shells(in Russ.)*. Nauka, Moscow, 1979.

[65] A. L. Gol'denveizer. *Theory of Thin Elastic Shells*. International Series of Monograph in Aeronautics and Astronautics. Pergamon Press, New York, 1961.

[66] A. L. Gol'denveizer. Qualitative analysis of free vibrations of an elastic thin shell (in Russ.). *Appl. Math. Mech*, 30(1):94–108, 1966.

[67] A. L. Gol'denveizer. *Theory of Thin Elastic Shells (in Russ.)*. Nauka, Moscow, 1976.

[68] A. L. Gol'denveizer. Bending of surfaces and the lowest natural frequencies of thin shells (in Russ.). *Mechanika tverdogo tela*, pages 106–117, 1977.

[69] A. L. Gol'denveizer. On the geometric theory of shells stability (in Russ.). *Mechanika Tverdogo Tela*, (6):143–154, 1983.

[70] A.L. Goldenveizer, J.D. Kaplunov, and E.V. Nolde. On Timoshenko-Reissner type theories of plates and shells. *International Journal of Solids and Structures*, 30(5):675–694, 1993.

[71] V. T. Golovchan, V. D. Kubenko, N. A. Shul'ga, A. N. Guz', and V. T. Grinchenko. *Dynamics of Elastic Bodies*, volume 5 of *Space Problems of Theory Elasticity and Plasticity*. Naukova Dumka, 1986.

[72] D. Gorman. An analytical solution for the free vibration analysis of rectangular plates resting on symmetrically distributed point supports. *J. Sound Vibr*, 79(4):561–574, 1981.

[73] D. Gorman and R. Singal. Analytical and experimental study of vibrating rectangular plates on rigid point supports. *AIAA J*, 29(5):838–844, 1991.

[74] A. G. Gorshkov, E. I. Starovoitov, and A. B. Yarovaya. *Mechanics of laminated viscoelastoplastic elements of constructions (in Russ.)*. Fizmatlit, Moscow, 2005.

[75] F.W. Grashof. Fairbairns Versuche tiber den Widerstand von Rohren gegen Zusammendruckung. *VDI-Zeitschrift*, 3(8–9):234–243, 18590.

[76] R.D. Gregory and I. Gladwell. The reflection of a symmetric Rayleigh-Lamb wave at the fixed or free edge of a plate. *Journal of Elasticity*, 13(2):185–206, 1983.

[77] E. I. Grigoluk and V.V. Kabanov. *Stability of Shells (in Russ.)*. Nauka, Moscov, 1978.

[78] E. I. Grigolyuk and A. G. Gorshkov. *Non-stationary Hydro-elasticity of Shells (in Russ.)*. Sudostroenie, Leningrad, 1974.

[79] E. I. Grigolyuk and G. M. Kulikov. General direction of development of the theory of multilayered shells. *Mechanics of Composite Materials*, 24(2):231–241, 1988.

[80] E.I. Grigolyuk and Kulikov G.M. *Multilayer Reinforced Shells: Calculation of Pneumatic Tires (in Russ.).* Mashinostroenie, Moscow, 1988.

[81] V.T. Grinchenko and M.M. Meleshko. On the resonance for a semi-infinite strip. *Prikladnaya Mekhanika*, 16(2):58–63, 1980 [English translation: Sov. Appl. Mech.].

[82] G.R. Gulgazaryan, L.G. Gulgazaryan, I.A. Miklashevich, and A.A. Pletezhov. Free interfacial vibrations of momentless infinite cylindrical shells with arbitrary smooth directing curve (in Russ.). *Vestnik of the Foundation for Fundamental Research*, 1(1):59–80, 2012.

[83] G.R. Gulgazaryan, L.G. Gulgazaryan, and R.D. Saakyan. The vibrations of a thin elastic orthotropic circular cylindrical shell with free and hinged edges (In Russ.). *Journal of Applied Mathematics and Mechanics*, 72:312–322, 2008.

[84] G.R. Gulgazaryan and R.G. Gulgazaryan. About free interfacial vibrations of thin elastic circular cylindrical shells (in Russ.). *Mechanics of machines, mechanisms and materials*, 4(25):12–19, 2013.

[85] A.N. Guz and V.D. Kubenko. *Theory of Nonstationary Aerohydroelasticity of Shells(in Russ.).* Naukova Dumka, Kiev, 1982.

[86] H. Hencky. Über die Berücksichtigung der Schubverzerrung in ebenen Platten. *Ingenieur-Archiv*, 16(1):72–76, 1947.

[87] C. H. Hodges. Confinement of vibration by structural irregularity. *Journal of Sound and Vibration*, 82:411–424, 1982.

[88] C.H. Hodges and J. Woodhouse. Vibration isolation from irregularity in a nearly periodic structure: Theory and measurements. *Journal of Acoustical Society of America*, 79(9):894–905, 1983.

[89] M. A. Il'gamov, V. A. Ivanov, and B. V. Gulin. *Stgength, Stability and Dynamics of Shells with an Elastic Filling (in Russ.).* Nauka, Moscow, 1977.

[90] E. Ince. Research into the characteristic numbers of Mathieu equation. *Proc. R. Soc. Edinburgh*, 46:20–29, 1927.

[91] A.Yu. Ishlinskii. On a limiting process in the theory of the stability of elastic rectangular plates (In Russ.). *Dokl. Akad. Nauk SSSR*, 95(3):477–479, 1954.

[92] A. Yu Ishlinsky. About one limiting transition in the theory of stability of rectangular plates (in Russ.). *Doklady Acad. Sci*, 95(5):129–132, 1954.

[93] D. N. Ivanov. Vibrations of cylindrical shells with identical radii (in Russ.). *Vestnik St. Petersburg Univ. Ser. 1. Mathematics. Mechanics. Astronomy.*, (1):111–116, 2000.

[94] D. N. Ivanov and S. B. Filippov. Vibrations of a cylindrical shell with a curvilinear edge (in Russ.). *Vestnik St. Petersburg Univ. Ser. 1. Mathematics. Mechanics. Astronomy.*, (4):95–100, 1998.

[95] E. Kamke. *Differentialgleichungen: Lösungsmetoden und Lösungen.* Verbesserte Auflage, Leipzig, 1959.

[96] R. K. Kapania. A review on the analysis of laminated shells. *J Press Vessel Technol*, 111(2):88–96, 1989.

[97] R. K. Kapania and S. Raciti. Recent advances in analysis of laminated beams and plates. Part I - Shear effects and buckling. *AIAA J*, 27(7):923–934, 1989.

[98] R. K. Kapania and S. Raciti. Recent advances in analysis of laminated beams and plates. Part II: Vibrations and Wave Propagation. *AIAA J*, 27(7):935–946, 1989.

[99] J. D. Kaplunov, G. A. Rogerson, and P.E. Tovstik. Localized vibration in elastic structures with slowly varying thickness. *QJ Mechanics Appl. Math.*, 58(4):645–664, 2005.

[100] J.D. Kaplunov, L.Yu. Kossovich, and E.V. Nolde. *Dynamics of Thin Walled Elastic Bodies.* Academic Press, San Diego, 1998.

[101] J.D. Kaplunov, L.Yu. Kossovich, and M.V. Wilde. Free localized vibrations of a semi-infinite cylindrical shell. *The Journal of the Acoustical Society of America*, 107(3):1383–1393, 2000.

[102] J.D. Kaplunov and M. V. Wilde. Resonances of Rayleigh-type waves in a semi-infinite strip (in Russ.). *Acustichesky Zhurnal*, 49(1):38–42, 2003.

[103] J.D. Kaplunov and M.V. Wilde. Edge and interfacial vibrations in elastic shells of revolution. *Zeitschrift fur Angewandte Mathematik und Physik*, 51:530–549, 2000.

[104] J.D. Kaplunov and M.V. Wilde. Free interfacial vibrations in cylindrical shells. *JASA*, 111(6):2692–2704, 2002.

[105] G.R. Kirchhoff. Über das Gleichgewicht und die Bewegung einer elastischen Scheibe . *Crelles Journal für die reine und angewandte Mathematik*, 40:51–88, 1850.

[106] A.P. Kiselev and M.V. Perel. Gaussian wave packets. *Opt. Spectrosc.*, 86(3):307–309, 1999.

[107] A.P. Kiselev and M.V. Perel. On the nature of quasiphotons. *J. Math. Sci.*, 96(4):3355–3358, 1999.

[108] A.P. Kiselev and M.V. Perel. Highly localized solutions of the wave equation. *J. Math. Phys.*, 41(4):1934–1955, 2000.

[109] R. Kochurov and K.V. Avramov. On effect of initial imperfections on parametric vibrations of cylindrical shells with geometrical nonlinearity. *International Journal of Solids and Structures*, 49(3):537–545, 2012.

[110] W. T. Koiter. On the nonlinear theory of thin elastic shells. *Proc. Koninkl. Acad.*, 69:1–54, 1966.

[111] Y.K. Konenkov. Rayleigh-type flexural wave. *Soviet Physics Acoustics*, 6:122–123, 1960.

[112] L.Yu. Kossovich and A. Parfenova. Flexural transient waves in shells of revolution: An asymptotic approach. *Zeitschrift für angewandte Mathematik und Physik ZAMP*, 51(4):611–628, 2000.

[113] L.R. Kumar, P.K. Datta, and D.L. Prabhakara. Tension buckling and parametric instability characteristics of doubly curved panels with circular cutout subjected to nonuniform tensile edge loading. *Thin-Walled Structures*, 42(7):947–962, 2004.

[114] S.P. Kuntsevich and G.I. Mikhasev. Local parametric vibrations of a noncircular conical shell subjected to nonuniform pulsating pressure. *Mechanics of Solids*, 37(3):134–139, 1996.

[115] S.P. Kuntsevich and G.I. Mikhasev. Parametric vibrations of viscoelastic cylindrical shell under static and periodic axial loads. *Technische Mechanik*, 19(3):187—195, 1999.

[116] D. Kurant, R. abd Gil'bert. *Methods of Mathematical Physics*, volume 1. Wiley, New York, 1966.

[117] R. E. Langer. The asymptotic solutions of ordinary linear differential equations of second order with special reference to a turning point. *Trans. Amer. Math. Soc.*, 67:461–490, 1949.

[118] J. Lawrie, J.B. Kaplunov. Edge waves and resonance on elastic structures: An overview. *Mathematics and Mechanics of Solids*, 17(1):4–16, 2012.

[119] M. Levinson. An accurate, simple theory of the static and dynamics of elastic plates. *Mech Res Comm*, 7(6):343–350, 1980.

[120] A. Libai and J.G. Simmonds. *The Nonlinear Theory of Elastic Shells.* Cambridge Univ. Press, Cambridge, 1998.

[121] L. Librescu, A. A. Khdeir, and J. N. Reddy. A comprehensive analysis of state of stress of elastic anisotropic flat plates using refined theories. *Acta Mechanica*, 70(1-4):57–81, 1987.

[122] R. Lorentz. Die nicht achsensummetrische Knickung dunnwandiger Hohlzylinder. *Physical Zeitschrift*, 12(7):10–16, 1911.

[123] R. Lorenz. Achsensymmetrische verzerrungen in dunnwandingen hohlzzylindern. *Zeitschrift des Vereines Deutscher Ingenieure*, 52(43):1706–1713, 1908.

[124] R. Lorenz. Die nicht achsensymmetrische knickung dunnwandiger hohlzylinder. *Physical Zeitschrift*, 12(7):241–260, 1911.

[125] A.E.H. Love. On the small free vibrations and deformations of the elastic shells. *Phil Trans Roy Soc London, Ser A*, 179:491–546, 1888.

[126] A.E.H. Love. *A Triatise on the Mathematical Theory of Elasticity, 2nd edn.* University Press, Cambridge, 1906.

[127] S. Lukasiewicz. *Local Loads in Plates and Shells.* Springer, 1979.

[128] A. Luongo. Mode localization in dynamics and buckling of linear imperfect continuous structures. *Nonlinear Dynamics*, 259(1-3):135–156, 2001.

[129] A.I. Lurie. *Statics of Thin Elastic Shells (in Russ.).* Gostekhizdat, Moscow, 1946.

[130] A.N. Markov. Dynamic stability of anisotropic Cylindrical shells (in Russ.). *Appl. Math. & Mech.*, 13(2):145–150, 1949.

[131] V. P. Maslov. *The Complex WKB Method for Nonlinear Equations I: Linear Theory.* Birkhöuser, Basel, Boston, 1994.

[132] V.P. Maslov. *The Complex WKB Method in Nonlinear Equations(in Russ.).* Nauka, Moscow, 1977.

[133] E. Mathieu. Memoire sur le mouvement vibratoire d'une membrane de forme elliptique. *J. Math. Pures Appl.*, 13:137–203, 1868.

[134] E. Mathieu. *Cours De Physique Mathematique (1873) (French Edition) (French).* Kessinger Publishing, 2010.

[135] I.K. McIvor. The elastic cylindrical shell under radial impulse. *Trans. ASME*, E33(4):831–837, 1966.

[136] G. Mikhasev and R. Atayev. Localized parametric vibrations of laminated cylindrical shell under non-uniform axial load periodically varying with time. In H. Altenbach, J. Chróścielewski, V. Eremeyev, and K. Wiśniewski, editors, *Recent Developments in the Theory of Shells. Advanced Structured Materials*, volume 110, pages 459–478. Springer, Cham, 2019.

[137] G. Mikhasev and I. Mlechka. Localized buckling of laminated cylindrical shells with low reduced shear modulus under non-uniform axial compression. *ZAMM*, 98(3):491–508, 2018.

[138] G. Mikhasev, I. Mlechka, and H. Altenbach. Soft suppression of travelling localized vibrations in medium-length thin sandwich-like cylindrical shells containing magnetorheological layers via nonstationary magnetic field. In J. Awrejcewicz, editor, *Dynamical Systems: Theoretical and Experimental Ananlysis*, volume 182 of *Springer Proceedings in Mathematics&Statistics*, pages 241–260. Springer, Singapore, 2016.

[139] G. Mikhasev, I. Mlechka, and S. Maevskaya. Effect of magnetic field on free and forced vibrations of laminated cylindrical shells containing magnetorheological elastomers. In H. Altenbach et al., editor, *Analysis and Modelling of Advanced Structures and Smart Systems*, volume 81 of *Advanced Structured Materials*, pages 119–148. Springer, Singapore, 2018.

[140] G. I. Mikhasev, H. Altenbach, and E. A. Korchevskaya. On the influence of the magnetic field on the eigenmodes of thin laminated cylindrical shells containing magnetorheological elastomer. *Composite Structures*, 113:186 196, 2014.

[141] G. I. Mikhasev and P. E. Tovstik. *Localized Vibrations and Waves in Thin Shells. Asymptotic Methods.* Fizmatlit, Moscow, 2009.

[142] G.I. Mikhasev. The propagation of flexural waves in a cylindrical shell (in Russian). *Vesnik Sankt. Peterburg. Univ. Ser. 1*, 2(8):99–103, 1993.

[143] G.I. Mikhasev. On propagation of bending waves in cylindrical shell (In Russ.). *Vestnik S. Peterburg. Univer. Ser. Math. Astronom.*, 8:99–103, 1993.

[144] G.I. Mikhasev. The propagation of flexural waves in a non-circular cylindrical shell (in Russian). *Izvestiya RAN: Mekhanika Tverdogo Tela*, 3:164–172, 1994.

[145] G.I. Mikhasev. On propagation of axisymmetric bending waves in an infinite cylindrical shell (In Russ.). *Vesti Akademii Nauk Belarusi: Ser. Fiz.-Mat. Navuk*, (1):39–45, 1994.

[146] G.I. Mikhasev. On the complex WKB-solution of the Cauchy problem for equations of motion of a thin cylindrical shell (In Russ.). *Dokl. Akad. Nauk Belarusi*, 38(4):24–27, 1994.

[147] G.I. Mikhasev. On the wave forms of motion of an infinite cylindrical shell with variable parameters (in Russ.). *Izvestiya RAN: Mekhanika Tverdogo Tela*, 6:129–137, 1995.

[148] G.I. Mikhasev. Investigation of flexural waves in an infinite shell of revolution. *International Applied Mechanics*, 32(7):545–549, 1996.

[149] G.I. Mikhasev. Localized wave forms of motion of an infinite shell of revolution . *Journal of Applied Mathematics and Mechanics*, 60(5):813–820, 1996.

[150] G.I. Mikhasev. Localized families of bending waves in a non-circular cylindrical shell with sloping edges. *Journal of Applied Mathematics and Mechanics*, 60(4):629–637, 1996.

[151] G.I. Mikhasev. Free and parametric vibrations of cylindrical shells under static and periodic axial loads . *Technische Mechanik*, 17(3):209—216, 1997.

[152] G.I. Mikhasev. Asymptotic solutions of system of equations for shallow shells in the form of two-dimensional wave packets (In Russ.). *Izvestiya Vuzov: Matematika*, 2:47–53, 1998.

[153] G.I. Mikhasev. Asymptotic solutions of a system of equations of shallow shells in the form of two-dimensional wave packets (In Russ.). *Russian Mathematics (Izvestiya VUZ. Matematika)*, 42(2):44–50, 1998.

[154] G.I. Mikhasev. Travelling wave packets in an infinite thin cylindrical shell under internal pressure. *Journal of Sound and Vibration*, 209(4):543–559, 1998.

[155] G.I. Mikhasev. Localized families of bending waves in a thin medium-length cylindrical shell under pressure. *Journal of Sound and Vibration*, 253(4):833–857, 2002.

[156] G.I. Mikhasev. Running two-dimensional wave packets in a thin non-uniform cylindrical shell. In *Proceedings of of the 8th International Conference on Mathematical and Numerical Aspects of Waves (Editors: Nick Biggs, et al.) University of Reading, 23rd-27th July 2007*, pages 340–341. University of Reading, 2007.

[157] G.I. Mikhasev. On localized modes of free vibrations of single-walled carbon nanotubes embedded in nonhomogeneous elastic medium. *ZAMM-Z. Angew. Math. Mech.*, 94(1–2):130–141, 2014.

[158] G.I. Mikhasev and H. Altenbvach. *Thin-walled Laminated Structures. Buckling, Vibrations, and Their Suppression.* Springer, 2019.

[159] G.I. Mikhasev and M.G. Botogova. Free localized vibrations of a long double-walled carbon nanotube introduced into an inhomogeneous elastic medium. *Vestnik St. Petersburg University. Mathematics*, 49(1):85–91, 2016.

[160] G.I. Mikhasev and S.P. Kuntsevich. Thermoparametric vibrations of noncircular cylindrical shell in nonstationary temperature field. *Technische Mechanik*, 17(2):113–120, 1997.

[161] R.D. Mindlin. Influence of rotatory inertia and shear on flexural motions of isotropic elastic plates. *Trans ASME J Appl Mech*, 18:31–38, 1951.

[162] R.D. Mindlin and H.H. Bleich. Response of an elastic cylindrical shell to a transverse step shock wave. *J. Appl. Mech.*, (2):189–195, 1953.

[163] E. N. Mnev and Pertzev A. K. *Hydro-elasticity of Shells (in Russ.)*. Sudostroenie, Leningrad, 1970.

[164] A. I. Molchanov. Free vibrations of non-circular cylindrical shells close to shells with the zero Gaussian curvature (in Russ.). *Vestnik Leningrad Univ. Serie 1*, (4):43–45, 1986.

[165] A. I. Molchanov. Asymptotic solution of a system of rquations of free vibrations of circular shells close to shells with the zero Gaussian curvature (in Russ.). *Vestnik Leningrad Univ. Serie 1*, (2):106–107, 1987.

[166] N.F. Morozov, P.E. Tovstik, and T.P. Tovstik. Generalized Timoshenko-Reissner model for a multilayer plate (English transl.). *Mechanics of Solids*, 51(5):527–537, 2016.

[167] N.F. Morozov, P.E. Tovstik, and T.P. Tovstik. The Timoshenko-Reissner generalized model of a plate highly nonuniform in thickness (English transl.). *Doklady Physics*, 61(8):394–398, 2016.

[168] K. Muschtari and K. Galimov. *Nonlinear Theory of Thin Elastic Shells*. NSF-NASA, Washington, 1961.

[169] M.A. Naimark. *Linear Differential Operator*. Ungar, New York, 1968.

[170] T.Y. Ng and K.Y. Lam. Effects of boundary conditions on the parametric resonance of cylindrical shells under axial loading. *Shok and Vibration*, 1(5):343–354, 1998.

[171] U.K. Nigul, J.A. Metsaveer, N.D. Veksler, and M.E. Kutser. *Echo-Signals from Elastic Objects (in Russ.)*. Valgus, Tallin, 1974.

[172] V. V. Novozhilov and R. M. Finkel'stein. On the error of the Kirchhoff hypotheses in the shell theory (in Russ.). *Appl. Math. & Mech.*, 17(5):331–340, 1960.

[173] V.V. Novozhilov. *Theory of Thin Shells (English Transl.)*. Wolters-Noordhoff, Groningen, 1970.

[174] W. Nowacki. Vibration and buckling of rectangular plates simply supported at the periphery and at several points inside. *Arch. Mech. Stosow*, 5(3):437–454, 1953.

[175] P.M. Ogibalov and V.F. Gribanov. *Thermostability of Plates and Shells (in Russ.).* Moskov. Universitet, Moscow, 1968.

[176] O.D. Oniashvili. On the dynamic stability of shells (in Russ.). *Issues of Georgian Acad. Sci*, 11(3):169–175, 1950.

[177] Maslov V. P. *Operator methods (in Russ.).* Mir, Moscow, 1976.

[178] N.J. Pagano. Exact solutions for composite laminates in cylindrical bending. *Journal of Composite Materials*, 3:398–411, 1969.

[179] N.J. Pagano. Exact solutions for rectangular bidirectional composites and sandwich plates bending. *Journal of Composite Materials*, 4:20–34, 1970.

[180] Y.-H. Pao. Elastic waves in solids. *J. Appl. Mech.*, 50:1152–1164, 1983.

[181] A.K. Pertzev and E.G. Platonov. *Dynamics of Shells and Plates (in Russ.).* Sudostroenie, Leningrad, 1987.

[182] C. Pierre. Mode localization and eigenvalue loci veering phenomena in disordered structures. *Journal of Sound and Vibration*, 126(3):485–502, 1988.

[183] C. Pierre. Weak and strong vibration localization in disordered structures: A statistical investigation. *Journal of Sound and Vibration*, 139(1):111–132, 1990.

[184] C. Pierre and P. D. Cha. Strong mode localization in nearly periodic disordered structures. *AIAA Journal*, 27(2):227–241, 1989.

[185] C. Pierre and E.H. Dowell. Localization of vibrations by structural irregularity. *Journal of Sound and Vibration*, 114(3):549 – 564, 1987.

[186] G.T. Piliposian, M.V. Belubekyan, and K.B. Ghazaryan. Localized bending waves in a transversely isotropic plate. *Journal of Sound and Vibration*, 329(2):3596–3605, 2010.

[187] G.S. Pisarenko and A.N. Chemeris. Dynamic stability of a cylindrical shell (in Russ.). In *Dissipation of Energy with the Vibrations of Mechanical Systems*, pages 107–114. Izd. Naukova Dumka, Kiev, 1968.

[188] M. S. Qatu. Recent research advances in the dynamic behaviour of shells: 1989-2000. Part 1: Laminated composite shells. *Applied Mechanics Review*, 55(4):325–350, 2002.

[189] M. S. Qatu. *Vibration of Laminated Shells and Plates.* Elsevier, San Diego, 2004.

[190] M.S. Qatu, R.W. Sullivan, and W. Wang. Recent research advances on the dynamic analysis of composite shells: 2000-2009. *Composite Structures*, 93(1):14–31, 2010.

[191] Y. Qu, X. Long, S. Wu, and G. Meng. A unified formulation for vibration analysis of composite laminated shells of revolution including shear deformation and rotatory inertia. *Composite Structures*, 98:169–191, 2013.

[192] Stoneley R. The elastic waves at the interface of two solids. *Proc. Roy Soc. London A.*, 106:416–429, 1924.

[193] I. Raju and C. Amba-Rao. Free vibrations of a square plate symmetrically supported at four points on the diagonals. *J. Sound Vibr*, 90(2):291–297, 1983.

[194] G. V. Rao, I. Raju, and C Amba-Rao. Vibrations of point supported plates. *J. Sound Vibr*, 29(3):387–391, 1973.

[195] J.W. Rayleigh. On waves propagated along the plate surface of an elastic solids. *Proc. London Math. Soc.*, 17:4–11, 2019.

[196] J. N. Reddy. An evaluation of equivalent-single-layer and layerwise theories of composite laminates. *Compos Struct*, 25:21–35, 1993.

[197] J. N. Reddy and C. M. Wang. An overview of the relationships between of the classical and shear deformation plate theories. *Compos. Sci. Technol.*, 60:2327–2335, 2000.

[198] J.N. Reddy. A simple higher-order theory for laminated composite plates. *Trans ASME J Appl Mech*, 51(4):745–752, 1984.

[199] J.N. Reddy. *Mechanics of Laminated Composite Plates and Shells: Theory and Analysis, 2nd edn.* CRC Press, Boca Raton, 2004.

[200] J.N. Reddy and C.F. Liu. A higher-order shear deformation theory of laminated elastic shells. *Int J Eng Sci*, 23(3):319–330, 1985.

[201] H. Reismann and P.S. Pawlik. Plane-strain dynamic response of a cylindrical shell-a comparison study of three different shell theories. *Trans. ASME*, 35(2):297–305, 1968.

[202] E. Reissner. On the theory of bending of elastic plates. *J Math Phys*, 23(11):184–194, 1944.

[203] E. Reissner. The effect of transverse shear deformation on the bending of elastic plates. *Trans. ASME J Appl Mech*, 12(11):A69–A77, 1945.

[204] I. Roitberg, D. Vassiliev, and T. Weidl. Edge resonance in an elastic semi-strip. *Q. J. Mech. Appl. Math.*, 51:1–13, 1998.

[205] J.L. Sanders. *An improved first approximation theory for thin shells.* NASA Report 24, NASA, Washington, DC, 1959.

[206] A.S. Sayyad and Y.M. Ghugal. On the free vibration analysis of laminated composite and sandwich plates: A review of recent literature with some numerical results. *Composite Structures*, 129:177 – 201, 2015.

[207] E. A. G. Shaw. On the resonant vibrations of thick barium titanate disks. *Journal of the Acoustical Society of America*, 28:38–50, 1956.

[208] J. Sheng. The response of a thin cylindrical shell to transient surface loading. *AIAA Journal*, 3(4):701–709, 1965.

[209] P. Shi, C. Dong, F. Sun, W. Liu, and Q. Hu. A new higher order shear deformation theory for static, vibration and buckling responses of laminated plates with the isogeometric analysis. *Composite Structures*, 204:342 – 358, 2018.

[210] B.K. Sinha. Some remarks on propagation characteristics of ridge guides for acoustic waves at low frequencies. *Journal of the Acoustical Society of Americ*, 56:16–18, 1974.

[211] E. Skudrzyk. *Simple and Complex Vibratory Systems.* Pennsylvania Univ. press., Univ. Park and London, 1968.

[212] I. Sneddon. *Fourier Transforms.* McGraw-Hill, New York, 1952.

[213] K. P. Soldatos and T. Timarci. A unified formulation of laminated composites, shear deformable, five-degrees-of-freedom cylindrical shell theories. *Composite Structures*, 25:165–171, 1993.

[214] S. R. Srinivas, C. V. Joga Rao, and A. K. Rao. An exact analysis for vibration of simply-supported homogeneous and laminated thick rectangular plates. *Journal of Sound and vibration*, 12:187–199, 1970.

[215] M.J.O. Strutt. Zur Wellenmechanik des Atomgitters. *Ann. Phys.*, 391(10):319–324, 1927.

[216] I. Thompson, I.D. Abrahams, and A.N. Norris. On the existence of flexural edge waves on this orthotropic plates. *Journal of the Acoustical Society of America*, 112:1756–1765, 2002.

[217] R.N. Thurston and J. McKenna. Flexural acoustic waves along the edge of a plate. *IEEE Transactions on Sonics and Ultrasonics*, 21:296–2978, 1974.

[218] P. S. Timoshenko and J. M. Gere. *Stability of Elastic Systems.* McGraw-Hill, New York, 1961.

[219] S. P. Timoshenko. On the correction for shear of the differential equation for transverse vibrations of prismatic bars. *Philos. Mag.*, 4(242):744–746, 1921.

[220] M. H. Toorani and A. A. Lakis. General equations of anisotropic plates and shells including transverse shear deformations, rotary inertia and initial curvature effects. *Journal of Sound and Vibration*, 237(4):561–615, 2000.

[221] P.J. Torvik. Reflection of wave trains in semi-infinite plates. *J. Acoust. Soc. Am.*, 41:346–353, 1967.

[222] P. E. Tovstik. *Stability of Thin Shells(in Russ.)*. Nauka, Moscow, 1995.

[223] P. E. Tovstik. Local stability of plates and elastic shells lying on an elastic foundation (in Russ.). *Mechanics of solids*, (1):147–160, 2005.

[224] P. E. Tovstik. Reaction of an elastic pre-stressed orthotropic foundation. *Vestnik St. Petersburg Univ. Serie 1.*, (4), 2006.

[225] P. E. Tovstik. On the asymptotic character of the approximate models of beams, plates and shells, English transl. *Vestnik St.Petersburg Univ. Mathematics*, (3):49–54, 2007.

[226] P. E. Tovstik and A. L. Smirnov. *Asymptotic Methods in the Buckling Theory of Elastic Shells* . World Scientific Singapore, New Jersey, London, Hong Kong, 2001.

[227] P. E. Tovstik and T. P. Tovstik. On the 2D models of plates and shells including the transversal shear. *Zeitschrift fur Angewandte Mathematik und Mechanik*, 87(2):160–171, 2007.

[228] P. E. Tovstik and T. P. Tovstik. Two-dimensional linear model of anisotropic shells. In *Shell Structures Theory and Applications*, volume 3, pages 85–96, London, 2014. CRC Press Balkema. Taylor & Francis Group.

[229] P. E. Tovstik and T.P. Tovstik. A thin-plate bending equation of second-order accuracy (English transl.). *Doklady Physics*, 59(3):389–392, 2014.

[230] P. E. Tovstik and T.P. Tovstik. An elastic plate bending equation of second-order accuracy. *Acta mechanica*, 228(10):3403–3419, 2017.

[231] P. E. Tovstik and T.P. Tovstik. Generalized Timoshenko-Reissner models for beams and plates, strongly heterogeneous in the thickness direction. *Zeitschrift fur Angewandte Mathematik und Mechanik*, 97(3):296–308, 2017.

[232] P.E. Tovstik. Some problems in cylindrical and conic shell buckling . *English transl. Appl. Math. Mech.*, 47:815–822, 1983.

[233] P.E. Tovstik. Two-dimensional problems of buckling and vibrations of the shells of zero Gaussian curvature . *English transl. Soviet Phys. Dokl.*, 28(7):593–594, 1983.

[234] P.E. Tovstik. Free high-frequency vibrations of anisotropic plates of variable thickness (English transl.). *Journal of Applied Mathematics and Mechanics*, 56(3):390–395, 1992.

[235] P.E. Tovstik. Free high-frequency vibrations of a plate with variable thickness (in Russ.). *Mechanika Tverdogo Tela*, (4):162–170, 1994.

[236] J. Utjes, G. S. Sarmiento, P. Laura, and R. Gelos. Vibrations of thin elastic plates with point supports: A comparative study. *Appl. Acoust.*, 19(1):17–24, 1986.

[237] N. V. Valishvili. *Computer methods of design of shells of revolution (in Russ.)*. Mashinostroenie, Moscow, 1976.

[238] A. Vijayaraghavan and R.M. Evan-Ivanowski. Parametric instability of circular cylindrical shells. *Trans ASME. E*, 34(4):985–990, 1967.

[239] M.V. Vilde, Yu.D. Kaplunov, and L.Yu. Kassovich. *Edge and interfacial resonance phenomenon in elastic solids (in Russ.)*. FIZMATLIT, Moscow, 2010.

[240] S. Vlachoutsis. Shear correction factors for plates and shells. *International Journal for Numerical Methods in Engineering*, 33:1537–1552, 1992.

[241] V.Z. Vlasov. The principal differential equations of the general shell theory (in Russ.). *Prikl Mat Mech*, 8(2):109–140, 1944.

[242] A.S. Vol'mir and L.N. Smetanina. Dynamic stability of glass-reinforced plastic shells. *Polymer Mechanics*, 4(1):79–82, 1968.

[243] W. Wenzke. Die dynamische Stabilitat der axialpulsierend belasteten Kreiszylinderschale. *Wiss. Z. Techn. Hochschule Otto von Querischke*, 7(1):93–124, 1963.

[244] W. H. Wittrick. Analytical three-dimensional elasticity solutions to some plate problems and some observations on Mindlin's plate theory. *International Journal of Solid and Structures*, 23(4):441–464, 1987.

[245] G. Wu. Free vibration of cyclic assemblies with a single disordered component. *Journal of Sound and Vibration*, 165(3):567–570, 1993.

[246] J.C. Yao. Dynamic stability of cylindrical shells and static and periodic axial and radial loads. *AIAA Journal*, 1(6):1391–1396, 1963.

[247] J.C. Yao. Nonlinear elastic buckling and parametric excitation of a cylinder under axial loads. *Trans ASME. E*, 32(1):109–115, 1965.

[248] D.D. Zakharov. Konenkov's waves in anisotropic layered plates. *Acoustical Physics*, 48:171–175, 2002.

[249] D.D. Zakharov and W. Becker. Rayleigh type bending waves in anisotropic media. *Journal of Sound and Vibration*, 261:805–818, 2003.

[250] D.D. Zakharov and W. Becker. Analysis of the acoustical edge flexural mode in a plate using refined asymptotics. *Journal of the Acoustical Society of America*, 116(2):872–878, 2004.

[251] A.V. Zelinskaya and P.E. Tovstik. Buckling of an axially compressed transversely isotropic cylindrical shell with a weakly supported curvilinear edge (English transl.). *Vestnik St. Petersburg University: Mathematics*, 48(2):109–118, 2015.

[252] A.S. Zilbergreit and E.B. Suslova. Contact waves of bend in thin plates(in Russ.). *Akoust. Journal.*, 29(2):186–191, 1985.

Index